Modern Radar for Automotive Applications

Other volumes in this series:

Volume 1 **Optimised Radar Processors** A. Farina (Editor)
Volume 3 **Weibull Radar Clutter** M. Sekine and Y. Mao
Volume 4 **Advanced Radar Techniques and Systems** G. Galati (Editor)
Volume 7 **Ultra-Wideband Radar Measurements: Analysis and processing** L. Yu. Astanin and A.A. Kostylev
Volume 8 **Aviation Weather Surveillance Systems: Advanced radar and surface sensors for flight safety and air traffic management** P.R. Mahapatra
Volume 10 **Radar Techniques Using Array Antennas** W. Wirth
Volume 11 **Air and Spaceborne Radar Systems: An introduction** P. Lacomme (Editor)
Volume 13 **Introduction to RF Stealth** D. Lynch
Volume 14 **Applications of Space-Time Adaptive Processing** R. Klemm (Editor)
Volume 15 **Ground Penetrating Radar, 2nd Edition** D. Daniels
Volume 16 **Target Detection by Marine Radar** J. Briggs
Volume 17 **Strapdown Inertial Navigation Technology, 2nd Edition** D. Titterton and J. Weston
Volume 18 **Introduction to Radar Target Recognition** P. Tait
Volume 19 **Radar Imaging and Holography** A. Pasmurov and S. Zinovjev
Volume 20 **Sea Clutter: Scattering, the K distribution and radar performance** K. Ward, R. Tough and S. Watts
Volume 21 **Principles of Space-Time Adaptive Processing, 3rd Edition** R. Klemm
Volume 22 **Waveform Design and Diversity for Advanced Radar Systems** F. Gini, A. De Maio and L.K. Patton
Volume 23 **Tracking Filter Engineering: The Gauss-Newton and Polynomial Filters** N. Morrison
Volume 25 **Sea Clutter: Scattering, the K distribution and radar performance, 2nd Edition** K. Ward, R. Tough and S. Watts
Volume 33 **Radar Automatic Target Recognition (ATR) and Non-Cooperative Target Recognition** D. Blacknell and H. Griffiths (Editor)
Volume 26 **Radar Techniques Using Array Antennas, 2nd Edition** W. Wirth
Volume 101 **Introduction to Airborne Radar, 2nd Edition** G.W. Stimson
Volume 530 **Radar Sea Clutter: Modelling and target detection** Luke Rosenburg and Simon Watts
Volume 534 **New Methodologies for Understanding Radar Data** Amit Kumar Mishra and Stefan Brüggenwirth
Volume 537 **Ocean Remote Sensing Technologies: High frequency, marine and GNSS-based radar** Weimin Huang and Eric W. Gill (Editor)

Modern Radar for Automotive Applications

Edited by
Zhengyu Peng, Changzhi Li and
Faruk Uysal

The Institution of Engineering and Technology

Published by SciTech Publishing, an imprint of The Institution of Engineering and Technology, London, United Kingdom

The Institution of Engineering and Technology is registered as a Charity in England & Wales (no. 211014) and Scotland (no. SC038698).

First published 2022

The Institution of Engineering and Technology
Futures Place
Six Hills Way, Stevenage
Herts, SG1 2UA, United Kingdom

www.theiet.org

British Library Cataloguing in Publication Data
A catalogue record for this product is available from the British Library

ISBN 978-1-83953-435-5 (hardback)
ISBN 978-1-83953-436-2 (PDF)

Typeset in India by Exeter Premedia Services Private Limited

Cover Image: Viaframe/Corbis via Getty Images

Contents

About the Editors

Zhengyu Peng is currently a radar systems engineer at Aptiv, USA; a global technology company that develops safer, greener and more connected solutions enabling the future of mobility. Dr Peng has expertise and experience in automotive radar, radar signal processing, mm wave and microwave systems, antenna array, and beamforming.

Changzhi Li is a professor in the Department of Electrical & Computer Engineering at Texas Tech University, USA. His research interest is microwave/mm-wave sensing for healthcare, security, and human-machine interfaces. Professor Li is a Microwave Theory and Techniques Society distinguished microwave lecturer. He was a recipient of the IEEE Microwave Society Outstanding Young Engineer Award, the IEEE Sensors Council Early Career Technical Achievement Award, and the ASEE Frederick Emmons Terman Award, amongst others.

Faruk Uysal is a scientist at the Netherlands Organisation for Applied Scientific Research (TNO), an affiliate member of the Advanced Radar Research Center (ARRC) at the University of Oklahoma USA, and a senior member of URSI and IEEE. He is a member of Dutch URSI committee and an elected member of the IEEE AESS Radar System Panel. Dr Uysal has expertise in beamforming theory, target detection, radar imaging, distributed radar networks, and automotive radar.

Chapter 1
Introduction

Peng Z[1], Li C[2], and Ran L[3]

The term "radar" stands for radio detection and ranging. A radar is an electromagnetic system used to detect, locate, track, and identify different objects within a certain area. A radar transmits electromagnetic energy in the direction of targets to observe the echoes from them. The targets could be ships, aircraft, astronomical bodies, automotive vehicles, etc. In early days, radar systems were only used in the military area due to their bulky size and high cost. Thanks to the advance of high-frequency integrated circuit (ICs) and monolithic microwave ICs, modern miniature radar systems can be realized on a printed circuit board or even on an IC [1–5]. Applications of radar systems have been extended to commercial areas, such as through-wall detection [6–9], indoor localization [10–13], biomedical applications [14, 15], and driver assistance [16, 17]. The topic of this book focuses on radar technology and its applications on automotive vehicles.

Radar is a key technology in the safety system of a modern vehicle. Automotive radars are the critical sensors in advanced driver-assistance systems, which are used for adaptive cruise control, collision avoidance, blind spot detection, lane change assistance, parking assistance, etc. Figure 1.1 shows a typical automotive radar configuration on a vehicle. A long-range radar (LRR) mounted in the front of the vehicle is usually for adaptive cruise control. The medium-range radars mounted in the front and rear provide wider coverage than the LRR, and they can be used for cross traffic alert and lane change assistance. The corner-mounted short-range radars support parking aid, obstacle/pedestrian detection, and blind spot monitoring. In real applications, these radars usually work together to provide more robust detection results. Regarding the operation frequency, the 24 GHz bands have been in use in legacy automotive sensors. Due to spectrum regulations and standards developed by the European Telecommunications Standards Institute and the Federal Communications Commission, the 77 GHz frequency band has been widely adopted for automotive radars. The mainstream automotive radar development has already

[1]Aptiv Corporation, Indiana Tech Center (ITC), Fort Wayne, IN, USA
[2]Department of Electrical and Computer Engineering, Texas Tech University, Lubbock, TX, USA
[3]Radar Technology Department, Netherlands Organisation for Applied Scientific Research (TNO), Mekelweg, Delft, Netherlands

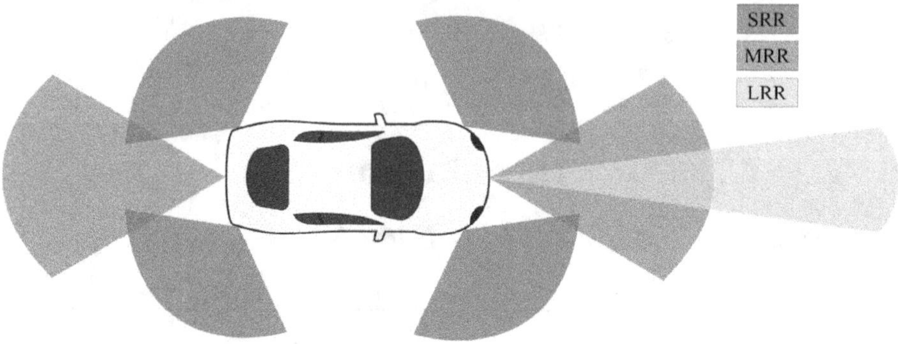

Figure 1.1 A typical radar configuration on a vehicle

moved to 77 GHz frequency band, which enables the radar a better resolution and a smaller form factor.

Evolving from driver assistance systems, a fully autonomous car has much higher demands for its sensors, especially for automotive radars. The Society of Automotive Engineers (SAE) has developed six generally accepted levels of self-driving vehicles based on the degree of human involvement in the act of driving [18], including level zero, which correlates to having no automation and instead complete human control of the vehicle. Figure 1.2 shows the definition of the levels of driving automation.

- **Level 1 (driver assistance):** A human driver is responsible for all tasks associated with operating the car, including accelerating, steering, braking, and monitoring of the surrounding environment. There is a driving automation system in the car that helps with either steering or accelerating, but not both.
- **Level 2 (partial automation):** At this level, the automation system in the car can assist with both steering and acceleration, while the driver is still responsible for most of the safety-critical functions and environment monitoring. Currently, the level 2 autonomous vehicles are the most common on the roads.
- **Level 3 (conditional automation):** Starting from level 3 and onward, the car itself monitors the environment by utilizing autonomous vehicle sensors and

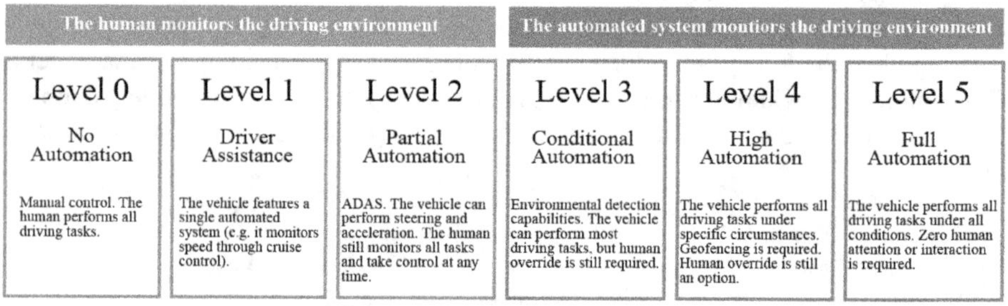

Figure 1.2 SAE levels of driving automation

performs other dynamic driving tasks such as braking. The human driver has to be prepared to intervene if a system failure occurs or other unexpected conditions arise while driving.

- **Level 4 (high automation):** Level 4 correlates to a high level of automation, where the car is capable of completing an entire journey without any intervention from the driver, even in extreme cases. However, there are some restrictions: the driver can switch the vehicle into this mode only when the system detects that the traffic conditions are safe and there are no traffic jams.
- **Level 5 (full automation):** Fully automated cars do not yet exist, but automakers are striving to achieve level 5 of autonomous driving, where the driver simply specifies their destination and the vehicle takes complete control and responsibility for all driving modes. Therefore, level 5 cars will have no provisions for any human control, such as steering wheels or pedals.

The future of autonomous vehicles looks fantastic; however, realizing a fully automated vehicle is still very challenging. Currently, the market remains dominated by level 2 partially autonomous vehicles.

Autonomous vehicles would be impossible without sensors, which allow the vehicle to see and sense everything on the road, as well as to collect the information needed in order to drive safely. Furthermore, this information is processed and analyzed in order to build a path from point A to point B and to send the appropriate instructions to the controllers of the car, such as steering, acceleration, and braking. Moreover, the information collected with the sensors in autonomous vehicles, including the actual path ahead, traffic jams, and any obstacles on the road, can also be shared between cars that are connected through vehicle-to-vehicle communication technology [19, 20], which can be an incredibly helpful resource for driving automation. The majority of today's automotive manufacturers most commonly use the following three types of sensors in autonomous vehicles: cameras, radars, and lidars. Compared with other sensors, radar provides targets' locations, as well as their velocities with relatively low cost. Moreover, radar is robust in harsh environments [21, 22], such as poor light, bad weather, and extreme temperatures. These features make radar the unique sensor for autonomous vehicles.

Current automotive radar technology is mostly based on the principle of frequency-modulated continuous-wave (FMCW) radar, which has been well known for several decades. In this book, the fundamentals of modern automotive radars are first introduced using the FMCW waveform as an example. The spatial dimension is obtained using modern multiple-input multiple-output techniques, which will be introduced in the chapter that follows. Together with an increase of hardware capabilities such as phase modulation, as well as a scaling up of simultaneously utilized transmit and receive channels with independent modulation features, new degrees of freedom have been added to traditional FMCW radar system design and signal processing. With more and more vehicles adopting automotive radar, it is inevitable that the radar will be interfered by other radars operating in the same frequency band. Interference and its mitigation methods will be introduced in Chapter 4. Chapter 5 introduces how radar sensors can be integrated together with other sensors to improve target detection. Chapter 6 provides several approaches to classify the commonly seen traffic targets with radar. Then, in Chapter 7, the application of using radar for road surface condition detection will be discussed.

In this book, we also recognize that the future of automotive radars should not only address conventional exterior applications but also play important roles for interior applications, such as gesture sensing for human–vehicle interaction, driver/passenger vital signs, and presence monitoring. Chapters 8 and 9 in this book introduce the details in these areas.

References

[1] Droitcour A.D., Boric-Lubecke O., Lubecke V.M., Lin J., Kovacs G.T.A. 'Range correlation and I/Q performance benefits in single-chip silicon Doppler radars for noncontact cardiopulmonary monitoring'. *IEEE Transactions on Microwave Theory and Techniques*. 2004, vol. 52(3), pp. 838–48.

[2] Yeap S.B., Qing X., Chen Z.N. '77-GHz dual-layer transmit-array for automotive radar applications'. *IEEE Transactions on Antennas and Propagation*. 2015, vol. 63(6), pp. 2833–7.

[3] Luo T.N., Wu C.H.E., Chen Y.J.E. 'A 77-GHz CMOs automotive radar transceiver with anti-interference function'. *IEEE Transactions on Circuits and Systems I: Regular Papers*. 2013, vol. 60(12), pp. 3247–55.

[4] Lee J., Li Y.-A., Hung M.-H., Huang S.-J. 'A fully-integrated 77-GHz FMCW radar transceiver in 65-nm CMOs technology'. *IEEE Journal of Solid-State Circuits*. 2010, vol. 45(12), pp. 2746–56.

[5] Li C., Yu X., Lee C.M., Li D., Ran L., Lin J. 'High-sensitivity software-configurable sensitivity software-configurable 5.8-ghz radar sensor receiver CHIP in 0.13-μM cmos for noncontact vital sign detection'. *IEEE Transactions on Microwave Theory and Techniques*. 2010, vol. 58(5), pp. 1410–19.

[6] Ahmad F., Amin M. 'Noncoherent approach to through-the-wall radar localization'. *IEEE Transactions on Aerospace and Electronic Systems*. 2006, vol. 42(4), pp. 1405–19.

[7] Lubecke V.M., Boric-Lubecke O., Host-Madsen A., *et al.* 'Through-the-wall radar life detection and monitoring'. *IEEE/MTT-S International Microwave Symposium*; 2007. pp. 769–72.

[8] Debes C., Amin M.G., Zoubir A.M. 'Target detection in single- and multiple-view through-the-wall radar imaging'. *IEEE Transactions on Geoscience and Remote Sensing*. 2009, vol. 47(5), pp. 1349–61.

[9] Yoon Y.-S., Amin M.G. 'High-resolution through-the-wall radar imaging using beamspace music'. *IEEE Transactions on Antennas and Propagation*. 2008, vol. 56(6), pp. 1763–74.

[10] Vossiek M., Wiebking L., Gulden P., Wieghardt J., Hoffmann C., Heide P. 'Wireless local positioning'. *IEEE Microwave Magazine*. 2003, vol. 4(4), pp. 77–86.

[11] Peng Z., Ran L., Li C. 'A 24-ghz low-cost continuous beam steering phased array for indoor smart radar'. *2015 IEEE 58th International Midwest Symposium on Circuits and Systems (MWSCAS)*; 2–5 August, Fort Collins, CO, USA, 2015. pp. 1–4.

[12] Peng Z., Li C. 'A portable 24-ghz FMCW radar based on six-port for short-range human tracking'. Presented at 2015 IEEE MTT-S 2015 International Microwave Workshop Series on RF and Wireless Technologies for Biomedical and Healthcare Applications (IMWS-BIO); Taipei, Taiwan, 21–23 September 2015.

[13] Peng Z., Li C. 'A portable K-Band 3-D MIMO radar with nonuniformly spaced array for short-range localization'. *IEEE Transactions on Microwave Theory and Techniques*. 2018, vol. 62(11), pp. 1–12.

[14] Li C., Lubecke V.M., Boric-Lubecke O., Lin J. 'A review on recent advances in Doppler radar sensors for noncontact healthcare monitoring'. *IEEE Transactions on Microwave Theory and Techniques*. 2013, vol. 61(5), pp. 2046–60.

[15] Li C., Peng Z., Huang T.-Y., *et al*. 'A review on recent progress of portable short-range noncontact microwave radar systems'. *IEEE Transactions on Microwave Theory and Techniques*. 2017, vol. 65(5), pp. 1692–706.

[16] Hasch J., Topak E., Schnabel R., Zwick T., Weigel R., Waldschmidt C. 'Millimeter-wave technology for automotive radar sensors in the 77 GHz frequency band'. *IEEE Transactions on Microwave Theory and Techniques*. 2012, vol. 60(3), pp. 845–60.

[17] Patole S.M., Torlak M., Wang D., Ali M. 'Automotive radars: a review of signal processing techniques'. *IEEE Signal Processing Magazine*. 2017, vol. 34(2), pp. 22–35.

[18] *J3016C: taxonomy and definitions for terms related to driving automation systems for on-road motor vehicles – SAE international[online]*. Available from https://www.sae.org/standards/content/j3016_202104/

[19] Vehicle-to-vehicle communication–NHTSA [online]. Available from https://www.nhtsa.gov/technology-innovation/vehicle-vehicle-communication

[20] Abbasi I., Shahid Khan A. 'A review of vehicle to vehicle communication protocols for VANETs in the urban environment'. *Future Internet*, vol. 10(2), p. 14.

[21] Hasirlioglu S., Doric I., Lauerer C., Brandmeier T. 'Modeling and simulation of rain for the test of automotive sensor systems'. Presented at 2016 IEEE Intelligent Vehicles Symposium (IV); Gotenburg, Sweden, 19–22 June 2016. IEEE,

[22] Gourova R., Krasnov O., Yarovoy A. 'Analysis of rain clutter detections in commercial 77 ghz automotive radar'. Presented at 2017 European Radar Conference (EURAD); Nuremberg, 11–13 October 2017. IEEE,

Chapter 2

Principles of automotive radar systems

Zhengyu Peng[1] and Changzhi Li[2]

2.1 Basic radar functions

A radar system radiates electromagnetic energy into space from an antenna or an antenna array. The radiated electromagnetic energy "illuminates" the surrounding targets. The "illuminated" targets intercept some of the radiated energy and reflect a portion back to the radar system. The radar system utilizes one or multiple receiver channels to detect the reflected energy to determine the targets' range, velocity, and relative angles.

Based on the different types of waveforms radiated by the radar transmitter, radar systems can be categorized as pulsed radars and continuous-wave (CW) radars. A pulsed radar consists of a repetitive train of short-duration pulses. The range of the target is measured based on the time delay between the transmitted pulse and the received pulse. Different from a pulsed radar, a CW radar usually transmits the electromagnetic wave continuously in a period of time. The properties of the targets are obtained by comparing between the received signal with a replica of the transmitted signal. For automotive applications, CW radar systems have been the dominant due to their advantages in multiple aspects. Compared with a pulsed radar, a CW radar features low peak transmit power, simpler, and highly integrated structure, which make its applications spread into various areas, especially for automotive applications. This chapter is trying to present a thorough and consistent description of the fundamentals of radar technology for automotive applications. Though many of the concepts are the same between pulsed radars and CW radars, CW radars are emphasized over pulsed radars in this book.

The functions of automotive radars can be classified as detection, tracking, and imaging. In this chapter, the emphasis is on detection, as well as the basic techniques of signal processing to perform the task. Tracking and imaging will be discussed in the following chapters. For target detection, the most fundamental problem is determining if the echo received by the receiver is from the reflection of an object or

[1]Aptiv Corporation, Indiana Tech Center (ITC), Carmel, IN, USA
[2]Department of Electrical and Computer Engineering, Texas Tech University, Lubbock, TX, USA

only noise. For a CW radar, the detection decisions are usually made by comparing the signal amplitude of the received echo to a threshold, which can be predefined or computed in realtime. For a robust radar system, the threshold is necessary to be computed adaptively from the real-time radar data.

In order to obtain the range of a target, certain types of modulations are required for a CW radar. The modulation is used to encode the range information into the echo signal, which needs to be extracted through signal processing. For example, a linear frequency modulated continuous-wave (FMCW) radar encodes the target's range into the frequency of the baseband signal. In a phase-modulated continuous-wave radar, the range information is encoded in the sequence of phase code, which can be extracted by calculating the correlation of the echo with the original code sequence. Despite the various type of modulation, the range resolution (ΔR) of a radar is inversely proportional to the bandwidth (BW) of the transmitted signal:

$$\Delta R \propto \frac{1}{BW} \tag{2.1}$$

A larger BW is usually favorable in automotive radar applications to achieve better range resolution.

A radar is also capable of obtaining the relative velocity of a target by utilizing the Doppler effect. This is one of the major advantages of an automotive radar compared with other automotive sensors, such as cameras and LiDAR, which is an acronym of "light detection and ranging". The Doppler effect is the change in frequency or phase of an electromagnetic wave in relation to a target that is moving relative to the radar. It is named after the Austrian physicist Christian Doppler, who described the phenomenon in 1842. The reason for the Doppler effect is that when the target is moving toward the radar, each successive electromagnetic wave crest is reflected from a position closer to the radar than the crest of the previous electromagnetic wave. Therefore, each electromagnetic wave takes slightly less time to reach the radar than the previous electromagnetic wave. Hence, the time between the arrivals of successive electromagnetic wave crests at the radar is reduced, causing a reduction in the phase in the received electromagnetic wave. On the contrary, if the target is moving away from the radar, each electromagnetic wave is reflected from a position farther from the radar than the previous electromagnetic wave, so the arrival time between successive electromagnetic waves is increased, increasing the phase. By calculating the phase evolving among a sequence of defections from a target, a radar is able to obtain the relative velocity of the target.

It is usually not sufficient to obtain the relative range and velocity of a target for automotive applications. In order to make a proper decision, such as performing an emergency brake, the vehicle also needs to know the location of targets in the 3D space. As shown in Figure 2.1, an automotive radar measures the location of a target (P) in a spherical coordinate system. The $+x$-axis is the boresight direction, which is usually perpendicular to the radar's antenna board. The angle ϕ on the $x-y$ plane is the azimuth angle, and angle θ is the elevation angle. There exist several techniques to obtain the azimuth and elevation angles of targets in a radar system. One of the most well-known methods is using a mechanical rotator [1]. In this method, a radar

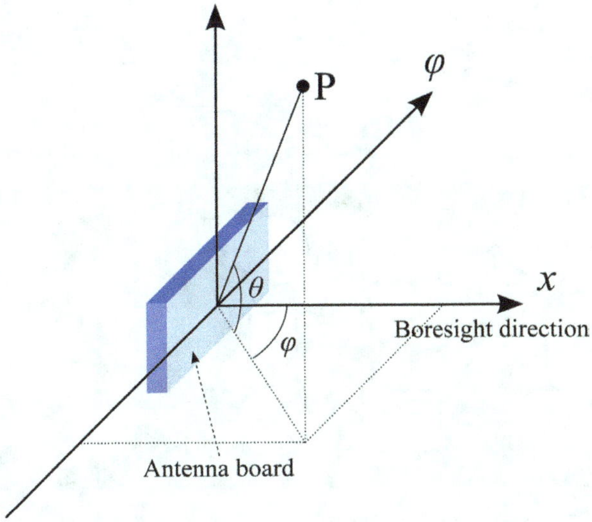

Figure 2.1 Spherical coordinate system for radar measurement

with a very narrow radiation beam is mounted on a rotator to mechanically scan the environment. The target's relative elevation and azimuth angles are given by the rotator's position. Mechanical scanning radar was widely used in military applications; however, due to its bulky size, it is not suitable for automotive applications, which require the radar to be compact and low cost. Other methods for angle measurements include phased-array, digital beamforming, as well as multiple-input and multiple-output techniques, which don't require any mechanical rotation structure and are highly integrated [1]. A brief introduction of phased-array and digital beamforming will be given in this chapter. Multiple-input and multiple-output techniques will be discussed in detail in the following chapter.

2.2 Automotive radar architecture

From a very top level, a basic automotive radar consists of a transmitter, a receiver, and the antennas. Figure 2.2 illustrates a simplified single-channel CW radar architecture [2]. This architecture applies to most of the CW radars. The actual hardware realization of the different parts of the radar could be different based on various types of waveforms. On the transmitter (Tx) side, a signal synthesizer is used to generate different waveforms. Then, the generated waveform is amplified and transmitted by the transmitter antenna. There is a portion of the generated signal fed into the receiver, serving as the local oscillator (LO). On the receiver (Rx) side, the receiver antenna picks up the reflected signals from targets. The received signals are first conditioned by a series of amplifying and filtering on the receiver channel. It is then followed by a quadrature down converter, which mixes the received signals with

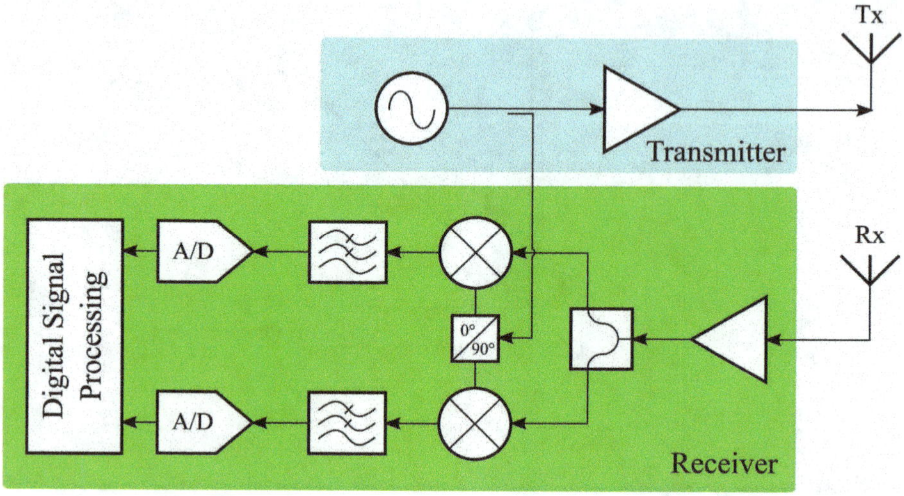

Figure 2.2 Simplified single-channel radar architecture

the LO to create the baseband signal. After down-conversion, the baseband signal is sampled into the digital domain for further processing.

As it has been mentioned, the simplified architecture in Figure 2.2 is not applicable to all types of CW radars due to their various waveforms. For example, in order to support multiple channels, many designs have an extra modulation in the transmitter to help the receiver separate the signals from different transmitter channels. Some designs have phase shifters in the transmitter channels to support beamforming. Another example is that on the receiver side, the quadrature down-converter is not always necessary, in some cases, a single-channel direct down-converter also works.

2.2.1 Transmitter

For an automotive radar, its transmitter plays a major role in determining the sensitivity and range resolution of the whole system. Using a higher transmitting power and high gain antennas improves the signal and increases radar's capability of detecting a smaller target at a longer range. Moreover, transmitting wider BW improves radar's capability of discriminate targets that are close in range. However, the frequency bands and maximum transmitted power used by the automotive radar transmitters are also highly regulated by the authorities. As electromagnetic waves are widely used in modern technology, particularly in telecommunication, to prevent interference between different users, the generation and transmission of electromagnetic waves are strictly regulated by national laws, coordinated by an international body, the International Telecommunication Union (ITU). For automotive radar applications, there are four dedicated frequency bands across the globe that include the 24 and 77 GHz bands. Table 2.1 shows the currently available frequency bands for automotive radars.

Table 2.1 Automotive radar frequency bands

Frequency	Bandwidth	Type of band
24–24.25 GHz	250 MHz	ISM Band
21–26 GHz	5 GHz	UWB
76–77 GHz	1 GHz	Automotive LRR Band
77–81 GHz	4 GHz	Automotive SRR Band

The 250 MHz BW from 24 to 24.25 GHz is the reserved industrial, scientific and medical (ISM) band, which is shared with unlicensed and licensed operations. Since it is a shared band, it suffers from the high likelihood of interference. The 5 GHz ultra-wide band (UWB) from 21 to 26 GHz has a very stringent requirement on a low transmit power, which limits its applications on short range. Two higher frequency bands, including the 76–77 GHz band for long range radar (LRR) and the 77–81 GHz band for short range radar (SRR), have been allocated by the authorities in most of the countries. These two higher frequency bands are the frequency bands of choices for automotive radars due to the following advantages. First, the 77 GHz frequency band ranges from 76 to 81 GHz, with a BW over 4 GHz. The wide BW increases range resolution of the radar, allowing it to discriminate against closely spaced targets. With the higher frequency, the resolution and accuracy of the velocity measurement are also improved due to the shorter wavelength. Moreover, the shorter wavelength also helps to reduce the antenna size of the automotive radars. Since the 77 GHz band is dedicated for automotive radar applications, the regulation also allows for a higher transmit power in this band. The authorized maximum effective isotropic radiated power (EIRP) for automotive radars operating at 79 GHz is 55 dBm and the worst case mean EIRP spectral density is below –3 dBm/MHz [3]. On the other hand, the 24 GHz band has a peak limitation of 20 dBm EIRP.

The signal synthesizer, which is also named the waveform generator, is one of the most important components for an automotive radar. The signal synthesizer is capable of generating various types of waveforms for different applications. Figure 2.3 illustrates four popular examples of the waveforms. The first waveform is a single tone signal, which is an un-modulated sinusoidal wave used in Doppler radars. The FMCW waveform is a linear modulated signal, whose frequency is changing linearly with the time. The stepped-frequency continuous-wave waveform has the frequency increases/decreases in a stair shape. The frequency-shift keying waveform has alternating frequencies. For automotive radars, FMCW waveform is currently the most widely used waveform due to its convenience in generation by a phase-locked loop (PLL), as well as the high efficiency in obtaining the range information from the baseband. In this chapter, the FMCW waveform will be used as the main example in the discussions, though most of the principles are applicable to the other waveforms.

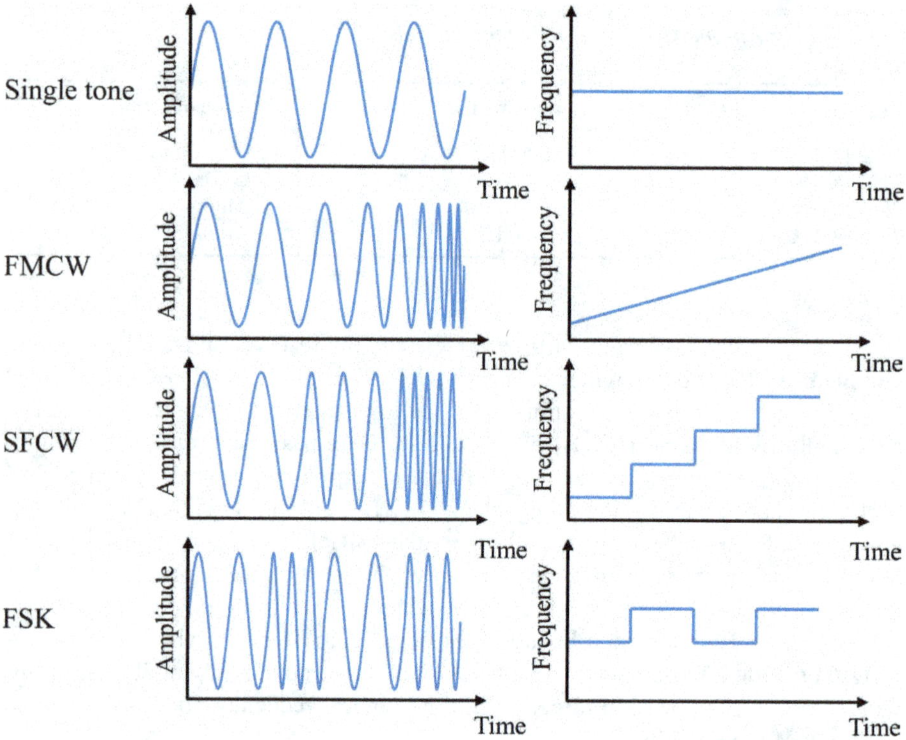

Figure 2.3 Different types of waveforms

2.2.2 Receiver

The main purposes of the receiver are coherently down-converting the received echo and obtaining the baseband signal. As shown in Figure 2.2, the received signal is split into two channels. One channel is mixed with the LO signal, which is a replica of the transmit signal, to obtain the in-phase or "I" channel of the baseband. The other channel is mixed with a 90° phase shifted LO signal to obtain the quadrature phase or "Q" channel of the baseband. Assume the transmit signal is $f(t)$, the received echo $r(t)$ can be written as

$$r(t) = A \cdot f(t - \delta_t) \tag{2.2}$$

where t is time, δ_t is the round-trip time delay of the electromagnetic wave between the radar and the target, and A is the signal amplitude change. After the mixer, the baseband signal r_b(t) can be simply expressed as

$$r_b(t) = A \cdot f(t) \cdot f^*(t - \delta_t) \tag{2.3}$$

where $*$ denotes the conjugate of the complex signal. In the following sections, the FMCW waveform will be used to discuss (2.2) and (2.3) in detail.

In a modern automotive radar, digital signal processing is commonly used to deal with the baseband signal. Thus, it is necessary to convert the analog baseband signal into a digital representation. The most fundamental question in analog-to-digital conversion is choosing a suitable sampling rate. The guidance has been provided by the Nyquist sampling theorem [4], which will not be discussed here.

2.2.3 Antenna and antenna array

In a radar system, its antenna or antenna array plays an essential role in determining the sensitivity and angular resolution. A wide variety of types of antennas have been used in radar systems. For automotive radars, patch antennas are the most widely used due to their low profile and ease of fabrication. Figure 2.4 demonstrates two examples of patch antennas [5]. Figure 2.4(a) is a 3-element series-fed patch antenna, and Figure 2.4(b) is a 5-element series-fed patch antenna. The most important properties of an antenna are its gain, beamwidth, and side lobe level. The antenna far-field radiation pattern $P(\theta, \phi)$ is usually used to describe antenna's radiation intensity in the direction (θ, ϕ) relative to the antenna boresight.

(a) (b)

Figure 2.4 Examples of patch antennas [5] (a) 3-element series-fed patch antenna and (b) 5-element series-fed patch antenna.

Figure 2.5 shows the far-field radiation pattern of the 3-element series-fed patch antenna in Figure 2.4(a). The E-plane corresponds to the plane parallel to the electrical field, which is parallel to the vertical edge of the antenna in Figure 2.4(a). The H-plane is the plane parallel to the magnetic field, which is perpendicular to the vertical edge of the antenna in Figure 2.4(a). The half-power beam width (HPBW) on E-plane is 48.6°, and the HPBW on H-plane is 81.5°. For the antenna in Figure 2.4(b), it has a higher gain but a narrower E-plane HPBW due to its longer length.

Besides the popular patch antennas, other types of antennas, such as substrate integrated waveguide antenna and lens antenna, are also widely used in automotive radars. Table 2.2 lists several 77 GHz low-profile automotive radar planar antennas in the literature.

An antenna array is a collection of cooperating antenna elements. An antenna array enables a radar's capability in finding the targets' angles related to the radar. By controlling the phase and amplitude of each array element, the array is able to steer the main beam to the angle of interest or even form multiple beams. On the other hand, the directions of the nulls in an array can also be adjusted, and they can be used to reject strong interference. A simple example of an eight-element linear array is shown in Figure 2.6. Assume that the array elements are isotropic radiation elements. The array factor of this linear array can be expressed as [1]:

$$AF(\varphi) = \sum_{n=1}^{N} \omega_n \cdot e^{jk\psi_n(\varphi)} \tag{2.4}$$

$$\psi_n(\varphi) = y_n \cdot \sin\varphi \tag{2.5}$$

where N is the number of elements and $k = 2\pi/\lambda$ is the wave number. λ is the wavelength, y_n is the location of the n-th element, and φ is the azimuth angle. ω_n is the weight of the n-th element, which corresponds to the properties, i.e., phase and amplitude, of the excitation.

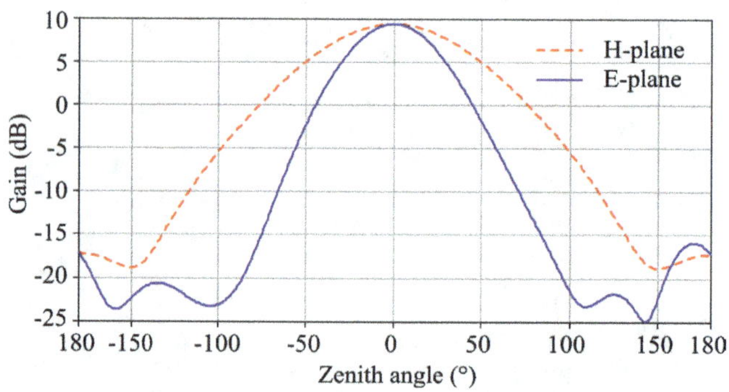

Figure 2.5 Simulated H- and E-plane patterns of the 3-element series-fed patch antenna [5]

Table 2.2 Automotive radar frequency bands

Reference	Antenna type	Frequency (GHz)	Gain (dBi)	HPBW° (*E*/*H* plane)
[6]	Folded reflector antenna	55–64, 76–77	34, 35	3.2/3.4, 2.9/3
[7]	Lens antenna	76.4–76.6	28	4/4.4
[8]	Low-temperature cofired ceramic (LTCC) patch array	77–81	13	7
[9]	Integrated lens antenna	73–94	20	18
[10]	Dual-layer transmit-array	75–78	18.5	15
[11]	Dielectric flat lens antenna	57–64, 76–80	15, 18.9	21, 20
[12]	Hybrid thin-film antenna	76–85	11	12/40
[13]	Series-fed patch array	75–80	15.9	4.5/12
[14]	Planar grid array	77–81	15	5
[15]	Microstrip antenna	75.6–78.5	21.7	15
[16]	Microstrip patch array	76–81.4	14.8	–
[17]	Circular-shaped patch array	76–79.5	21	19/9.5
[18]	Microstrip planar array	76.5–81.5	10	–
[19]	Patch antenna array	76–81	11.5	8.1

Consider a common case when the spacing between adjacent elements $d = \lambda/2$, (2.4) can be simplified to

$$AF(\varphi) = \sum_{n=1}^{N} \omega_n \cdot e^{j(n-1)\pi \sin\varphi} \tag{2.6}$$

With a uniform excitation, where,

$$\boldsymbol{\omega} = [\omega_1, \omega_2, ..., \omega_8] = [1, 1, ..., 1] \tag{2.7}$$

The corresponding array factor is illustrated in Figure 2.7. The main lobe of the pattern directs to 0°, and the sidelobe level is about −11 dB. If the phases of the elements are tuned, e.g.,

$$\omega_n = e^{-jn\pi \sin(-20^\circ)} \tag{2.8}$$

The corresponding array factor is illustrated in Figure 2.8, where the main lobe of the pattern is steered to −20°.

A more complex beamforming can be realized by tuning both the phase and amplitude of each element. An example shown in Figure 2.9 is the weighting in (2.8) multiplied with a 40 dB sidelobe level Dolph–Chebyshev window [20].

Figure 2.6 Eight-element linear array

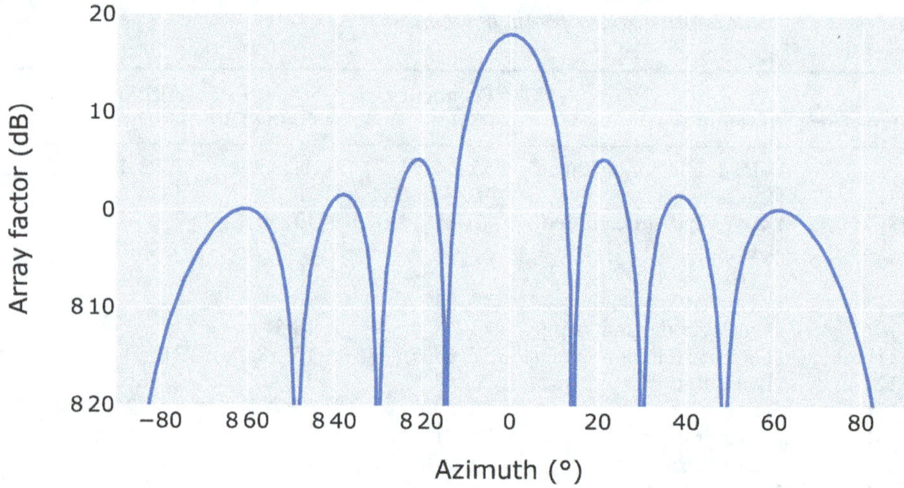

Figure 2.7 Array factor with a uniform excitation

The array factor assumes that the array elements are isotropic radiation elements. For an array with realistic radiation elements, its full radiation $P_{array}(\varphi)$ is the multiplication between the array factor and the element's radiation pattern $P(\varphi)$ [1]:

$$P_{array}(\varphi) = P(\varphi) \cdot AF(\varphi) \tag{2.9}$$

2.3 Signal models

A radar's transmitter usually sends out a carefully designed and well-defined signal. However, the received return signal is the superposition of several components,

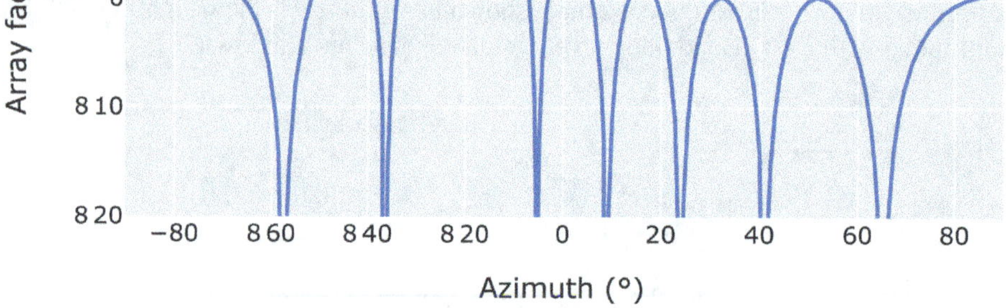

Figure 2.8 Array factor with the weighting of (2.8)

including targets' reflections, clutters, noises, and in some cases, interference. None of these components is entirely under the control of the radar's designer. The ultimate goal of radar signal processing is to extract useful information regarding the presence of targets and their characteristics. The presence of noise and interference degrades the ability or accuracy of measuring targets' characteristics. Various figures of merits, such as detection probability and signal-to-noise ratio (SNR), can be used to evaluate the effectiveness of radar system design and its signal processing.

For a conventional CW radar, assume it transmits a well-designed waveform, whose frequency can be expressed as $f(t)$ with $-T_0/2 \leqslant t < T_0/2$, where T_0 is the duration of the waveform. It is easy to know that for a single tone signal

$$f(t) = f_c \tag{2.10}$$

where f_c is the center frequency.

An FMCW waveform can be written as

$$f(t) = f_c + \frac{BW}{T_0} t \tag{2.11}$$

The instantaneous phase of the transmitted signal $\phi(t)$ is the integration of $f(t)$,

$$\phi(t) = 2\pi \cdot \int_{T_0/2}^{t} f(t)dt \tag{2.12}$$

and the time domain transmitted signal is

$$x(t) = a(t) \cdot e^{j[\phi(t)+\theta(t)]} \tag{2.13}$$

where $a(t)$ is the amplitude of the transmitted signal and $\theta(t)$ is the additional phase of the transmitted signal. Based on the design of the waveform, $a(t)$ and $\theta(t)$ can be either constant values or time-varying modulations.

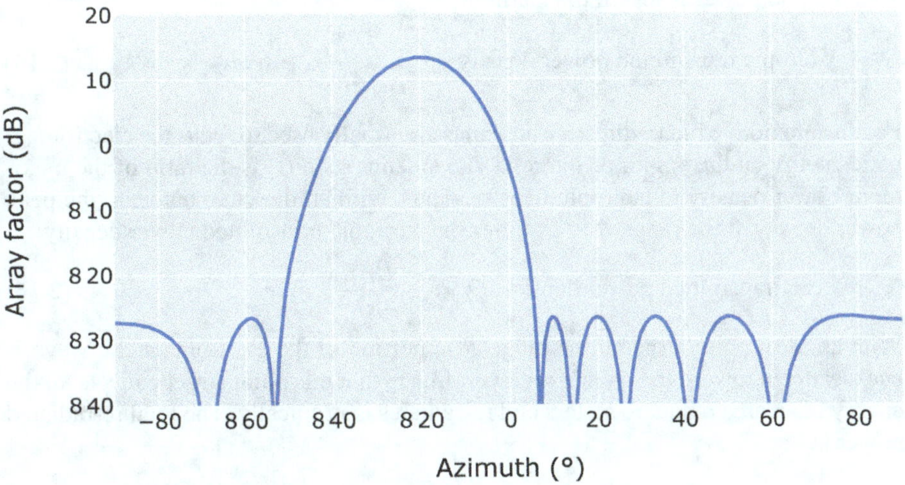

Figure 2.9 Array factor with the weighting of (2.8) multiplied with a Dolph–Chebyshev window

The received signals are delayed echoes of the transmitted waveform, whose amplitude and phase modulation could be altered by propagation loss and Doppler shift. Receiver noise can be treated as an additive random signal. The full receive signal can be modeled as

$$y(t) = \sum_i b_i(t) \cdot e^{j[\phi(t-t_i)+\varphi_i(t)]} + n(t) \tag{2.14}$$

where $t_i = 2R_i/c$ is the round-trip time delay of the i-th reflector, R_i is the range of the i-th reflector, $n(t)$ is receiver noise, $b_i(t)$ is the amplitude of the i-th reflector, and $\varphi_i(t)$ is the phase modulation of the i-th reflector. The important parameters in $y(t)$ are the delay time t_i, signal amplitude $b_i(t)$, and the phase modulation $\varphi_i(t)$. The target's range can be estimated from t_i. The target's size determines the signal amplitude $b_i(t)$, and its motion status varies the phase modulation term $\varphi_i(t)$. It should be noted that i doesn't represent the i-th target. In practice, one target consists of multiple reflection surfaces.

In order to design effective signal processing algorithms, it is necessary to have good models of the signals to be processed. In this chapter, several models of radar signal characteristics will be introduced for a better understanding of radar measurements.

2.3.1 Amplitude models

2.3.1.1 Point target radar range equation

The radar range equation is one of the most widely used tools for basic radar system design analysis. The radar range equation describes the physical dependence of electromagnetic power from the transmitter to the receiver.

To derive the range equation, assume an isotropic radiating element transmits a waveform with the power P_t watts into a lossless medium. The power density at a range R is the total power P_t divided by the surface area of a sphere of radius R assuming no power is lost in the medium:

$$\text{Isotropic transmitted power density: } Q_{iso} = \frac{P_t}{4\pi R^2} \quad W/m^2 \tag{2.15}$$

For an automotive radar, directive antennas are usually used to focus the electromagnetic energy on the angles of interest. The antenna gain G_t is the ratio of the maximum power density to isotropic density. Thus, with an directive antenna, the peak power density at the range R is G_t times the isotropic transmitted power density:

$$\text{Peak transmitted power density: } Q_{peak} = \frac{P_t G_t}{4\pi R^2} \quad W/m^2 \tag{2.16}$$

With an ideal point target at the range R, a portion of the electromagnetic wave is backscattered toward the radar's receiver. Image that this point target collects all the energy of an area σ square meters and re-radiates isotropically. The total re-radiated power is

$$\text{Total re-radiated power: } P_{back} = \frac{P_t G_t \sigma}{4\pi R^2} \quad W \tag{2.17}$$

The quantity σ is called the radar cross section (RCS) of the target. RCS is the measure of a target's ability to reflect radar signals in the direction of the radar receiver. It is the ratio of backscattered density in the direction of the radar (from the target) to the power density that is intercepted by the target. Thus, RCS is usually not equal to the physical cross sectional area of the target. In addition, RCS is also defined under the assumption that the backscattered power is re-radiated isotropically. Similar to (2.15), the power density on the radar's receiver at a range R can be derived by dividing the power of (2.17) by the surface of a sphere with a radius R,

$$\text{Backscattered power density: } Q_{back} = \frac{P_t G_t \sigma}{(4\pi)^2 R^4} \quad W/m^2 \tag{2.18}$$

If the aperture of the radar's receiver antenna is A_e, the total backscattered power collected by the receiver antenna will be

$$\text{Received power: } P_r = \frac{P_t G_t A_e \sigma}{(4\pi)^2 R^4} \quad W \tag{2.19}$$

According to the definition of the effective aperture size, the relation between the effective aperture size A_e and receiver antenna gain G_r is

$$A_e = \frac{\lambda^2 G_r}{4\pi} \tag{2.20}$$

where λ is the wavelength. Equation (2.19) can thus been re-written as

$$P_r = \frac{P_t G_t G_r \lambda^2 \sigma}{(4\pi)^3 R^4} \quad W \tag{2.21}$$

which describes an ideal case that the radar's electromagnetic wave doesn't suffer any additional loss in the atmosphere. Moreover, the radar itself here is also an ideal system without additional loss from the components or gain from signal processing.

The radar range equation is a fundamental tool for radar system design and analysis. For example, through (2.21), it can be seen that the received power decreases as the fourth power of range between the radar and the target. Thus, in order to increase the detection range of a given RCS target, the transmit power P_t or the antenna gains G_t, G_r needs to be increased. About 12 dB increment from $P_t G_t G_r$ can only double the detection range of the given RCS target. It is always challenging for a radar system designer to make such improvements.

As mentioned above, (2.21) doesn't include the gain contributed by signal processing. A well-designed signal processing algorithm can increase the effective received power, e.g., improving the SNR, and therefore increase the detection range. The effect of signal processing will be discussed in the following sections.

2.3.1.2 Radar cross section

The concept of RCS has been intuitively introduced above, which is a representative of the amount of power re-radiated toward the radar's receiver.

Assume the incident power density at the target's location is Q_{peak}, as shown in (2.16) and the re-radiated power density at radar's receiver with the range R is Q_{back}, as shown in (2.18). RCS σ is a portion of the area over the incident power density. The relation among Q_{peak}, Q_{back}, and σ should satisfy:

$$Q_{back} = \frac{\sigma Q_{peak}}{4\pi R^2} \tag{2.22}$$

It is assumed that the backscattered power density is from isotropic radiation of the target. Equation (2.22) can thus been re-written as

$$\sigma = 4\pi R^2 \frac{Q_{back}}{Q_{peak}} \tag{2.23}$$

The definition of RCS is usually written in terms of electric fields. The range R tends to infinity to make the definition depends only on the target's characteristics.

$$\sigma = 4\pi \lim_{R\to\infty} \left(R^2 \frac{|\mathbf{E}^r|^2}{|\mathbf{E}^t|^2} \right) \tag{2.24}$$

where $|\mathbf{E}^r|^2$ and $|\mathbf{E}^t|^2$ are the backscattered and incident electric field squared magnitudes, respectively. Typical values of RCS range from 0.01 m^2 to hundreds of square meters. It is common to use decibels (dBsm, dB square meter) for RCS values:

$$1\ m^2 = 10\log_{10}(1)\ \text{dBsm} \tag{2.25}$$

The RCS defined above is a real scalar number. A more generalized definition of RCS is the polarization scattering matrix (PSM) S, which describes the polarization state of the backscattered electromagnetic field. The relation between the incident electromagnetic field and backscattered electromagnetic field can be described by the PSM:

$$\begin{bmatrix} E_H^s \\ E_V^s \end{bmatrix} = \begin{bmatrix} S_{HH} & S_{HV} \\ S_{VH} & S_{VV} \end{bmatrix} \begin{bmatrix} E_H^i \\ E_V^i \end{bmatrix} \tag{2.26}$$

where the subscript H represents horizontal polarization, and subscript V represents vertical polarization. The four elements of the scattering matrix are complex and can be obtained from the magnitudes and phases measured by the four channels of a polarimetric radar. The discussion of polarimetric techniques is beyond the scope of this book. Hence, in this book, it is assumed that only a single fixed polarization is transmitted and a single fixed polarization received.

A real target's RCS is a function of aspect angle, frequency, and polarization. It can't be simply modeled as a constant. In theory, the RCS of a target can be determined by solving Maxwell's equations with proper boundary conditions. However, only objects with simple geometries can be determined in this way. The approximate formulas for the RCS of some simple objects are shown in Table 2.3.

The RCS of a large complex target is highly dependent on the aspect angle and frequency. Figure 2.10 shows a simple example of a trihedral corner reflector, whose edge length is $a = 0.1$ m. Figure 2.11 illustrates its RCS versus the frequency when the radar is facing the center of the reflector. The RCS of this

Table 2.3 Approximate forms for the RCS of simple objects [21]

Object	Aspect	RCS	Symbol
Sphere	Any	πa^2	a: Radius
Cone	Axial	$\frac{\lambda^2}{16\pi}\tan^4\theta$	θ: Cone half angle
Paraboloidal	Axial	πa^2	a: Apex radius of curvature
Cylinder	Normal to axis	$\frac{2\pi a L^2}{\lambda}$	a: Radius, L: Length
Large flat plate	Normal	$\frac{4\pi A^2}{\lambda^2}$	A: Plate area
Square plate	Angle to normal	$\frac{4\pi a^4}{\lambda^2}\left[\frac{\sin(ka\sin\theta)}{ka\sin\theta}\right]$	a: Length of side, k:$\frac{2\pi}{\lambda}$
Circular plate	Angle to normal	$\frac{\pi a^2}{\tan^2\theta}J_1^2(2ka\sin\theta)$	a: Radius, J_1: Bessel function of the first order
Dihedral	Maximum direction	$\frac{8\pi a^2 b^2}{\lambda^2}$	a, b: Edge length
Trihedral	Maximum direction	$\frac{4\pi a^4}{3\lambda^2}$	a: Edge length
Square trihedral	Maximum direction	$\frac{12\pi a^4}{\lambda^2}$	a: Edge length

reflector is –3 dBsm at 10 GHz frequency, and its RCS reaches to 14 dBsm at 77 GHz.

The observation angle is another important factor to affect the RCS of a complex object. As shown in Figure 2.12, or the same trihedral reflector, its RCS reaches to the maximum when observed at the center of the reflector and degrades with the

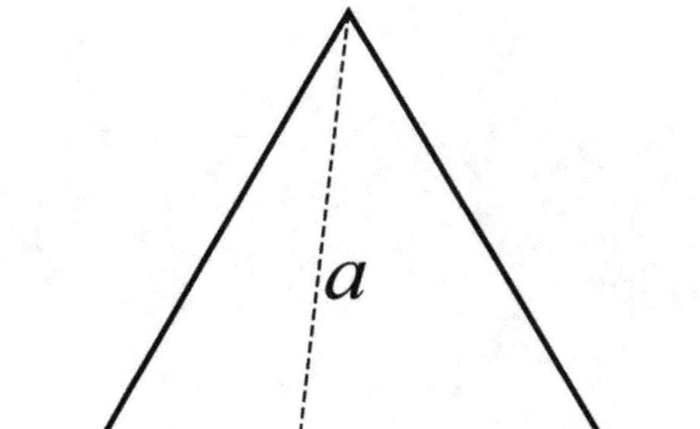

Figure 2.10 Trihedral corner reflector

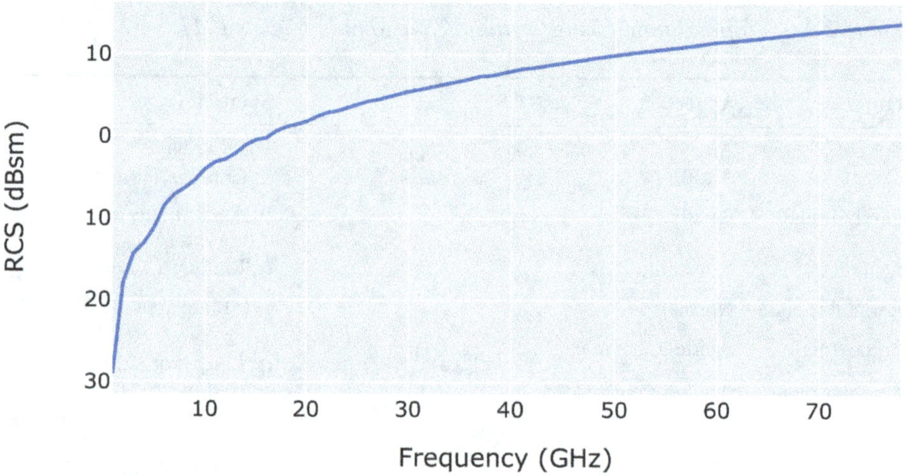

Figure 2.11 *Trihedral corner reflector's RCS versus frequency*

change of the observation angle. The RCS reduces by more than 70 dB when looking at the side of the reflector.

For an even more complex target, such as a passenger car for automotive applications, its RCS varies more rapidly from angle to angle. Figure 2.13 is an example of a passenger car's RCS from different observation angles at 77 GHz. 0° means observing from the front of the car, and 180° is the back of the car. The largest RCS is at 90°, which is from the side of the car.

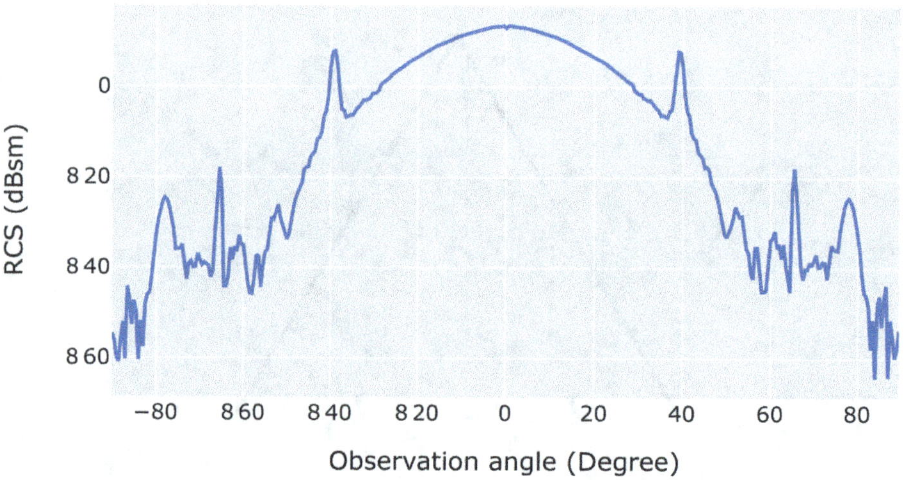

Figure 2.12 *Trihedral corner reflector's RCS versus observation angle at 77 GHz*

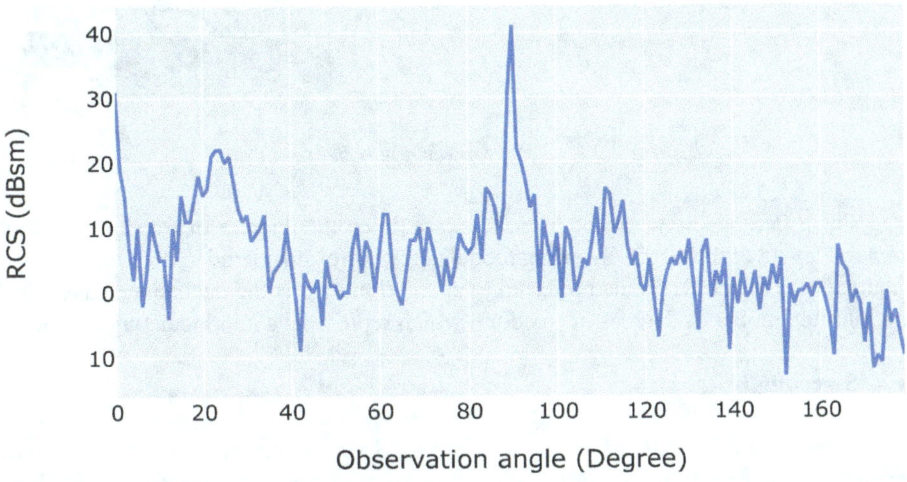

Figure 2.13 *RCS of a passenger car at 77 GHz*

2.3.1.3 Swerling models

A single point target in space with a given radar RCS has been used in the radar equation models above for the basic analysis. It has also been discussed that the RCS of a real object is difficult to estimate except for simple shapes, as shown in Table 2.3. Before the introduction of detailed computer modeling, the RCS for real-world objects was generally measured instead of calculated. However, the measured RCS in a controlled environment fails to account for real-world effects due to the radar signal reflecting off multiple points on the target. Figure 2.14 illustrates a simple example of a passenger car with two reflection paths r and r'. Due to the relative angles of r and r' to the normal of the radar, the difference between r and r' will change when the range of the car changes. This results the received signal being amplified or diminished depending on the difference of r and r'. As the target moves in relation to the radar, these distances change and create a constantly changing signal. Fluctuation loss is an effect seen in radar systems as the target object moves or changes its orientation relative to the radar system. It was extensively studied during the 1950s by Peter Swerling, who introduced the Swerling models to allow the effect to be simulated [22, 23].

The Swerling target models treat a target as a number of individual radiators and consider the RCS σ using the chi-squared distribution

$$p(\sigma) = \frac{o}{2\Gamma(o/2)\sigma_{mean}} \left(\frac{o\sigma}{2\sigma_{mean}}\right)^{o/2-1} e^{-\frac{o\sigma}{2\sigma_{mean}}} I_{[0,\infty)}(\sigma) \tag{2.27}$$

where σ_{mean} is mean value of σ, $\Gamma(o/2)$ denotes the gamma function, and $I_{[0,\infty)}(\sigma)$ is the modified Bessel function of the first kind. o is the degree of freedom. The standard deviation of chi-squared distribution is $\sqrt{2o}$, and the mean is o. The ratio of the standard deviation to the mean value is equal to $1/\sqrt{o/2}$, which means larger values of o will result in smaller fluctuations. If $o = \infty$, the target's RCS is nonfluctuating.

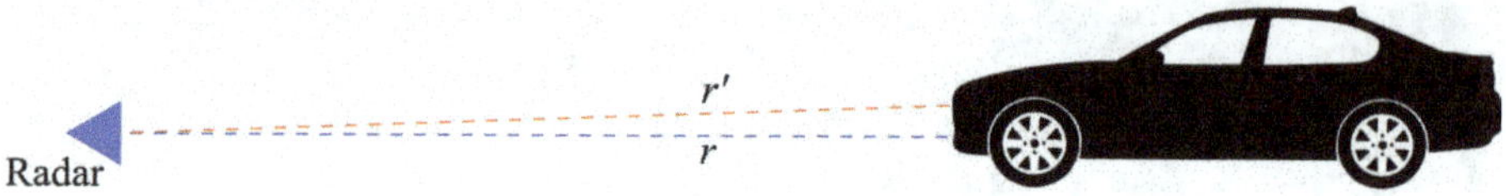

Figure 2.14 Passenger car reflection paths

The difference between the models is largely on the degrees of freedom and the general layout of the target. Swerling models I–IV were considered in Swerling's original paper. The model V, also referred to as the 0 model, is the degenerate case with an infinite number of degrees of freedom, which represents a nonfluctuating target.

- **Swerling I**

Swerling I describes the case where the RCS varies according to a chi-squared probability density function (pdf) with two degrees of freedom ($o = 2$). This applies to a target that is made up of many independent reflectors with similar intensity. For this model, the target's velocity is low compared to the observation time. The RCS is constant from pulse-to-pulse but varies independently from scan to scan. The pdf of the RCS is reduced to the Rayleigh function:

$$p(\sigma) = \frac{1}{\sigma_{mean}} e^{-\frac{\sigma}{\sigma_{mean}}} \tag{2.28}$$

- **Swerling II**

Swerling II model is similar to Swerling I, except that the RCS value varies faster, from pulse-to-pulse, instead of scan-to-scan.

- **Swerling III**

Swerling III describes the case where the RCS varies according to a chi-squared pdf with four degrees of freedom ($o = 4$). This applies to a target that is made up of many independent reflectors with one large dominant reflector. For this model, the

Table 2.4 Swerling target models

Model	Degree of freedom	Fluctuation rate	Nature of scattering
I	2	Slow	Several independent reflectors of similar intensity
II	2	Fast	
III	4	Slow	Several independent reflectors and one dominate
IV	4	Fast	

target's velocity is low compared to the observation time. The RCS is constant from pulse-to-pulse but varies independently from scan-to-scan. The pdf of the RCS is

$$p(\sigma) = \frac{4\sigma}{\sigma_{mean}^2} e^{-\frac{2\sigma}{\sigma_{mean}}} \tag{2.29}$$

- **Swerling IV**

Swerling IV model is similar to Swerling III, except that the RCS value varies faster, from pulse-to-pulse, instead of scan-to-scan.

- **Swerling 0/V**

Swerling V (also known as Swerling 0) describes a constant RCS, corresponding to infinite degrees of freedom ($o = \infty$).

Table 2.4 provides a summary of the Swerling target models from Swerling models I to IV. For automotive radar applications, passenger cars, trucks, and motorcycles are usually considered as Swerling III targets. Cyclists and pedestrians are usually considered as Swerling I targets.

2.3.2 Noise model

Noise is inevitable for any type of receiver. In a radar system, the echo signal received from a target needs to compete with noise to get identified. The sources of noise in an automotive radar include external noise received by the receiver antenna, as well as internal noise generated in the radar receiver itself.

External noise could be classified into atmospheric noise, extraterrestrial noise, and manmade noise. Atmospheric noise caused by natural atmospheric processes, primarily lightning discharges in thunderstorms. The energy of this kind of noise is usually spread over the spectrum. Extraterrestrial noise includes cosmic and solar noise. Manmade noise is the noise that comes from human activities.

Internal noise includes thermal noise, shot noise, partition noise, and flicker noise. Thermal noise is due to ohmic losses. Shot noise is produced by the random arrival of electrons or holes at the collector or drain in a transistor. It is also caused by the random movement of electrons or holes across a PN junction. Partition noise occurs whenever current has to divide between two or more paths and results from the random fluctuations in the division. Flicker noise is a type of electronic noise with a $1/f$ power spectral density, and it is usually related to a direct current (dc), as resistance fluctuations are transformed to voltage or current fluctuations by Ohm's law [24].

Among the above-mentioned various noise sources, thermal noise is usually dominant. Thermal noise can be approximated as a white noise [25]. The power spectrum S_n of the thermal noise is

$$S_n = k_B T \quad \text{W/Hz} \tag{2.30}$$

where $k_B = 1.38 \times 10^{-23}$ J/K is Boltzmann's constant, and T is the temperature of the noise source in Kelvin (K). In many practical systems, a "standard" temperature $T_0 = 290$ K is used.

A radar receiver doesn't have an infinite BW, and a real filter doesn't have a perfect rectangular pass-band. To analyze noise power, the noise-equivalent BW β_n of a filter is used. The noise equivalent BW is the width of an ideal rectangular filter, with gain or loss G equal to the peak gain or loss of the actual filter. The total noise power P_N can be written as

$$P_N = k_B T \beta_n G \tag{2.31}$$

For the whole receiver chain, the power spectral density of the noise at the final output can also be described as the product of Boltzmann's constant and an equivalent temperature T'. The total noise power at the output of the whole receiver chain will be

$$P_N = k_B T' \beta_n G_s \tag{2.32}$$

where G_s is the gain of the whole receiver chain. In (2.32), the total noise power can be separated into two parts: the external thermal noise, which is $k_B T_0 \beta_n G_s$, and the internal noise generated by the receiver itself, which can be written as $k_B T_e \beta_n G_s$ that is the effective temperature of the system.

$$P_N = k_B T_0 \beta_n G_s + k_B T_e \beta_n G_s \tag{2.33}$$

The noise temperature description of noise power is useful for a low-noise receiver. A more common description in radar is using noise figure NF. Noise figure is the ratio of the noise power at the output of the receiver to the noise power contributed by the external noise $k_B T_0 \beta_n G_s$:

$$NF = \frac{P_N}{k_B T_0 \beta_n G_s} \tag{2.34}$$

Noise figure is one of the most important parameters to characterize a radar's performance. Detailed discussions on using noise figures to analyze the radar's performance will be presented in the following sections.

2.4 Radar waveforms and signal processing

For a CW radar, the waveform determines its basic signal processing flow as well as some of the key capabilities. In this section, the FMCW waveform will be used as an example to discuss the signal model and basic signal processing flow.

Figure 2.15 plots an FMCW waveform example. This waveform consists of a series of chirps. For each chirp, its frequency versus time relation is defined as

$$f(t) = f_c + \frac{BW}{T_0} t \tag{2.35}$$

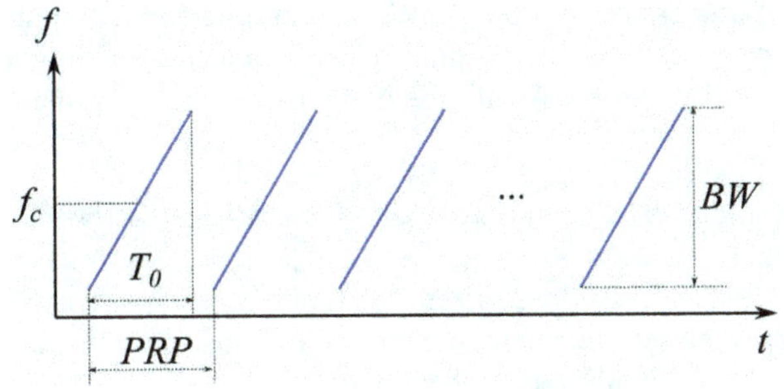

Figure 2.15 A typical FMCW waveform

where $-T_0/2 \leqslant t < T_0/2$. T_0 is the chirps length, f_c is the center frequency, and BW is the bandwidth.

Based on (2.12), the instantaneous phase of the transmitted signal $\phi(t)$ is the integration of $f(t)$:

$$\phi(t) = 2\pi \int_{-T_0/2}^{t} f(t)dt = 2\pi(f_c t + \frac{BW}{2T_0}t^2) + C \tag{2.36}$$

and,

$$-T_0/2 \leqslant t < T_0/2 \tag{2.37}$$

where C is a constant from the indefinite integral and can be ignored in the following calculations. The time domain transmitted signal is

$$s_{tx}(t) = A_{tx}e^{j\left[2\pi(f_c t + \frac{K}{2}t^2) + \theta\right]} \tag{2.38}$$

where,

$$K = \frac{BW}{T_0} \tag{2.39}$$

is the slope of the chirp.

2.4.1 Range processing

Assume a stationary target located at range R. The signal reflected from the target has a time-domain representation of

$$s_{rx}(t) = A_{rx} \cdot w_R(t) \cdot e^{j\left\{2\pi\left[f_c(t-\delta t_0) + \frac{K}{2}(t-\delta t_0)^2\right] + \varphi\right\}} \tag{2.40}$$

where $\delta t_0 = 2R/c$ is the round-trip delay of the electromagnetic wave, and φ is the phase. $w_R(t)$ is a rectangular window function:

$$w_R(t) = \begin{cases} 1, & -T_0/2 \leqslant t < T_0/2 \\ 0, & t < -T_0/2 \text{ or } t \geqslant T_0/2 \end{cases} \tag{2.41}$$

The de-chirping processing in the FMCW radar architecture is mixing the received signal with the original transmitted signal. The mathematical expression of the baseband signal is multiplying $s_{tx}(t)$ with the complex conjugate of the received signal $s_{rx}^*(t)$:

$$s_b(t) = s_{tx}(t) \cdot s_{rx}^*(t) = A_b \cdot w_R(t) \cdot e^{j(2\pi f_c \delta t_0 + 2\pi K t \delta t_0 - \pi K \delta t_0^2 + \varphi_b)} \tag{2.42}$$

Since $\pi K \delta t_0^2 = 4\pi K R^2/c^2$ is close to zero for typical automotive applications, it can be discarded. φ_b is the residue phase, which is not of interest in this case and can thus be discarded too. The simplified baseband signal is

$$s_b(t) = s_{tx}(t) \cdot s_{rx}^*(t) = A_b \cdot w_R(t) \cdot e^{j(2\pi f_c \delta t_0 + 2\pi K t \delta t_0)} \tag{2.43}$$

After performing Fourier transform to the baseband signal, the frequency domain representation of the baseband signal is

$$S_b(f) = A_b \cdot T_0 \cdot e^{j(2\pi f_c \delta t_0)} \cdot \mathrm{sinc}\left[T_0(f - K\delta t_0)\right] \tag{2.44}$$

$$= A_b \cdot T_0 \cdot e^{j(2\pi f_c \delta t_0)} \cdot \mathrm{sinc}\left[T_0\left(f - \frac{2BWR}{T_0 c}\right)\right] \tag{2.45}$$

For a sinc function, its peak is at the location of $f_p - K\delta t_0 = 0$, where f_p is the peak frequency. Thus, the range of the target can be obtained by

$$R = \frac{f_p T_0 c}{2BW} \tag{2.46}$$

2.4.2 Doppler processing

In (2.45), there is a phase term $e^{j(2\pi f_c \delta t_0)}$, which is also related to the location of the target. With a single chirp, this phase term is a constant. However, for a second chirp following the first chirp, its phase term $e^{j(2\pi f_c \delta t_1)}$ will be different as $\delta t_1 \neq \delta t_0$ if the target is moving. Thus, the simplest way to obtain the Doppler information is comparing the phase difference between two consecutive chirps

$$\Delta\varphi = 2\pi f_c(\delta t_1 - \delta t_0) + 2k_a\pi \tag{2.47}$$

$$= 2\pi f_c\left[\frac{2(R_0 + v\Delta t)}{c} - \frac{2(R_0)}{c}\right] + 2k_a\pi \tag{2.48}$$

$$v = \frac{(\Delta\varphi - 2k_a\pi)c}{4\pi f_c \Delta t} \tag{2.49}$$

where R_0 is the initial range of the target, v is the target's speed relative to the radar, Δt is the time difference between the two consecutive chirps, and k_a is an integer. The $2k_a\pi$ represents the phase aliasing. A correct speed can be obtained only when $k_a = 0$. Besides aliasing, one of the other major issues of using (2.49) is that if there are more than one target at the same range, it will be very difficult to obtain the correct speed for each of those targets.

A more popular and robust approach to find a target's range and Doppler is using a sequence of chirps, as shown in Figure 2.15. First, for (2.43), replace δt_0 by δt

$$s_b(t) = A_b \cdot w_R(t) \cdot e^{j(2\pi f_c \delta t + 2\pi K t \delta t)} \tag{2.50}$$

where,

$$\delta t = \frac{2(R_0 + v\tau)}{c} \tag{2.51}$$

τ is the center time of each chirp, and $0 \leqslant \tau < M \cdot PRP$, where M is the total number of chirps. Equation (2.50) can be re-written as a function of t and τ, which are commonly called "fast-time" and "slow-time," respectively.

$$s_b(t, \tau) = A_b \cdot w_R(t) \cdot w_D(\tau) \cdot e^{j\left[\frac{4\pi f_c(R_0+v\tau)}{c} + \frac{4\pi K t(R_0+v\tau)}{c}\right]} \tag{2.52}$$

and,

$$w_D(\tau) = \begin{cases} 1, & 0 \leqslant \tau < M \cdot PRP \\ 0, & \tau < 0 \; or \; \tau \geqslant M \cdot PRP \end{cases} \tag{2.53}$$

After some arrangement,

$$s_b(t, \tau) = A_b \cdot w_R(t) \cdot w_D(\tau) \cdot e^{j\left(\frac{4\pi f_c R_0}{c}\right)} \cdot e^{j\left(\frac{4\pi R_0 K t}{c}\right)} \cdot e^{j\left(\frac{4\pi f_c v \tau}{c}\right)} \cdot e^{j\left(\frac{4\pi K v t \tau}{c}\right)} \tag{2.54}$$

The signal processing part is operated in digital domain, and $t = n/f_s$, where n is the n-th sample in a chirp and f_s is the sampling rate. For the m-th chirp,

$$\tau_m = mPRP + t = mPRP + \frac{n}{f_s}, \quad -\frac{N}{2} \leqslant n < \frac{N}{2} \tag{2.55}$$

The baseband signal for the m-th chirp is

$$s_b(n, m) = A_b \cdot w_R(t) \cdot w_D(\tau) \tag{2.56}$$

$$\cdot e^{j\left[\frac{4\pi f_c R_0}{c}\right]} \quad \text{Constant phase} \tag{2.57}$$

$$\cdot e^{j\left[\frac{(4\pi R_0 K + 4\pi f_c v)n}{f_s c}\right]} \quad \text{Ranging and intra-chirp Doppler phase} \tag{2.58}$$

$$\cdot e^{j\left[\frac{4\pi f_c v PRP m}{c}\right]} \quad \text{Inter-chirp Doppler phase} \tag{2.59}$$

$$\cdot e^{j\left[\frac{4\pi K v PRP mn}{f_s c}\right]} \quad \text{Range Doppler coupling} \tag{2.60}$$

$$\cdot e^{j\left[\frac{4\pi K v n^2}{f_s^2 c}\right]} \quad \text{Higher order term} \tag{2.61}$$

For easier analysis, the higher order term is ignored here as the value is very close to one in a common FMCW radar system. There are two time-related parameters: n and m. A discrete Fourier transform (DFT) can be applied to n first, which results in

$$S_b(\kappa, m) = A_b \cdot N \cdot w_D(\tau) \cdot e^{j\left[\frac{4\pi f_c R_0}{c}\right]} \tag{2.62}$$

$$\cdot e^{j\left[\frac{4\pi f_c vPRPm}{c}\right]} \tag{2.63}$$

$$\cdot f_{sinc}(\kappa - N\frac{2R_0K + 2f_c v + 2KvPRPm}{f_s c}) \tag{2.64}$$

where κ is the range bin, f_{sinc} is a function similar to a sinc function defined as

$$f_{sinc}(n) = \frac{\sin(\pi n)}{\sin\left(\frac{\pi n}{N}\right)} \tag{2.65}$$

where N is the length of the samples, and $f_{sinc}(0) = N$ is the peak. $S_b(\kappa, m)$ is also named as the range profile for the m-th chirp. Similar to (2.45), the peak of the range profile is obtained when,

$$\kappa_p = N \cdot \frac{2R_0K + 2f_c v + 2KvPRPm}{f_s c} \tag{2.66}$$

when the target's speed $v = 0$, (2.66) can be simplify into

$$\frac{\kappa_p f_s}{N} = f_p = \frac{2R_0K}{c} \tag{2.67}$$

which matches with (2.46). With $v \neq 0$, the target's speed is assumed as a constant during the sequence of chirps to simplify the problem. Let $R_m = R_0 + PRPmv$ be the target's range at the center of the m-th chirp. The range bin of the target is

$$\bar{\kappa_p} = N \cdot \frac{2KR_m + 2f_c v}{f_s c} \tag{2.68}$$

and,

$$R_m = \frac{\bar{\kappa_p} f_s c}{2KN} - \frac{f_c v}{K} \tag{2.69}$$

With the target's range, R_m, and range bin, $\bar{\kappa}_p$, obtained, the next step is trying to derive the target's Doppler property. In order to find the speed of the target, the location of the Doppler bin needs to be identified. With the known $\kappa = \bar{\kappa}_p$, let

$$f_{sinc}(\bar{\kappa}_p - N \cdot \frac{2R_0K + 2f_c v + 2K_v PRPm}{f_s c}) = N \cdot \beta(m) \tag{2.70}$$

where $\beta(m)$ is the amplitude variation with a peak of

$$\beta\left[\frac{(M-1)}{2}\right] = 1 \tag{2.71}$$

Based on the assumption that the target's speed is a constant, $\beta(m)$ is symmetric referenced to $(M-1)/2$:

$$\beta(m) = \beta(M-1-m) \tag{2.72}$$

Thus, the amplitude of the DFT results of $\beta(m)$ is also symmetric:

$$|B(\xi)| = |B(M-1-\xi)| \tag{2.73}$$

Putting $\kappa = \bar{\kappa}_p$ and (2.72) to (2.62), the equation can be simplified to:

$$S_b(\bar{\kappa_p}, m) = A_b \cdot N^2 \cdot w_D(m) \cdot \beta(m) \cdot e^{j\left[\frac{4\pi f_c R_0}{c}\right]} \cdot e^{j\left[\frac{4\pi f_c vPRPm}{c}\right]} \tag{2.74}$$

Apply DFT over m, the baseband signal can be further derived as

$$S_b(\bar{\kappa_p}, \xi) = A_b \cdot N^2 \cdot M \cdot e^{j\left[\frac{4\pi f_c R_0}{c}\right]} \tag{2.75}$$

$$\cdot(e^{-j\frac{2\pi\xi}{M}(M-1)/2} \cdot f_{sinc}(\xi - M\frac{2f_c PRPv}{c})) \tag{2.76}$$

$$*B(\xi) \tag{2.77}$$

where $*$ is convolution and ξ is the Doppler bin. As discussed above, the amplitude of $B(\xi)$ is symmetric, which is similar to a window function. The convolution between f_{sinc} and $B(\xi)$ will not change the peak location of f_{sinc}. Thus, the Doppler bin of the target is

$$\xi_p = M\frac{2f_c PRPv}{c} \tag{2.78}$$

and the target's speed is

$$v = \frac{c}{2Mf_c PRP}\xi_p \tag{2.79}$$

The target's range and Doppler can thus be extracted from (2.69) and (2.79).

In actual algorithm realization of the range-Doppler processing, a 2D fast Fourier transform can be directly applied to $s_b(n, m)$ to obtain $S_b(\kappa, \xi)$ with high efficiency, and targets' range and Doppler information can be measured within milliseconds. This is essential for automotive applications that usually require high update rate and low latency.

2.4.3 Typical waveform parameters for FMCW automotive radar applications

The automotive radar systems operating in the 76–81 GHz range can be separated into three main groups based on the radar applications. Each group has different technical waveform parameters to satisfy requirements.

The first group, uses frequency band 76–77 GHz, is known as LRR systems. LRR consists of applications such as adaptive cruise control (ACC) and (high speed) collision avoidance (CA). These applications assist drivers in driving and aim to avoid accidents and collisions at high speeds. In LRR applications, the radar detection range can be up to 250 meters. A 1 GHz continuous BW is usually required to have enough range resolution.

The other two groups use frequency band 76–81 GHz and cover short- and medium-range radar (SRR and MRR) applications. In general, these applications need higher range resolution and use up to 4 GHz BW. Short-range radar applications, such as in-cabin occupancy detection and obstacle detection in close proximity to a vehicle, do not need a long detection range. Some SRR applications, such as pedestrian detection, parking-aid, and low speed (<30 km/h) emergency braking, may need a higher field of view. For MRR applications, such as rear-traffic-crossing-alert, blind-spot detection, and lane-change detection, the detection range can be up to 100 m with a narrower field of view.

Sweep time shows differences based on the application and is usually selected between 10 and 40 μs. Even though automotive radar waveform parameters show differences based on the radar hardware, radar waveform parameters for a classical FMCW are summarized in Table 2.5 for selected applications. Interested readers are further referred to [26] for recommendation to detailed technical parameters of radar systems operating in the frequency bands of 76–77 and 77–81 GHz.

2.4.4 Window taper function

During the range and Doppler processing*, the time-limited baseband signal results in a signal in the form of sinc function after Fourier transform, e.g., (2.45). While the sinc function makes a peak at the target location, it creates a signature consisting of "mainlobe" and "sidelobes." The mainlobe contains high power and makes the peak at the intended target location (such as range and Doppler). The

Table 2.5 Example automotive radar waveform parameters

		LRR	**MRR**	**SRR**
Parameter	**Unit**			
Sub-band usage	GHz	76–77	76–81	76–81
Unambiguous range	m	250	100	50
Range resolution	m	1	0.5	0.05
Unambiguous velocity	m/s	60	60	10
Velocity resolution	m/s	0.2	0.2	0.2
Sweep time	μs	10–40	10–40	10–40
Applications		ACC CA	Blind spot detection Lane-change assist Cross-traffic alert	Parking-aid Obstacle detection Emergency braking

*Window taper function (may also know weight function) can also be applied to angle measurements.

mainlobe width at the half-power beamwidth, which is ~ 3 dB lower than the peak power, defines the range resolution. The other lobes generated by the sinc function are called "sidelobes," and usually represent unwanted signal returns related to the detected target. These unwanted sidelobes may hinder weak targets because the first side lobe of sinc function appears ~ −13.26 dB lower than the mainlobe peak.

To mitigate sidelobes, alternative window taper functions were proposed [27]. Many of them are used in current systems to adjust the trade-off between mainlobe and sidelobes. There are some metrics for selecting the window function, which can also be used while selecting an alternative waveform other than linear FMCW.

Mainlobe width: The lobe containing the higher (peak) power of the target response. The theoretical width of the mainlobe is related by the BW of the waveform. Using different window functions widens the mainlobe.

SNR loss: Using different window tapers results in a loss in the peak power of the target response, thus the SNR decreases. The SNR loss for the sinc function is zero and is assumed to be the benchmark.

Peak sidelobe ratio (PSLR): The ratio of peak power to the maximum sidelobe power. A larger PSLR is preferred to detect weaker targets. For sinc function, PSLR is −13.26 dB. The nulls occur at $\kappa(1/T_0)$, where $\kappa = \pm 1, \pm 2, \cdots, \pm N/2$.

Integrated sidelobe level: The ratio of the sidelobe energy to the mainlobe energy. The integrated sidelobe level ratio is −10.12 dB for sinc function (from the first null to the end of the sidelobe, excluding the mainlobe).

To address the high sidelobe issue, a window function is used to increase the PSLR with a trade-off of wider mainlobe width. For the interested readers, a catalog of the most used window taper functions for sidelobe control can be found at [27]. When predefined window functions do not satisfy the requirements of the user, it is possible to optimize the window function for a given template as well as the desired properties by using modern optimization techniques [28].

2.5 Detection fundamentals

For a radar measurement that is to be tested for the presence of a target, one of the two following hypotheses can be assumed to be true

- H_0 – The measurement is the result of noise only.
- H_1 – The measurement is the combined result of noise and echo from a target.

The radar system's detection logic examines each measurement and selects one of the hypotheses as the best accounting for the measurement. If H_0 is selected, the system declares the measurement is solely from noise and the target is not present. On the other hand, if H_1 is selected, the radar system declares that a target is present.

The pdf can be used to analyze the signals. Two pdfs, $p_y(y|H_0)$ and $p_y(y|H_1)$, are defined H_0 and H_1 hypotheses, respectively. $\boldsymbol{y}$ denotes the measurement samples to be tested. $p_y(y|H_0)$ is the pdf of $\boldsymbol{y}$ that a target is not present, and $p_y(y|H_1)$ is the pdf of $\boldsymbol{y}$ that a target is present. Based on these two pdfs, the following probabilities can be defined:

- Probability of detection, P_D: The probability that the radar system declares a target is detected (H_1 is chosen) when a target is actually present.
- Probability of false alarm, P_{FA}: The probability that the radar system declares a target is detected (H_1 is chosen) when a target is actually not present.

To further discuss how P_D and P_{FA} are used in radar system characterization, consider the noise of the radar system as white noise, and $p_y(y|H_0)$ is a normal distribution. In addition, consider the target as nonfluctuation, and thus $p_y(y|H_1)$ is also a normal distribution as H_1 is the combination of noise and the echo from a target.

$$H_0 : y \sim N(a_n, \sigma_n^2) \tag{2.80}$$

$$H_1 : y \sim N(a_s, \sigma_s^2) \tag{2.81}$$

where a_n and a_s are the mean amplitude for hypothesis H_0 and H_1, respectively. σ_n^2 and σ_s^2 are the variance for hypothesis H_0 and H_1, respectively. Therefore, the pdfs are

$$p_y(y|H_0) = \frac{1}{\sqrt{2\pi\sigma_n^2}} e^{-\frac{(y-a_n)^2}{2\sigma_n^2}} \tag{2.82}$$

$$p_y(y|H_1) = \frac{1}{\sqrt{2\pi\sigma_s^2}} e^{-\frac{(y-a_s)^2}{2\sigma_s^2}} \tag{2.83}$$

Figure 2.16 illustrates the two $p_y(y|H_0)$ and $p_y(y|H_1)$. In a radar system, it is more common to use the Neyman–Pearson criterion, where a threshold T is selected to maximize the probability of detection P_D under the constraint that the probability of false alarm P_{FA} doesn't exceed a constant. As shown in Figure 2.16, P_{FA} is the area above threshold T in $p_y(y|H_0)$ (blue shade), which can be calculated through

$$P_{FA} = \int_T^{+\infty} p_y(y|H_0)dy \tag{2.84}$$

P_D is the area above threshold T in $p_y(y|H_1)$ (red shade):

$$P_D = \int_T^{+\infty} p_y(y|H_1)dy \tag{2.85}$$

It needs to be noted that as long as the threshold T is selected, the P_{FA} stays unchanged. However, P_D varies with the SNR, which is $(a_s - a_n)/a_n$ in Figure 2.16. P_D increases with the increase in SNR, as shown in Figure 2.17, where $P_{FA} = 1 \times 10^{-6}$. Coherent processing means taking the phase information into consideration, while

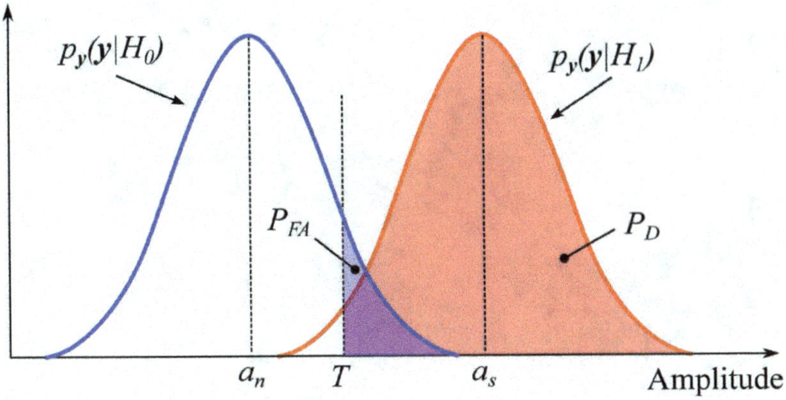

Figure 2.16 The pdfs of hypotheses H_0 *and* H_1

noncoherent processing discards the phase information. Figure 2.18 shows the P_D versus P_{FA} plots with 10 dB SNR.

2.5.1 Coherent and noncoherent integration

Integration is one of the common techniques in radar signal processing to improve the SNR. Both coherent integration and noncoherent integration are widely used. Coherent integration refers to integration of complex data, and noncoherent integration means integration based only on the magnitude of the data. It can be seen in Figures 2.17 and 2.18 that there will be a few SNR loss if the phase is discarded in the processing even without any integration.

Coherent integration is favorable for SNR improvement. However, it also requires more processing as it needs to deal with complex data. Assume a radar

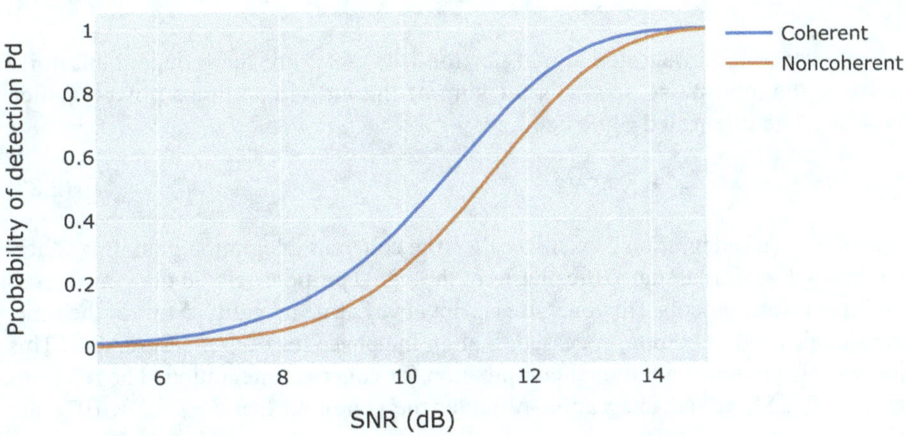

Figure 2.17 P_D *versus SNR with* $P_{FA} = 1 \times 10^{-6}$

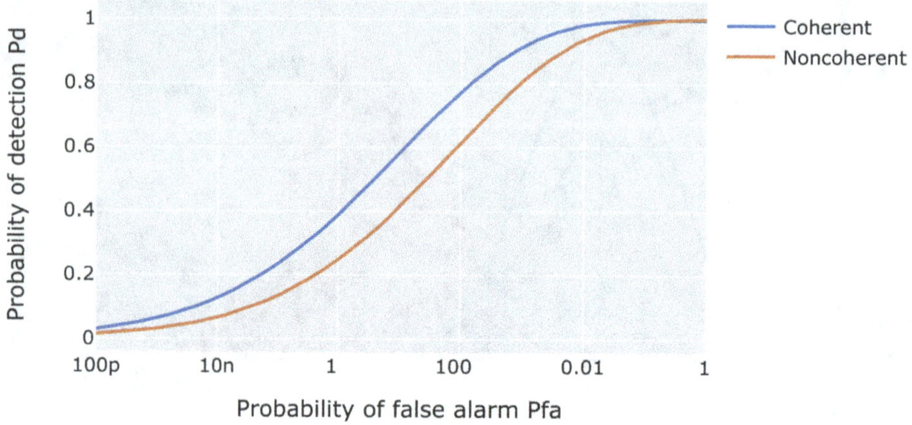

Figure 2.18 P_D *versus* P_{FA} *with 10 dB SNR*

received a measurement $A \cdot e^{j\varphi} + \omega$ from the echo of a target, where A is the magnitude, φ is the phase and ω is the additive noise with the power of σ^2. The SNR is defined as

$$SNR_1 = \frac{A^2}{\sigma^2} \tag{2.86}$$

Suppose the measurement has been taken N times, it could be that there are N channels taking the same measurement or repeating the measurement N times. For the repeated measurement data, it is expected that the echo response is the same, but the noise is independent each time. Thus, all the measurement results can be added with the phase information.

$$\sum_{n=0}^{N-1} \{A \cdot e^{j\varphi} + \omega[n]\} = N \cdot A \cdot e^{j\varphi} + \sum_{n=0}^{N-1} \omega[n] \tag{2.87}$$

The power of the integrated signal portion is $N^2 \cdot A^2$. For the independent noise portion, the total noise power is the sum of the individual noise power, which is $N \cdot \sigma^2$. The integrated SNR becomes:

$$SNR_N = \frac{N^2 \cdot A^2}{N \cdot \sigma^2} = N \cdot SNR_1 \tag{2.88}$$

The number of integration N is also called the coherent integration gain. For coherent integration, it is assumed the phases of the signal portions remain the same across multiple measurements. However, this is not always true in reality. Most of the time, the signals need to be pre-processed to align the phase for each measurement. This process also requires additional computation for coherent integration. The relations among P_D, SNR, and integration N under the condition that $P_{FA} = 1 \times 10^{-6}$ are shown in Figure 2.19.

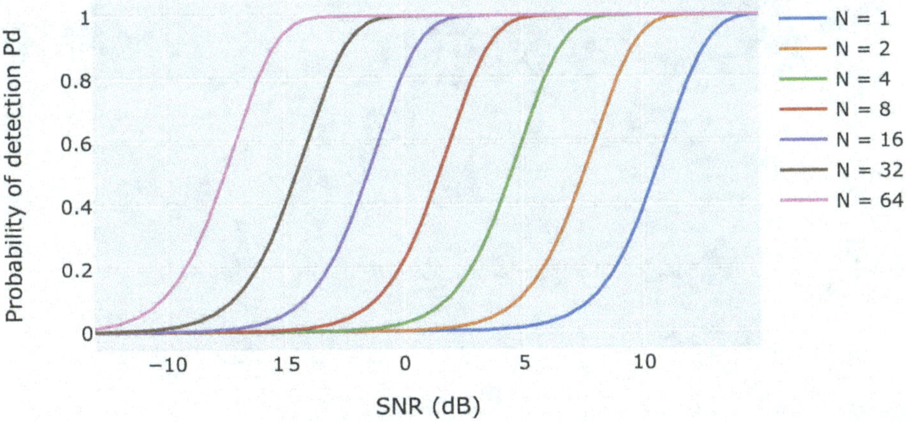

Figure 2.19 P_D *versus SNR with coherent integration under* $P_{FA} = 1 \times 10^{-6}$

Consider noncoherent integration of a nonfluctuating target, which is also known as "Swerling 0" or "Swerling V" target, with white Gaussian noise ω. For simplicity, assume the total power of ω is 1. Under hypothesis H_0, the target is absent, the pdf of $y = |\omega|$ follows a Rayleigh distribution.

$$p_y(y|H_0) = \begin{cases} 2ye^{-y^2} & y \geqslant 0 \\ 0, & y < 0 \end{cases} \tag{2.89}$$

Under hypothesis H_1, $y = |Ae^{j\varphi} + \omega|$ is a Rice distribution

$$p_y(y|H_1) = \begin{cases} 2ye^{-(y^2+A^2)}I_0\left(2Ay\right) & y \geqslant 0 \\ 0, & y < 0 \end{cases} \tag{2.90}$$

where $I_0(y)$ is the modified Bessel function of the first kind with order zero.

By noncoherently integrating N ($N > 1$) samples, the pdf under H_0 hypothesis becomes the Erlang distribution.

$$p_{y'}(y'|H_0) = \frac{(y')^{N-1}}{(N-1)!}e^{-y'} \tag{2.91}$$

and the pdf under H_1 hypothesis is

$$p_{y'}(y'|H_1) = (\frac{y'}{NA^2})^{(N-1)/2}e^{-y'-NA^2}I_{N-1}(2\sqrt{NA^2y'}), \tag{2.92}$$

where $y' = \sum y^2$ as a square law detector is used. The detailed derivation of these equations is not shown in this chapter, as it is out of the scope of this book. For better understanding of noncoherent integration, Figures 2.20 and 2.21 provide simple illustrations of the effect of noncoherent integration. The original SNR without

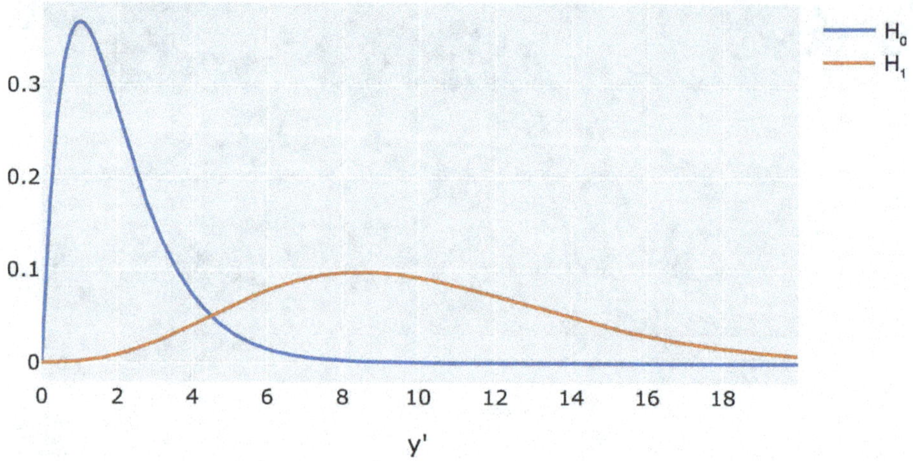

Figure 2.20 PDFs of H_0 *and* H_1 *with noncoherent integration of two measurements*

integration is 4 (6 dB) in both Figure 2.20 and Figure 2.21. Figure 2.20 is the case that $N = 2$, and $N = 10$ in Figure 2.21. In order to calculate the SNR after integration, the means for all the PDFs need to be calculated. For the Erlang distribution in (2.91), the mean is $\mu_{H_0} = N$. For the distribution in (2.93), the mean can be evaluated through the numerical method by calculating

$$\mu_{H_1} = \int_0^{\infty} y' p_{y'}(y'|H_1) dy' \tag{2.93}$$

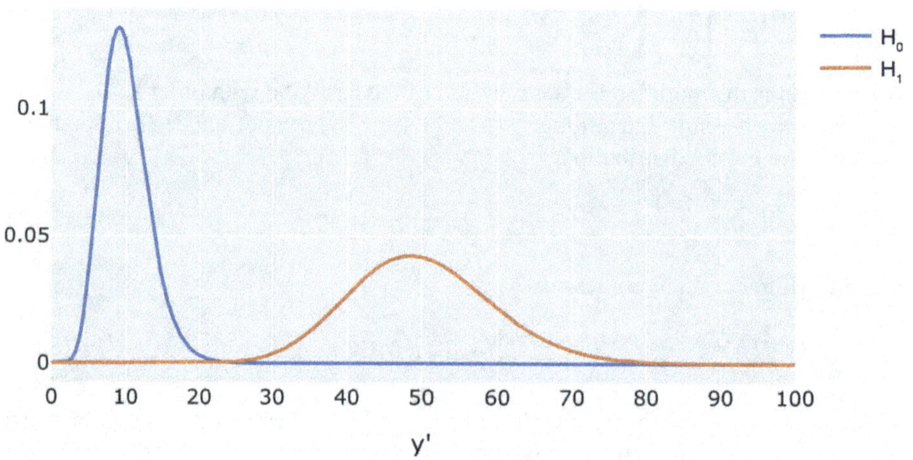

Figure 2.21 PDFs of H_0 *and* H_1 *with noncoherent integration of 10 measurements*

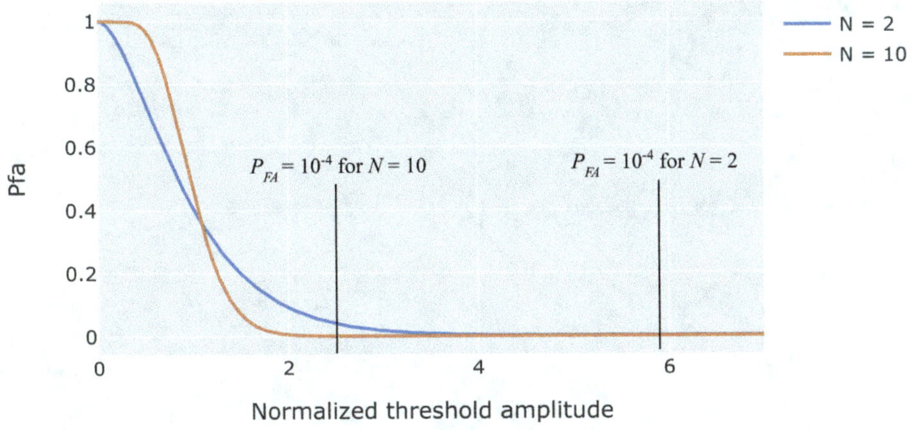

Figure 2.22 P_{FA} *versus normalized threshold with different integration values*

For $N = 2$ in Figure 2.20, $\mu_{H_0,N=2} = 2$ and $\mu_{H_1,N=2} = 10$,

$$SNR_{N=2} = \frac{\mu_{H_1,N=2} - \mu_{H_0,N=2}}{\mu_{H_0,N=2}} = 4 \tag{2.94}$$

For the case in Figure 2.21 that $N = 10$, $\mu_{H_0,N=10} = 10$ and $\mu_{H_1,N=10} = 50$,

$$SNR_{N=10} = \frac{\mu_{H_1,N=10} - \mu_{H_0,N=10}}{\mu_{H_0,N=10}} = 4 \tag{2.95}$$

Thus, noncoherent integration doesn't actually improve the SNR.

Now consider P_{FA} with different N values. It is difficult to analytically derive the threshold T based on P_{FA} and (2.91). An simpler approach is using numerical method to calculate the integration of $p_{y'}(y'|H_0)$ as shown below:

$$P_{FA}(T) = \int_T^{\infty} p_{y'}(y'|H_0)dy' = 1 - \int_0^T p_{y'}(y'|H_0)dy' \tag{2.96}$$

Figure 2.22 shows the P_{FA} versus the normalized threshold T' with $N = 2$ and $N = 10$. The threshold is normalized to the mean of the PDF in (2.89) for better comparison. For $N = 2$, $T' = 5.9$ to obtain 10^{-4}, on the other hand, to achieve the same P_{FA}, $T' = 2.6$ for $N = 10$ case. Thus, with noncoherent integration, the effective gain is achieved by lowering the threshold for certain P_{FA}. In other words, with more integrations, it is possible to achieve the same P_{FA} with a lower threshold, and the required minimal SNR for a target to be detected is also reduced.

The previous analysis only considered nonfluctuating targets, which are referring to the "Swerling 0" or "Swerling V" case. A more realistic case would be for fluctuating targets mentioned above, including "Swerling I, II, III, and IV" models. The derivations of pdfs for fluctuating targets are out of the scope of this book.

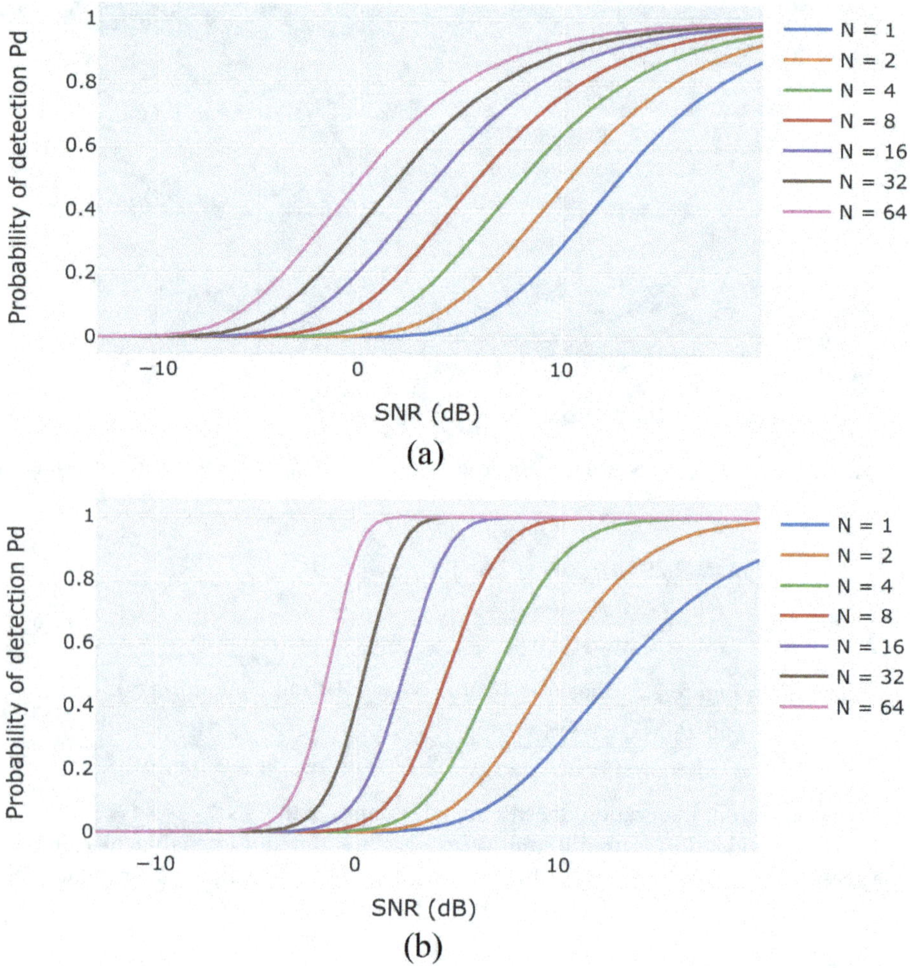

Figure 2.23 P_D *versus SNR with noncoherent integration under* $P_{FA} = 1 \times 10^{-6}$

Popular signal processing tools, such as MATLAB®, have handy features to do this kind of analysis. Figures 2.23 and 2.24 show the P_D versus SNR with noncoherent integration under $P_{FA} = 1 \times 10^{-6}$ for Swerling I, II, III, IV, and V models. (Table 2.6)

2.5.2 *Minimal SNR for certain P_D and P_{FA}*

The relation of P_D and P_{FA} has been discussed in the previous contents. For a certain P_{FA}, a threshold can be obtained from $p_{y'}(y'|H_0)$, which is the probability density function (PDF) of the noise without echoes from the target. Another threshold can also be derived from $p_{y'}(y'|H_1)$, which is the PDF of the echoes from the target with noise, based on a certain P_D. Usually, these two thresholds are separated, and their difference decreases with the reduction of SNR. At a certain SNR, these two

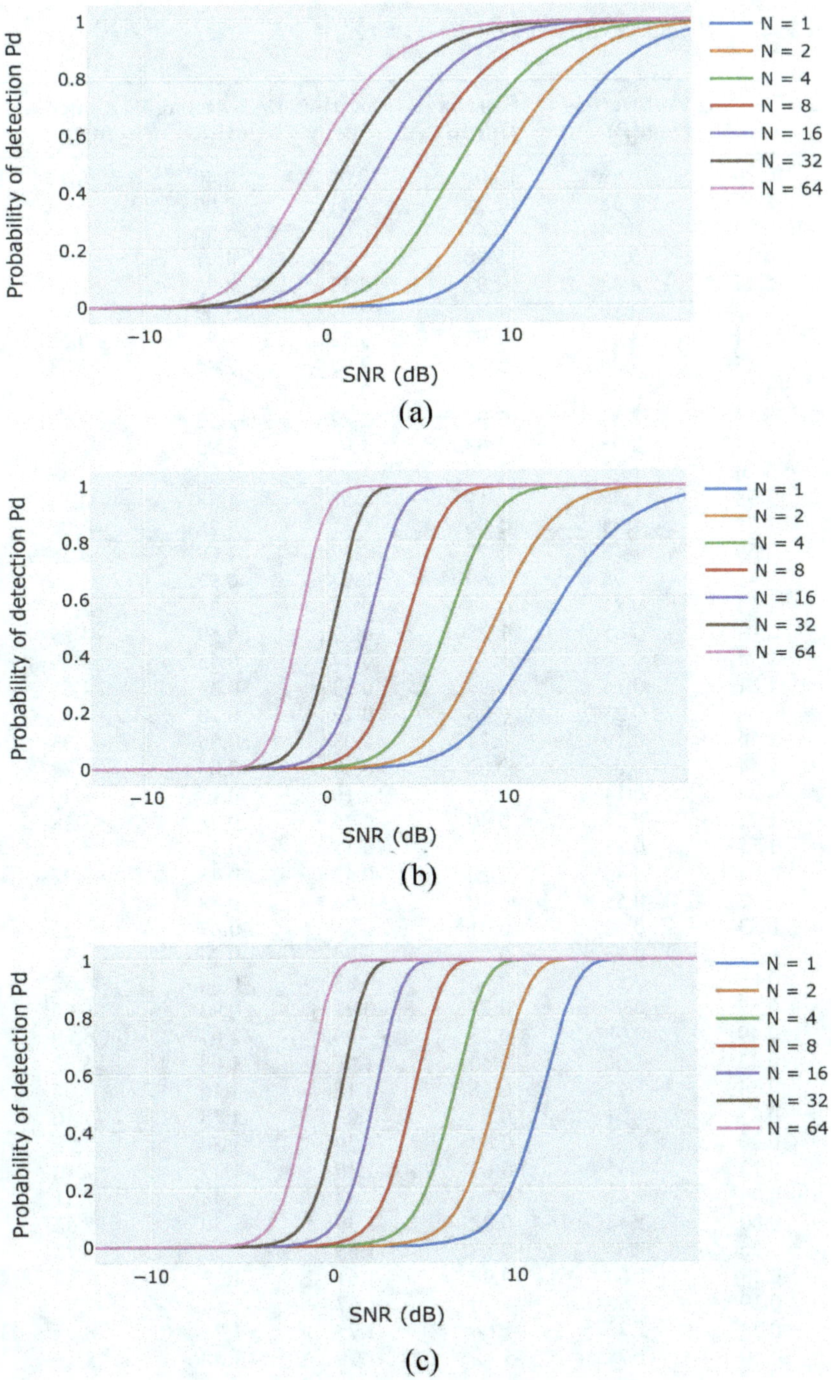

Figure 2.24 P_D *versus SNR with noncoherent integration under* $P_{FA} = 1 \times 10^{-6}$

Table 2.6 Minimal SNR for $P_{FA} = 10^{-4}$ and $P_D = 0.5$.

N	Swerling I (dB)	Swerling II (dB)	Swerling III (dB)	Swerling IV (dB)	Swerling V (dB)	Coherent (dB)
1	10.89	10.89	10.09	10.09	9.40	8.40
2	8.59	7.78	7.78	7.41	7.09	5.39
3	7.30	6.24	6.49	6.01	5.80	3.63
4	6.41	5.23	5.60	5.06	4.91	2.38
5	5.74	4.48	4.92	4.35	4.23	1.41
6	5.19	3.89	4.38	3.79	3.69	0.62
7	4.74	3.41	3.93	3.32	3.24	–0.05
8	4.35	3.00	3.54	2.92	2.85	–0.63
9	4.02	2.64	3.20	2.57	2.51	–1.14
10	3.72	2.33	2.90	2.27	2.22	–1.60
11	3.45	2.05	2.63	1.99	1.95	–2.02
12	3.20	1.79	2.39	1.75	1.70	–2.39
13	2.98	1.56	2.17	1.52	1.48	–2.74
14	2.78	1.35	1.96	1.31	1.28	–3.06
15	2.59	1.16	1.77	1.12	1.09	–3.36
16	2.41	0.98	1.60	0.94	0.91	–3.64
17	2.25	0.81	1.43	0.78	0.75	–3.91
18	2.09	0.65	1.28	0.62	0.59	–4.15
19	1.95	0.50	1.13	0.47	0.45	–4.39
20	1.81	0.36	0.99	0.33	0.31	–4.61
21	1.68	0.23	0.86	0.20	0.18	–4.82
22	1.56	0.10	0.74	0.08	0.06	–5.03
23	1.44	–0.02	0.62	–0.04	–0.06	–5.22
24	1.32	–0.14	0.51	–0.16	–0.17	–5.40
25	1.22	–0.24	0.40	–0.26	–0.28	–5.58
26	1.11	–0.35	0.30	–0.37	–0.39	–5.75
27	1.01	–0.45	0.20	–0.47	–0.48	–5.92
28	0.92	–0.55	0.10	–0.56	–0.58	–6.07
29	0.83	–0.64	0.01	–0.66	–0.67	–6.23
30	0.74	–0.73	–0.08	–0.74	–0.76	–6.37
31	0.65	–0.82	–0.16	–0.83	–0.84	–6.52
32	0.57	–0.90	–0.25	–0.91	–0.93	–6.65
33	0.49	–0.98	–0.33	–0.99	–1.01	–6.79
34	0.42	–1.06	–0.40	–1.07	–1.08	–6.92
35	0.34	–1.13	–0.48	–1.15	–1.16	–7.04
36	0.27	–1.21	–0.55	–1.22	–1.23	–7.16
37	0.20	–1.28	–0.62	–1.29	–1.30	–7.28
38	0.13	–1.35	–0.69	–1.36	–1.37	–7.40
39	0.06	–1.41	–0.76	–1.42	–1.43	–7.51
40	0.00	–1.48	–0.82	–1.49	–1.50	–7.62
41	–0.06	–1.54	–0.88	–1.55	–1.56	–7.73
42	–0.13	–1.60	–0.94	–1.61	–1.62	–7.83
43	–0.19	–1.66	–1.00	–1.67	–1.68	–7.94
44	–0.24	–1.72	–1.06	–1.73	–1.74	–8.04
45	–0.30	–1.78	–1.12	–1.79	–1.80	–8.13

(Continues)

Table 2.6 Continued

N	Swerling I (dB)	Swerling II (dB)	Swerling III (dB)	Swerling IV (dB)	Swerling V (dB)	Coherent (dB)
46	–0.36	–1.84	–1.18	–1.85	–1.85	–8.23
47	–0.41	–1.89	–1.23	–1.90	–1.91	–8.32
48	–0.46	–1.94	–1.28	–1.95	–1.96	–8.41
49	–0.52	–2.00	–1.34	–2.01	–2.01	–8.50
50	–0.57	–2.05	–1.39	–2.19	–2.06	–8.59
51	–0.62	–2.10	–1.44	–2.24	–2.11	–8.68
52	–0.67	–2.15	–1.49	–2.29	–2.16	–8.76
53	–0.71	–2.20	–1.53	–2.33	–2.21	–8.84
54	–0.76	–2.24	–1.58	–2.38	–2.26	–8.93
55	–0.81	–2.29	–1.63	–2.42	–2.30	–9.01
56	–0.85	–2.33	–1.67	–2.47	–2.35	–9.08
57	–0.90	–2.38	–1.72	–2.51	–2.39	–9.16
58	–0.94	–2.42	–1.76	–2.55	–2.44	–9.24
59	–0.98	–2.47	–1.80	–2.60	–2.48	–9.31
60	–1.02	–2.51	–1.84	–2.64	–2.52	–9.38
61	–1.07	–2.55	–1.88	–2.68	–2.56	–9.46
62	–1.11	–2.59	–1.93	–2.72	–2.60	–9.53
63	–1.15	–2.63	–1.97	–2.75	–2.64	–9.60
64	–1.18	–2.67	–2.00	–2.79	–2.68	–9.66
65	–1.22	–2.71	–2.04	–2.83	–2.72	–9.73
66	–1.26	–2.75	–2.08	–2.87	–2.76	–9.80
67	–1.30	–2.78	–2.12	–2.90	–2.79	–9.86
68	–1.33	–2.82	–2.15	–2.94	–2.83	–9.93
69	–1.37	–2.86	–2.19	–2.97	–2.87	–9.99
70	–1.41	–2.89	–2.23	–3.01	–2.90	–10.05
71	–1.44	–2.93	–2.26	–3.04	–2.94	–10.11
72	–1.48	–2.96	–2.30	–3.08	–2.97	–10.18
73	–1.51	–3.00	–2.33	–3.11	–3.01	–10.23
74	–1.54	–3.03	–2.36	–3.14	–3.04	–10.29
75	–1.58	–3.06	–2.40	–3.18	–3.07	–10.35
76	–1.61	–3.09	–2.43	–3.21	–3.10	–10.41
77	–1.64	–3.13	–2.46	–3.24	–3.14	–10.47
78	–1.67	–3.16	–2.49	–3.27	–3.17	–10.52
79	–1.70	–3.19	–2.52	–3.30	–3.20	–10.58
80	–1.74	–3.22	–2.56	–3.33	–3.23	–10.63
81	–1.77	–3.25	–2.59	–3.36	–3.26	–10.69
82	–1.80	–3.28	–2.62	–3.39	–3.29	–10.74
83	–1.83	–3.31	–2.65	–3.42	–3.32	–10.79
84	–1.85	–3.34	–2.68	–3.45	–3.35	–10.84
85	–1.88	–3.37	–2.70	–3.48	–3.38	–10.90
86	–1.91	–3.40	–2.73	–3.50	–3.41	–10.95
87	–1.94	–3.43	–2.76	–3.53	–3.44	–11.00
88	–1.97	–3.46	–2.79	–3.56	–3.46	–11.05
89	–2.00	–3.48	–2.82	–3.58	–3.49	–11.10
90	–2.02	–3.51	–2.84	–3.61	–3.52	–11.14
91	–2.05	–3.54	–2.87	–3.64	–3.54	–11.19
92	–2.08	–3.56	–2.90	–3.66	–3.57	–11.24

(Continues)

Table 2.6 Continued

N	Swerling I (dB)	Swerling II (dB)	Swerling III (dB)	Swerling IV (dB)	Swerling V (dB)	Coherent (dB)
93	–2.10	–3.59	–2.92	–3.69	–3.60	–11.29
94	–2.13	–3.62	–2.95	–3.72	–3.62	–11.33
95	–2.16	–3.64	–2.98	–3.74	–3.65	–11.38
96	–2.18	–3.67	–3.00	–3.77	–3.67	–11.42
97	–2.21	–3.69	–3.03	–3.79	–3.70	–11.47
98	–2.23	–3.72	–3.05	–3.81	–3.72	–11.51
99	–2.26	–3.74	–3.08	–3.84	–3.75	–11.56
100	–2.28	–3.77	–3.10	–3.86	–3.77	–11.60

thresholds are equal to each other. This SNR is the minimal SNR required to satisfy the predefined P_D and P_{FA}. In the case of integration, the difference of minimal SNR after integration equals to the integration gain. Table 2.6 shows the minimal SNR required for $P_{FA} = 10^{-4}$ and $P_D = 0.5$ with different target models and integration types.

For automotive radar applications, the minimal SNR is useful for evaluating the detection capability, or sensitivity, of a radar. For example, a radar is used in an emergency brake system and it is required to detect a pedestrian with $P_D = 0.5$ and $P_{FA} = 10^{-4}$ at range R_0. A pedestrian is usually classified as a "Swerling I" target, and its typical RCS is around –10 dBsm. From Table 2.6, the required minimal SNR is 10.89 dB. Thus, the radar needs to be designed to have at least 10.89 dB SNR at range R_0 for a –10 dBsm target. If the designed radar noncoherently integrates eight channels in the processing, the required minimal SNR can be relaxed to 4.35 dB. The SNR difference, 10.89 – 4.35 = 6.54 dB is the noncoherent integration gain for Swerling I targets.

2.5.3 *Constant false alarm rate detection*

As discussed above, standard radar threshold detection assumes that the noise level on the receiver is known and stable. An accurate threshold can thus be obtained to guarantee a specified P_{FA}. However, in the real world, the noise levels vary with multiple factors, including environment, temperature, and component aging. A radar system needs to estimate the noise level from the measurements and derive the threshold based on the estimation. Temperature compensation and periodic recalibration can partially resolve the problem of varying noise power; however, these methods are not practical with large volume production in automotive applications. In addition, automotive radars are often exposed to complex electromagnetic environments with interference from radars in surrounding vehicles. The strong interference could significantly increase the noise power. Thus, adaptive calculation of the noise power is always necessary for automotive radars.

Constant false alarm rate (CFAR) detection, also referred to as "adaptive threshold detection" or "automatic detection," is a set of techniques designed to provide

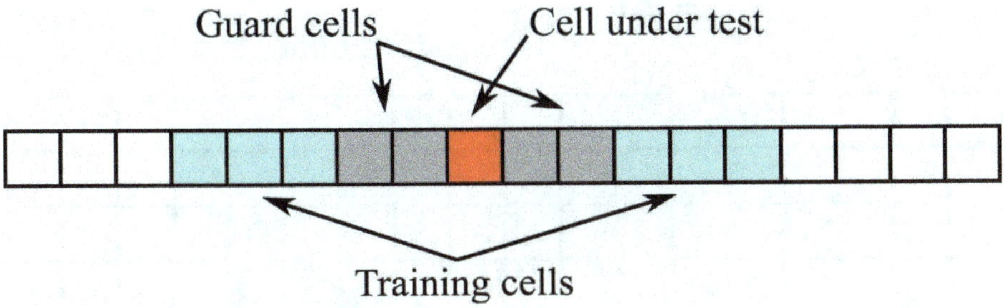

Figure 2.25 1-D CA-CFAR processing

predictable detection and false alarm behavior in realistic noise scenarios. For CFAR detection, the actual noise power must be estimated from the data in real time, so that the detector threshold can be adjusted to maintain a desired P_{FA}. Various CFAR algorithms from simple to sophisticated have been reported. To provide a basic impression on CFAR detection, this section will give a brief discussion on two basic CFAR algorithms, namely cell-averaging CFAR (CA-CFAR) and ordered-statistic CFAR (OS-CFAR).

2.5.3.1 CA-CFAR

CA-CFAR is one of the most basic CFAR algorithms. It is also used as a baseline comparison for other CFAR techniques. Figure 2.25 illustrates the sliding window processing structure of 1-D CA-CFAR, which can also be calculated in a 2-D range-Doppler image, as shown in Figure 2.26. Noise samples are extracted from both leading and lagging cells (training cells) around the cell under test (CUT). Guard cells are placed adjacent to the CUT. The purpose of these guard cells is to avoid signal components from leaking into the training cell, which could adversely affect the noise estimation. The noise power P_n can be computed as

$$P_n = \frac{1}{M}\sum_{m=1}^{M} y_m \tag{2.97}$$

where M is the number of training cells and y_m is the power of the m-th training cell. The numbers of training cells and guard cells can be adjusted as needed. The threshold T is given in

$$T = \alpha P_n \tag{2.98}$$

where α is the scaling factor called the threshold factor. The value of α is calculated based on the desired false alarm rate P_{FA} and (2.96). If there is no pulse integration, α can be obtained through:

$$\alpha = M \cdot \left(P_{FA}^{-1/M} - 1\right) \tag{2.99}$$

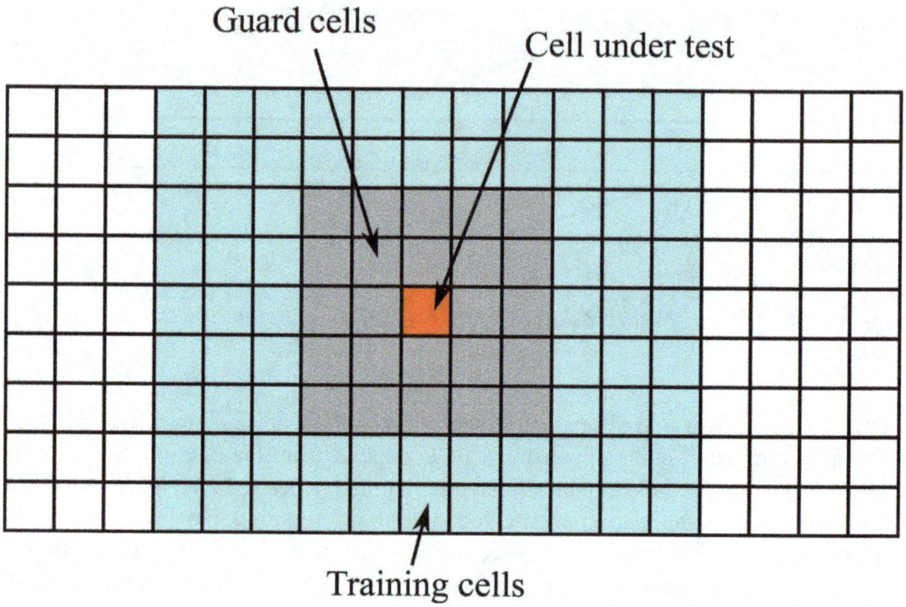

Figure 2.26 2-D CA-CFAR processing

Figure 2.27 shows the operation of CA-CFAR in an FMCW radar. There are two targets at 95 m and 30 m in the test. The CA-CFAR uses two guard cells and 10 trailing cells with $P_{FA} = 10^{-2}$. It can be seen in Figure 2.27 that the CA-CFAR threshold has high shoulders beside a target peak, which means a smaller target besides a large target can be masked by the threshold. Thus, CA-CFAR requires the assumption that targets need to be isolated, and targets need to be separated far enough so that they are not masked by the shoulders of the others. Another

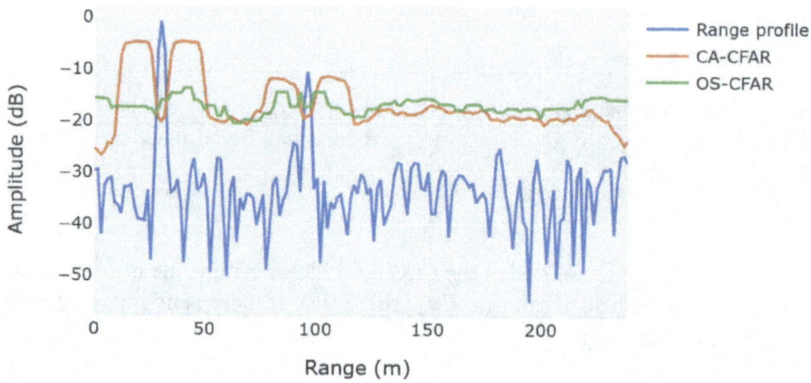

Figure 2.27 Example of CA-CFAR and OS-CFAR behaviors

assumption of CA-CFAR is that the noise is independent and identically distributed for the training cells.

The performance limitations of CA-CFAR have led to the development of many extensions to the CA-CFAR, such as smallest-of-cell-averaging CFAR and greater-of-cell-averaging CFAR.

2.5.3.2 OS-CFAR

OS-CFAR is an alternative range-based approach for CFAR threshold. OS-CFAR uses the similar 1-D or 2-D sliding window structures as CA-CFAR shown in Figures 2.25 and 2.26. The guard cells can be used if necessary. Different from CA-CFAR, OS-CFAR rank orders the data samples in the training cells $[y_1, y_2, ..., y_M]$ to form a new sequence in ascending order, denoted by $[y_{(1)}, y_{(2)}, ..., y_{(M)}]$. Then, the k -th element in the ordered list is selected as representative of the noise level and a threshold is set as a multiple of this value:

$$T_{OS} = \alpha_{OS} y_{(k)} \tag{2.100}$$

The calculation of α_{OS} can be found in Reference [29]. The noise is estimated from only one data sample, instead of the average of all the data samples. However, since the used sample is determined based on all the data samples, the threshold actually utilizes the information of all the samples.

Figure 2.27 shows the operation of OS-CFAR in an FMCW radar; $M = 20$, $k = 15$, and $P_{FA} = 10^{-2}$. It can be seen that OS-CFAR doesn't have the high shoulders around the target peaks. The major disadvantage of the OS-CFAR is the high processing power required for performing the sorting algorithm, and this processing power must be provided during the real-time part of radar signal processing.

2.6 Radar design considerations

The above sections give a brief overview of the fundamentals of an automotive radar. In designing a radar system, several aspects are essential and determine the key performances of a radar system. In this section, an FMCW radar will be used as an example to discuss these design considerations.

2.6.1 Sensitivity

A radar's sensitivity defines the weakest echo from a target that can be successfully detected at certain P_{FA} and P_D. In the above sections, both the signal models and the noise model have been introduced. With these models, a radar system designer will be able to predict the radar's sensitivity at the very early stage of the development.

The analysis of the sensitivity starts from the radar range (2.21). Assume the total gain in the receiver channel is G_s, which includes gains from amplifiers, loss/gain from the down-conversion mixer, and baseband amplifier. The signal power on the baseband is

$$P_s = P_r \cdot G_s = \frac{P_t G_t G_r \lambda^2 \sigma}{(4\pi)^3 R^4} \cdot G_s \tag{2.101}$$

On the other hand, from (2.34), the noise power on the baseband is

$$P_n = NF \cdot kT_0 \beta_n G_s \tag{2.102}$$

Thus, the SNR can be obtained as

$$SNR = \frac{P_s}{P_n} = \frac{P_t G_t G_r \lambda^2 \sigma}{(4\pi)^3 R^4 \cdot NF \cdot kT_0 \beta_n} \cdot \frac{G_{SP}}{L_{SP}}, \tag{2.103}$$

where G_{SP} is the signal processing gain, and L_{SP} is the signal processing loss. β_n is approximate to the BW of the anti-aliasing filter before the analog-to-digital converter. For an FMCW radar, G_{SP} mainly comes from the DFT on "fast-time" and "slow-time." With the range-Doppler processing, the total signal processing gain $G_{SP} = N \cdot M$, where N is the DFT size on "fast-time" and M is the DFT size on "slow-time," which are also referred to as the numbers of range and Doppler bins, respectively. Regarding L_{SP}, one of the causes is the windowing function applied to the data before DFT.

For the design of an automotive radar, the requirements are usually provided for specific cases. For example, an automotive manufacturer needs a forward-looking radar to support its automatic emergency braking feature, which requires the radar to detect a pedestrian at certain range with more than 50% of probability of detection ($P_D > 0.5$) and 1×10^{-4} a false alarm rate of $P_{FA} = 1 \times 10^{-4}$. A pedestrian is usually recognized as a Swerling I target. From Table 2.6, the minimal SNR for a Swerling I target to achieve $P_D = 0.5$ and $P_{FA} = 1 \times 10^{-4}$ is $SNR_{min} = 10.89$ dB without the integration. Thus, the maximum detection range is

$$R_{max} = \sqrt[4]{\frac{P_t G_t G_r \lambda^2 \sigma}{(4\pi)^3 \cdot SNR_{min} \cdot NF \cdot kT_0 \beta_n} \cdot \frac{G_{SP}}{L_{SP}}} \tag{2.104}$$

One of the major tasks for a radar designer is finding a good balance among the parameters in (2.105). kT_0 is a constant. λ is determined by the operating frequency, which can also be treated as a constant. Transmit power P_t and noise figure NF are usually determined by the radar transceiver chip, which are usually limited by the fabrication process. G_t and G_r are dependent on antenna design, which also need to consider the trade-off between the maximum antenna gain and the beam coverage (antenna beamwidth). β_n, G_{SP}, and L_{SP} are the parameters related to radar's waveform design and digital signal processing. SNR_{min} could be reduced by using integration and thus improving the maximum detection range. However, integration also requires more processing.

2.6.2 Range/Doppler coverage

Radar's range coverage can be determined from (2.46) to (2.105). From (2.46), the maximum range is obtained when $f_p = f_s$ and I/Q baseband is used:

$$R_{max} = \frac{f_s T_0 c}{2BW} \tag{2.105}$$

This maximum range is based on waveform and sampling rate, regardless of the sensitivity. With a single channel baseband (without I/Q baseband), the maximum range is reduced by half:

$$R_{max} = \frac{f_s T_0 c}{4BW} \tag{2.106}$$

Thus, the actual range coverage of a radar should be

$$R_{max} = \min\left[\sqrt[4]{\frac{P_t G_t G_r \lambda^2 \sigma}{(4\pi)^3 \cdot SNR_{min} \cdot NF \cdot kT_0 \beta_n} \cdot \frac{G_{SP}}{L_{SP}}}, \frac{f_s T_0 c}{2BW}\right] \tag{2.107}$$

or

$$R_{max} = \min\left[\sqrt[4]{\frac{P_t G_t G_r \lambda^2 \sigma}{(4\pi)^3 \cdot SNR_{min} \cdot NF \cdot kT_0 \beta_n} \cdot \frac{G_{SP}}{L_{SP}}}, \frac{f_s T_0 c}{4BW}\right] \tag{2.108}$$

for single channel baseband.

For Doppler coverage, the maximum velocity can be calculated from (2.79) when,$\xi_p = M$

$$v_{max} = \frac{c}{2f_c PRP} \tag{2.109}$$

v_{max} is also named as the maximum unambiguous velocity, as velocities outside v_{max} are folded into the interval.

2.6.3 Range/Doppler resolution

The target resolution of a radar is its ability to distinguish between targets that are very close in either range or Doppler. For an FMCW radar, the range of a target can be obtained from (2.69). The range resolution ΔR is the difference between two adjacent range bins:

$$\Delta R = \frac{(\kappa + 1) f_s c}{2KN} - \frac{\kappa f_s c}{2KN} \tag{2.110}$$

$$= \frac{f_s c}{2KN} \tag{2.111}$$

Since $N/f_s = T_0$, the range resolution can be derived as

$$\Delta R = \frac{c}{2BW} \tag{2.112}$$

Similar to the range resolution, the Doppler (velocity) resolution Δv can be obtained from (2.79):

$$\Delta v = \frac{c}{2Mf_c PRP}(\xi + 1) - \frac{c}{2Mf_c PRP}\xi \tag{2.113}$$

$$= \frac{c}{2Mf_c PRP} \tag{2.114}$$

It is worth noting that ΔR and Δv derived from (2.112) and (2.114) are for an ideal condition, including point targets with the same size, high SNR, and rectangular window. In a real scenario, the range and Doppler resolutions will be impacted by various factors. For example, if the sizes of the targets are different, it is more difficult to distinguish the smaller target from the larger ones. Another example is that certain window functions can be used in the range-Doppler processing to achieve a reasonable sidelobe level, which usually increases the mainlobe width and degrades the range and Doppler resolution [30].

2.6.4 Phase noise

A perfect sinusoidal wave is only available in textbooks. Phase and frequency instabilities are available in all natural and manmade oscillators. These instabilities are known as the phase noise. Phase noise is related to the materials, architectural design, as well as the random noise phenomena in the oscillator. Figure 2.28(A) illustrates the spectrum of a general oscillator, where f_0 is the center or carrier frequency. The power of the phase noise decreases as the frequency offset f_a increases. The phase noise portion of the spectrum can be divided into two parts, i.e., the near-carrier phase noise and the far-from carrier phase noise. For automotive radars, a PLL is usually used to synthesize the waveform. Figure 2.28(B) illustrates a typical spectrum of a PLL. A pedestal can be seen in Figure 2.28(B), which is due to a PLL-based synthesizer having a finite loop BW, or due to the finite BW of the frequency multiplier chain being employed in the system to frequency multiplication. A detailed analysis of the behavior of the noise pedestal under frequency multiplication can be found in References [31–33].

In a radar's receiver channels, mixers could add or cancel the phase noise in the received signals. Cancellation of phase noise happens when the two input signals are coherent, i.e., they have a defined phase relationship with each other. In other works, the input signals of the mixer are derived from the same reference source. In an

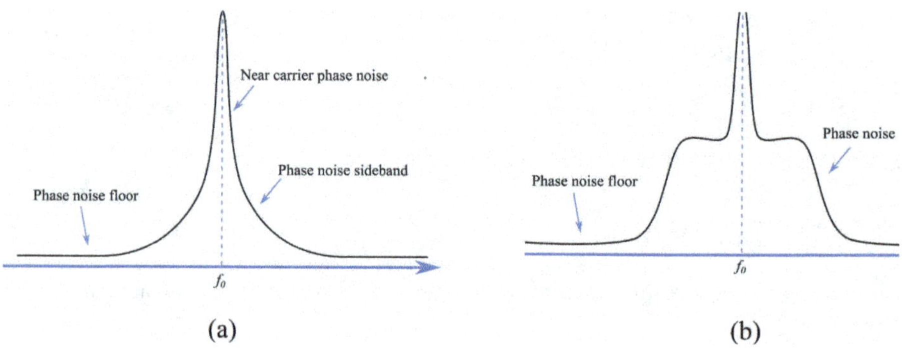

Figure 2.28 Spectrum of radar signal sources (a) Spectrum of a general oscillator (b) Spectrum of a synthesized signal generator including the noise pedestal

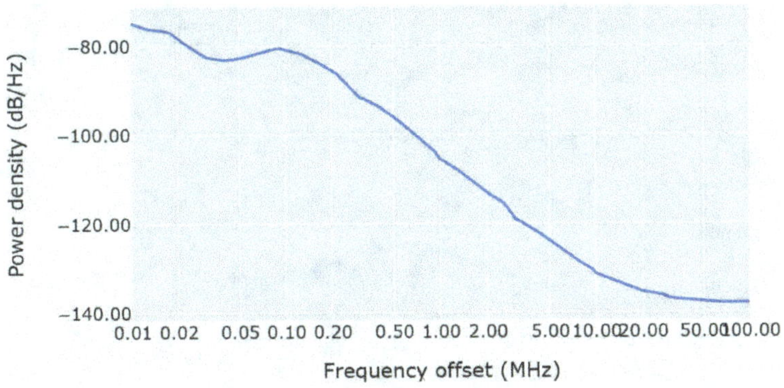

Figure 2.29 An example of measured phase noise of a radar transmitter

automotive radar system, the mixers mix the transmitted signal with a time-delayed replica, which causes phase noise de-correlation as follows [34, 35]:

$$PN_{baseband}(f_a) = PN_{Tx}(f_a) \cdot 4\sin^2(\pi f_a \delta t) \tag{2.115}$$

The phase noise de-correlation factor is $4\sin^2(\pi f_a \delta t)$, as shown in (2.115). It can be seen that the phase noise can be fully canceled when $\delta t = 0$. With the increase in δt, $PN_{baseband}(f_a)$ shows the property of periodicity.

In order to better illustrate the effect of phase noise in an automotive radar, the measured phase noise data of a radar transceiver is used in a simulation. Figure 2.29 is the measured phase noise. The phase noise pedestal can be clearly seen in this figure. The simulated range profiles are shown in Figures 2.30 and 2.31, which have a target at 80 m and 150 m, respectively. The range profiles are the averages of 128 chirps to obtain the shapes of the noise floor. As indicated in (2.115), the phase noise introduces shoulders in the range profiles. The shoulders are higher with the increased range of the target as the de-correlation increases with longer δt. In automotive radar

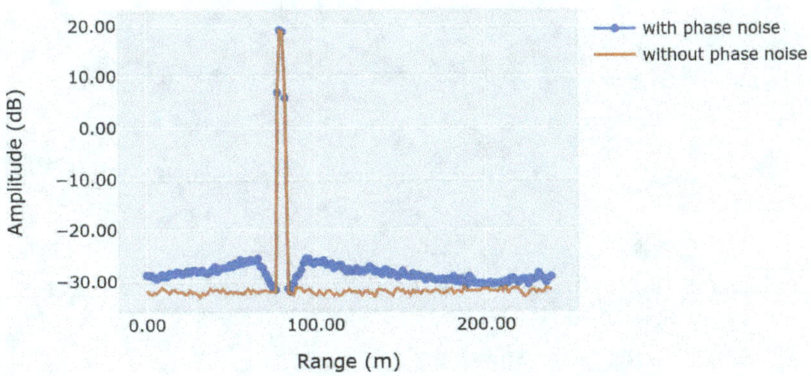

Figure 2.30 Range profile of a target at 80 m

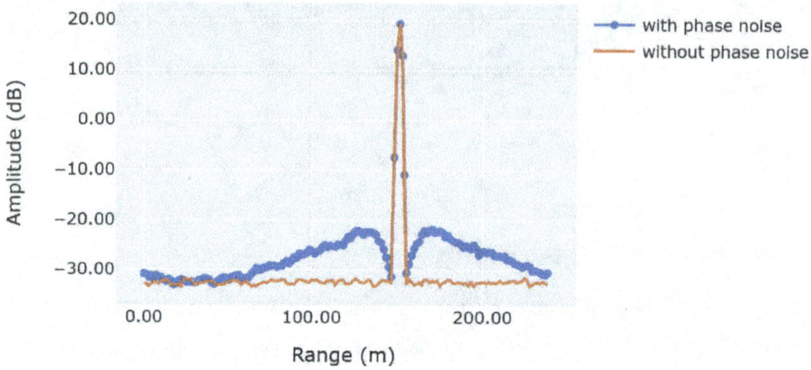

Figure 2.31 *Range profile of a target at 150 m*

applications, a higher shoulder means a reduced dynamic range. In this case, a smaller target beside a large one may not be detected. For example, a pedestrian in front of a car could be masked by the phase noise shoulder. Thus, it is essential to have low phase noise for the radar signal generator to maintain sufficient dynamic range.

2.6.5 Chirp non-linearity

The other major disturbance in an automotive radar for range measurement is the chirp non-linearity. An ideal FMCW radar has a perfect linear chirp, which can be expressed as

$$f(t) = f_c + Kt \tag{2.116}$$

where K is the slope of the chirp. However, there always exist frequency deviations in a real signal synthesizer. Figure 2.32 shows an example of the chirp at 24 GHz that has a slight deviation compared with an ideal linear chirp.

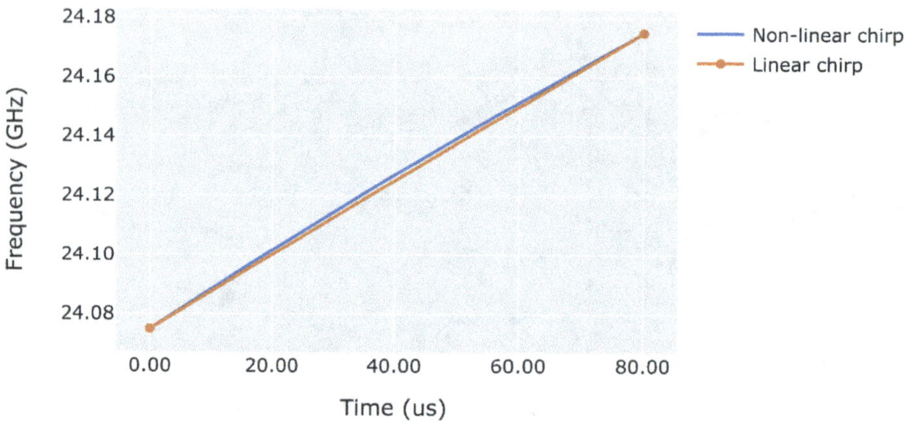

Figure 2.32 *A non-linear chirp*

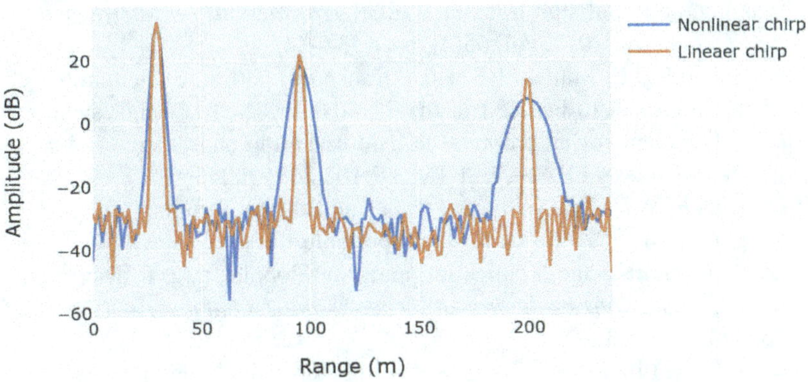

Figure 2.33 Range profile of targets with non-linear chirps

The baseband frequency and the target's range accuracy are dependent on the type of the non-linearity in the frequency ramp. Thus, the impacts introduced by chirp non-linearity should be evaluated case by case. For instance, The influence of a nonlinear frequency ramp with a sinusoidal deviation is described in Reference [36]. In Reference [37], square deviation in a free running voltage-controlled oscillator is evaluated in an FMCW radar. Due to the difficulty of analytically specifying the impact of the chirp non-linearity, radar designers usually combine measurement and simulation to analyze the impact of chirp non-linearity. Modern advanced signal and spectrum analysis instruments, such as Keysight 89,601BHPC and Rohde &Schwarz FSW signal and spectrum analyzer, provide the capability to measure the frequency versus time transient waveform directly. Radar designers can thus utilize the measured waveform in a radar simulation to evaluate the impact due to the non-linearity.

For the measured waveform in Figure 2.32, the general CW transmit signal (2.13) can be used to create the time domain transmit signal from the frequency versus time transient waveform. Based on the interested range, certain time delays can be applied into (2.42) to obtain the baseband signal for evaluation. The example of the simulated range profiles is shown in Figure 2.33, where three targets are at 30, 95, and 200 m with different RCS values. The effects of the non-linearity are obvious in Figure 2.33, where the widths of the target beams are broadened as the range increases, which degrades both the range discrimination and range accuracy. The SNR also has more degradation with a longer range. Thus, in automotive radar design, it is essential to have a good linearity in the chirp to maintain radar's performance across the range coverage.

References

[1] Visser H.J. *Array and phased array antenna basics*. Chichester: John Wiley & Sons; 2005.

[2] Peng Z., Muñoz-Ferreras J.M., Tang Y., *et al*. 'A portable FMCW interferometry radar with programmable low-IF architecture for localization, ISAR

imaging, and vital sign tracking'. *IEEE Transactions on Microwave Theory and Techniques*. 2017, vol. 65(4), pp. 1334–44.

[3] Compatibility E, Matters RS. Road transport and traffic telematics (RTTT); radio equipment to be used in the 77 GHz to 81 GHz band; system reference document for automotive collision warning short range radar. Sophia Antipolis Cedex - FRANCE: ETSI TR 102 263; 2004.

[4] Oppenheim A.V. *Signals and systems*. 2nd Edition. Pearson; 1997.

[5] Peng Z., Ran L., Li C. 'A K -band portable fmcw radar with beamforming array for short-range localization and vital-doppler targets discrimination'. *IEEE Transactions on Microwave Theory and Techniques*. 2017, vol. 65(9), pp. 3443–52.

[6] Menzel W., Pilz D., Al-Tikriti M. 'Millimeter-wave folded reflector antennas with high gain, low loss, and low profile'. *IEEE Antennas and Propagation Magazine*. 2002, vol. 44(3), pp. 24–9.

[7] Hallbjorner P., Zhongxia S.H., Bruce S., Chengs, *et al.* 'Low-profile 77-ghz lens antenna with array feeder'. *IEEE Antennas and Wireless Propagation Letters*. 2012, vol. 11, pp. 205–07.

[8] Wang X., Stelzer A. 'A 79-GHz LTCC patch array antenna using a laminated waveguide-based vertical parallel feed'. *IEEE Antennas and Wireless Propagation Letters*. 2013, vol. 12, pp. 987–90.

[9] Artemenko A., Mozharovskiy A., Maltsev A., Maslennikov R., Sevastyanov A., Ssorin V. 'Experimental characterization of E-band two-dimensional electronically beam-steerable integrated lens antennas'. *IEEE Antennas and Wireless Propagation Letters*. 2013, vol. 12, pp. 1188–91.

[10] Yeap S.B., Qing X., Chen Z.N. '77-GHz dual-layer transmit-array for automotive radar applications'. *IEEE Transactions on Antennas and Propagation*. 2015, vol. 63(6), pp. 2833–7.

[11] Imbert M., Papió A., De Flaviis F., Jofre L., Romeu J. 'Design and performance evaluation of a dielectric flat lens antenna for millimeter-wave applications'. *IEEE Antennas and Wireless Propagation Letters*. 2014, vol. 14, pp. 342–5.

[12] Khan O., Meyer J., Baur K., Waldschmidt C. 'Hybrid thin film antenna for automotive radar at 79 GHz'. *IEEE Transactions on Antennas and Propagation*. 2017, vol. 65(10), pp. 5076–85.

[13] Xu J., Hong W., Zhang H., Wang G., Yu Y., Jiang Z.H. 'An array antenna for both long- and medium-range 77 GHz automotive radar applications'. *IEEE Transactions on Antennas and Propagation*, vol. 65(12), pp. 7207–16.

[14] Arnieri E., Greco F., Boccia L., Amendola G. 'A reduced size planar grid array antenna for automotive radar sensors'. *IEEE Antennas and Wireless Propagation Letters*. 2018, vol. 17(12), pp. 2389–93.

[15] Yu Y., Hong W., Zhang H., Xu J., Jiang Z.H. 'Optimization and implementation of SIW slot array for both medium- and long-range 77 GHz automotive

radar application'. *IEEE Transactions on Antennas and Propagation*. 2018, vol. 66(7), pp. 3769–74.

[16] Tian H., Liu C., Gu X. 'Proximity-coupled feed patch antenna array for 79 GHz automotive radar'. *The Journal of Engineering*. 2019, vol. 2019(19), pp. 6244–6.

[17] Xu J., Hong W., Jiang Z.H., Zhang H. 'Low-profile circular patch array fed by Slotted substrate integrated waveguide'. *IEEE Transactions on Antennas and Propagation*. 2019, vol. 67(2), pp. 960–70.

[18] Hamberger G.F., Späth S., Siart U., Eibert T.F. 'A mixed circular/linear dual-polarized phased array concept for automotive radar—planar antenna designs and system evaluation at 78 GHz'. *IEEE Transactions on Antennas and Propagation*. 2018, vol. 67(3), pp. 1562–72.

[19] Yoo S., Milyakh Y., Kim H., Hong C., Choo H. 'Patch array antenna using a dual coupled feeding structure for 79 GHz automotive radar applications'. *IEEE Antennas and Wireless Propagation Letters*. 2020, vol. 19(4), pp. 676–9.

[20] Oppenheim A.V., Schafer R.W. *Discrete-time signal processing*. 3rd Edition. Pearson; 2010.

[21] Bourlier C. *Radar propagation modeling in a complex maritime environment*. Elsevier; 2018.

[22] Swerling P. 'Probability of detection for fluctuating targets'. *IEEE Transactions on Information Theory*. 1960, vol. 6(2), pp. 269–308.

[23] Meyers D. *Radar target detection: handbook of theory and practice*. 1st Edition. Elsevier; 1973.

[24] Voss R.F., Clarke J. 'Flicker (1/f) noise: equilibrium temperature and resistance fluctuations'. *Physical review. B*. 1976, vol. 13(2), pp. 556–73.

[25] Curlander J.C., McDonough R.N. *Synthetic aperture radar: Systems and signal processing*. Wiley; 1992.

[26] Series M. *Systems characteristics of automotive radars operating in the frequency band 76-81 GHz for intelligent transport systems applications*. 2014.

[27] Doerry A.W. *Catalog of window taper functions for Sidelobe control*. Albuquerque, NM: Sandia National Lab. (SNL-NM); 2017. SAND2017-4042 652561

[28] Coutino M., Uysal F., Anitori L. Waveform-Aware Optimal Window Function Design for Mismatched Filtering. 2022 IEEE Radar Conference (RadarConf22); 2022.

[29] Richards M.A. *Fundamentals of radar signal processing*. McGraw-Hill Education; 2005.

[30] Harris F.J. 'On the use of windows for harmonic analysis with the discrete Fourier transform'. *Proceedings of the IEEE*. 1978, vol. 66(1), pp. 51–83.

[31] Walls F.L., Demarchi A. 'Rf spectrum of a signal after frequency multiplication; measurement and comparison with a simple calculation'. *IEEE Transactions on Instrumentation and Measurement*. 1975, vol. 24(3), pp. 210–7.

[32] Schlecht E., Mehdi I. 'Effects of local oscillator phase noise on submillimeter-wave spectrometer performance effects of local oscillator phase noise on submillimeter-wave spectrometer performance'. *39th International Conference on Infrared, Millimeter, and Terahertz waves (IRMMW-THz)*; IEEE, Tucson, AZ, 2014. pp. 1–2.

[33] Bava E., De Marchi A., Godone A. 'Spectral analysis of synthesized signals in the MM wavelength region'. *IEEE Transactions on Instrumentation and Measurement*. 1977, vol. 26(2), pp. 128–32.

[34] Raven R.S. 'Requirements on master oscillators for coherent radar'. *Proceedings of the IEEE*. 1966, vol. 54(2), pp. 237–43.

[35] Stove A.G. 'Linear FMCW radar techniques'. *IEEE Proceedings F Radar and Signal Processing*. 1992, vol. 139(5), pp. 343–50.

[36] Piper S.O. 'Homodyne FMCW radar range resolution effects with sinusoidal nonlinearities in the frequency sweep'. *Proceedings International Radar Conference*; IEEE, Alexandria, VA, 1995. pp. 563–67.

[37] Brennan P.V., Huang Y., Ash M., Chetty K. 'Determination of sweep linearity requirements in FMCW radar systems based on simple voltage-Controlled oscillator sources'. *IEEE Transactions on Aerospace and Electronic Systems*. 2011, vol. 47(3), pp. 1594–604.

Chapter 3
MIMO radar technology

Shunqiao Sun[1]

Automotive radar with a small number of antennas has been used for advanced driver-assistance system (ADAS) purposes since the late 1990s. These early automotive radars mostly provided target detection and velocity information. However, the current generation of automotive radar for ADAS has rather limited ability to resolve closely spaced targets. LiDAR systems have a better angular resolution (less than 1 degree) and have been introduced in Level 4 and Level 5 autonomous driving systems. LiDAR can provide point clouds. Via the use of deep neural networks, such as PointNet [1] and PointNet++ [2], the point clouds can lead to target identification. However, due to its use of light spectrum wavelength, LiDAR is susceptible to bad weather conditions, such as fog, rain, snow, and dust in the air. In addition, the cost of LiDAR is high. On the other hand, automotive radar with millimeter-waveform technology has the potential to provide point clouds at a much lower cost than LiDAR and with more robustness to weather conditions. Such radar is referred to as a "high end radar" or *imaging radar* [3]. Computer vision techniques [1, 2] that were previously reserved for high-resolution camera sensors and LiDAR systems can be applied to imaging radar data to identify targets. For example, a car can be identified based on two-dimensional (2D) radar points of an imaging radar using PointNet [4]. Imaging radars have been attracting the interest of those developing fully autonomous vehicles, major Tier-1 suppliers, and automotive radar startups.

In addition to sensitivity, the important requirements for automotive radar are high resolution, low hardware cost, and small size. Multiple-input multiple-output (MIMO) radar technology has been receiving considerable attention from the automotive radar community because it can achieve high angular resolution with relatively small numbers of antennas and receivers. For that ability, it has been exploited in current generation automotive radar for ADAS as well as in next-generation high-resolution imaging radar for autonomous driving. For autonomous driving, information in both azimuth and elevation is crucial. In particular, the

[1]Department of Electrical and Computer Engineering, The University of Alabama, Tuscaloosa, AL, USA

Figure 3.1 *Automotive radars need to provide elevation resolution to enable drive-over and drive-under functions. © [2021] IEEE. Reprinted with permission from [5].*

height information of targets is required to enable drive-over and drive-under functions. Two typical scenarios are shown in Figure 3.1. It is safe to drive over a metal beverage can on the road and to drive under a steel pedestrian bridge over the road. To meet such requirement, the array is required to have a large aperture in both azimuth and elevation. The MIMO radar is a good candidate for high-resolution imaging radar for autonomous driving. In the MIMO radar, the targets are first distinguished in range and Doppler domains. Then, large virtual arrays with hundreds of elements can be synthesized to provide high resolution in both azimuth and elevation. As a result, point clouds with similar performance to LiDAR can be generated at a much lower cost.

In this chapter, we introduce the concept of imaging radars using MIMO technology, present some examples for synthesizing hundreds of virtual array elements by cascading multiple radar transceivers with each supporting a small number of antennas, and discuss design challenges.

3.1 Virtual array synthesis via MIMO radar

In the state-of-the-art automotive FMCW radar, the range and Doppler parameters of targets can be estimated by using a single receive antenna. However, to estimate the angle parameter of targets, a receive antenna array is needed. In the MIMO radar, the antennas transmit waveforms in a way that guarantees their orthogonality. At each receive antenna, the contribution of each transmit antenna is extracted by exploiting waveform orthogonality. For a uniform transmit antenna array with M_t transmit antennas and element spacing of d_t and a uniform receive antenna array with M_r receive antennas and element spacing of d_r, a virtual array with M_tM_r elements can be synthesized. The array response of the *synthesized* array, i.e., the array corresponding to a MIMO radar can be expressed as $\mathbf{a}_t(\theta_t) \otimes \mathbf{a}_r(\theta_r)$. Here, $\otimes$ denotes the Kronecker product, and $\mathbf{a}_t(\theta_t)$ and $\mathbf{a}_r(\theta_r)$ are, respectively, the transmit and receive array steering vectors, corresponding to the direction of departure (DOD) θ_t and direction of arrival (DOA) θ_r, defined in the following:

$$\mathbf{a}_t(\theta_t) = \left[1, e^{j2\pi \frac{d_t}{\lambda} \sin(\theta_t)}, \cdots, e^{j2\pi \frac{(M_t-1)d_t}{\lambda} \sin(\theta_t)}\right]^T \tag{3.1}$$

$$\mathbf{a}_r(\theta_r) = \left[1, e^{j2\pi \frac{d_r}{\lambda} \sin(\theta_r)}, \cdots, e^{j2\pi \frac{(M_r-1)d_r}{\lambda} \sin(\theta_r)}\right]^T \tag{3.2}$$

Figure 3.2 shows two MIMO radar configurations with $M_t = 2$ transmit and M_r receive antennas [6]. One is in interleaved mode with $d_r = 2d_t$ and $d_t = \frac{\lambda}{2}$, and the other is in stacked mode with $d_t = M_r d_r$ and $d_r = \frac{\lambda}{2}$. Here, $\lambda = \frac{c}{f_c}$ is the wavelength. It can be verified that for both array configurations, the synthetic virtual arrays are equivalent to uniform linear arrays (ULAs) with $M_t M_r$ elements and spacing of d_r when $\theta_t = \theta_r$.

It should be noted that because the MIMO radar transmits orthogonality waveform, when isotropic array elements are used, the array beampattern—also referred to as the MIMO radar array factor—is omnidirectional. Thus, the MIMO radar loses the coherent array processing gain advantage enjoyed by traditional phased-array radar systems, which is $10 \log N$ for a phased array with N elements [7]; the SNR of the array response at a given angular direction is less than that of the phased-array radar with transmit beamforming. Still, in the automotive application scenario, the high-resolution angle-finding ability of the MIMO radar coupled with its low cost is viewed as more important than the loss of coherent processing gain.

3.2 Waveform orthogonality strategies in automotive MIMO radar

Virtual array synthesis in automotive radar using MIMO radar technology relies on the separability of the transmit signals of the different antennas. The separation is easier when the transmit signals of different antennas are orthogonal. In the following, we review the techniques to achieve waveform orthogonality while transmitting FMCW, such as time-division multiplexing (TDM), Doppler-division multiplexing (DDM), and frequency-division multiplexing (FDM).

3.2.1 Waveform orthogonality via TDM

In the TDM MIMO radar [8–12], only one transmit antenna is scheduled to transmit at each time slot. In Figure 3.3, a signal processing example of an MIMO radar in TDM is given by NXP Semiconductors [13], where $M_t = 2$ transmit antennas emit FMCW chirps alternatively. The switch delay between transmit antennas is $\Delta t = T_{\text{PRI}}$. At each receive antenna, range FFTs of length N_r are conducted for each chirp and the FFT outputs of $2N_d$ chirps are assembled in two matrices corresponding to odd

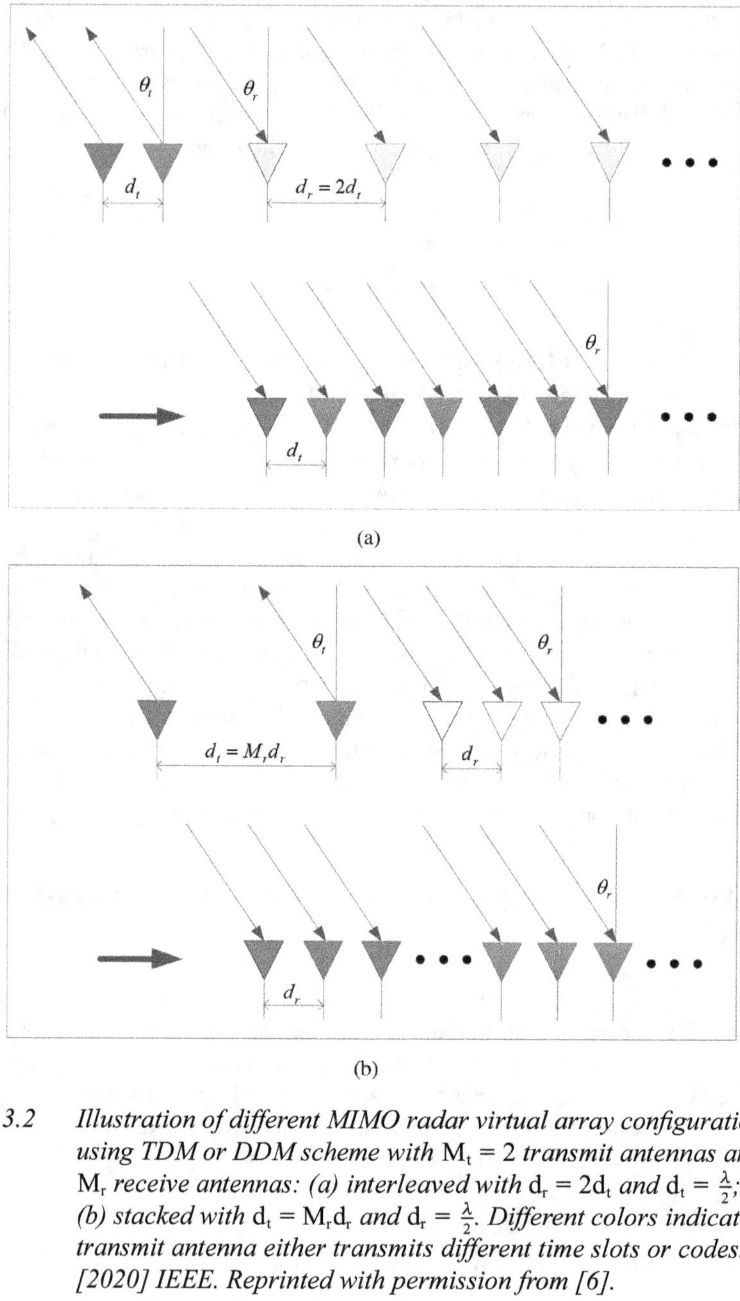

Figure 3.2 *Illustration of different MIMO radar virtual array configurations using TDM or DDM scheme with* $M_t = 2$ *transmit antennas and* M_r *receive antennas: (a) interleaved with* $d_r = 2d_t$ *and* $d_t = \frac{\lambda}{2}$*; (b) stacked with* $d_t = M_r d_r$ *and* $d_r = \frac{\lambda}{2}$*. Different colors indicate the transmit antenna either transmits different time slots or codes. © [2020] IEEE. Reprinted with permission from [6].*

and even chirp sequences, respectively. The receive array corresponding to odd and even chirp sequences forms two subarrays, which can be used to synthesize a virtual array according to interleaved or stacked configurations.

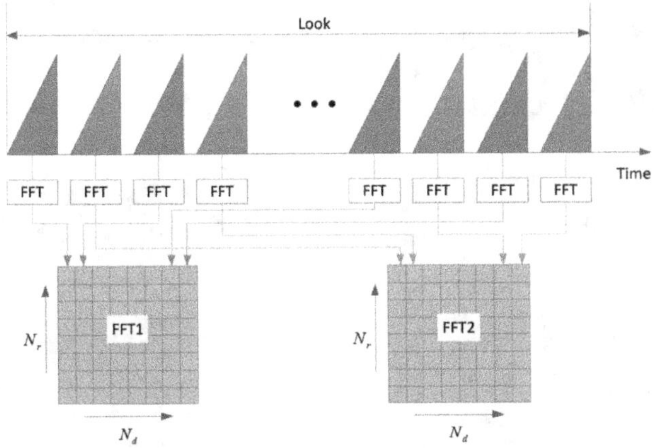

Figure 3.3 *Example of radar signal processing with TDM scheme, where $M_t = 2$ transmit antennas transmit FMCW chirp sequences alternatively. The red and green colors denote the odd and even echo chirp sequences, respectively. The range FFTs are conducted for each chirp, and the FFT outputs are stored in two matrices corresponding to odd and even sequences, respectively, for further processing. © [2020] IEEE. Reprinted with permission from [6].*

For a moving target with a velocity of v, the switching delays of transmit antennas introduce a target phase migration from chirp to chirp, which is defined as $\phi = 2\pi f_D \Delta t = \frac{4\pi}{\lambda} v \Delta t$. As a result, the virtual array pattern would be distorted [10]. In Figure 3.3, the phase difference between corresponding columns in the two matrices is $\phi = \frac{4\pi}{\lambda} v T_{\mathrm{PRI}}$. If $v = -0.5 v_{\max}$ and $v = -v_{\max}$, where $v_{\max}$ is the maximum unambiguous detectable radial speed and $v_{\max} = \frac{c}{4 f_c T_{\mathrm{PRI}}}$, the phase shifts are $\phi = -\pi/2$ and $\phi = -\pi$, respectively. The array beampattern distortion is demonstrated in Figure 3.4 for a moving target with a range of 35 m and an azimuth angle of $\theta = 0°$ [10]. Here, the MIMO radar with $M_t = 2, M_r = 8$ operates in a TDM fashion. As stated in Reference 10, and can also be seen in Figure 3.4, for interleaved MIMO array configuration, when the target velocity increases, the grating lobes at the edge of field of view (FOV) show up, while the peak at the target direction decreases and totally disappears when $v = -v_{\max}$. For stacked MIMO array configuration, as the target velocity increases, the peak is slightly off the boresight with a mirror grating lobe in the opposite direction.

The phase migration introduced by every moving target in the virtual array response needs to be compensated for before angle finding. The phase shift estimate $\hat{\phi}$ can be obtained after each target velocity has been estimated based on the 2D-FFT of a single receive antenna or non-coherent 2D-FFT integration of the same subarray. For example, in Figure 3.3, the phase in the beam vector of the subarray obtained from the even chirps needs to be compensated by multiplying them with

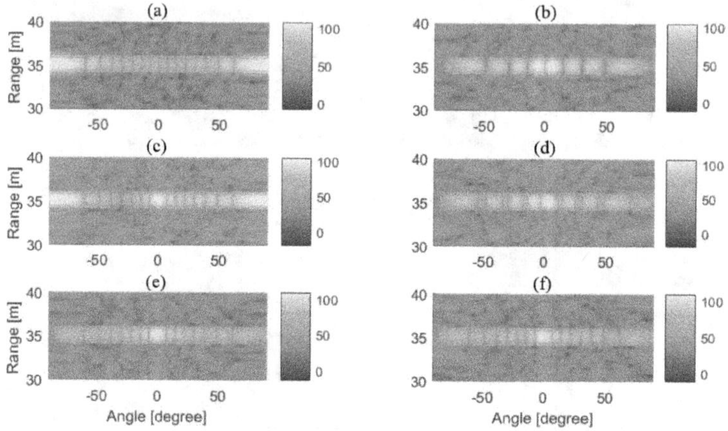

Figure 3.4 *Examples of MIMO radar range and azimuth images using TDM with $M_t = 2$, $M_r = 8$ for a single target with a range of 35 m and an azimuth angle of $\theta = 0°$. Two MIMO array configurations, interleaved (left column) and stacked (right column), are considered. The radial velocity of (a-b) $v = -v_{\max}$, (c-d) $v = -0.5v_{\max}$, and (e-f) $v = 0$. © [2020] IEEE. Reprinted with permission from [6].*

$e^{-j\hat{\phi}}$, while the phase in the beam vector of the subarray corresponding to odd chirps is kept unchanged.

It should be noted that, in the TDM MIMO radar, the pulse repetition interval is enlarged by the transmit antenna number M_t. As a result, the maximum unambiguous detectable velocity will be reduced by a factor of M_t [14].

3.2.2 Waveform orthogonality via DDM

In one look, a total of N chirps (i.e., pulses) are transmitted sequentially, with pulse repetition interval T_{PRI}. All transmit antennas simultaneously transmit the same FMCW waveform, after multiplying it with a phase code that is different for each antenna and changes between pulses, i.e., $x_m(n) = e^{j2\pi\alpha_m(n)}, m = 1, \cdots, M_t, n = 1, \cdots, N$ [15]. To separate the h-th transmit signal at the l-th receiver, after range FFT, a slow-time Doppler demodulation is applied to all range bins corresponding to the same chirp. The Doppler demodulated outputs of N chirps are assembled into a vector $\mathbf{s}_l^h$. Then, the Doppler FFT is applied to the vector $\mathbf{s}_l^h$. To separate the transmit signals in the Doppler domain, one of the two methods described in the following can be applied.

The first approach is to design phase codes such that the Doppler FFT of the interference $e^{j2\pi(\alpha_m(n)-\alpha_h(n))}$ is shifted to a frequency that is higher than the maximum detectable Doppler frequency $f_D^{\max}$. Therefore, a low-pass filter (LPF) can be applied to remove the interference [15]. One example of such phase codes is

$$\alpha_m(n) = \alpha_m n, m = 1, ...M_t, n = 1, ...N, \tag{3.3}$$

where the starting phase α_m is linear across different transmit antennas, i.e., $\alpha_m = a_0 m$. Figure 3.5a shows the range and Doppler spectrum of a target with a range of 75 m and a velocity of 10 m/s. The automotive MIMO radar has two transmit antennas and transmit phase codes given in (3.3) with $a_0 = 1$ and $N = 512$. It can be seen that signals from different transmit antennas are shifted to a higher Doppler spectrum, which can be removed via an LPF in the Doppler domain. Under this approach, the radar pulse repetition frequency f_{PRF} should be larger than $M_t f_D^{\max}$ [16]. Thus, if the f_{PRF} is kept unchanged, the maximum detectable unambiguous Doppler frequency is reduced by a factor of M_t. In practice, a Doppler unfolding or de-aliasing algorithm needs to be developed with different f_{PRF} in different looks.

The second approach is to design phase codes so that the Doppler FFT of the interference can be distributed into the entire Doppler spectrum as pseudo-noise. It is desired to minimize the peak interference residual (PIR) in the Doppler spectrum [17] calculated using the discrete-time Fourier transform (DTFT) for $m = 1, \cdots, M_t$, i.e.,

$$\mathrm{PIR} = \max_{f, m \neq h} \left| \sum_{n=1}^{N} e^{j2\pi(\alpha_m(n) - \alpha_h(n))} e^{j2\pi fn} \right| \tag{3.4}$$

where $f \in \left[-\frac{1}{2} f_{\mathrm{PRF}}, \frac{1}{2} f_{\mathrm{PRF}}\right]$. Following (3.4), the cross-correlation of the spectra of two codes needs to be flat [17], since the Fourier transform of multiplication of two codes in the time domain is equivalent to the convolution of the spectrum of one code with time-reversed and complex conjugate of the other. The maximum auto-correlation value of a unimodular sequence of length N is N. The ideal cross-correlation of two unimodular sequences of length N has a magnitude of $\sqrt{N}$. Thus, in the ideal case, according to [17], the maximum power gain of the currently transmitted signal over other signals is $\sqrt{N}$. For example, the maximum achievable waveform attenuation is about 27.1 dB for the unimodular sequence set with $N = 512$.

Constant amplitude zero auto-correlation codes are good candidates for DDM. The discrete Fourier transform (DFT) of a constant amplitude zero auto-correlation code has also constant amplitude and zero auto-correlation [18]. One of such examples is the Chu sequence [19], which is defined as $x_m(n) = e^{\frac{j\pi}{N} m(n+1)n}, m = 1, \ldots M_t, n = 1, \ldots, N$, where N is a prime number. In practice, the Chu sequence of prime length is first generated and then truncated into a length for efficient FFT. For example, we generate Chu codes of prime length 521 and truncate them to length $N = 512$. By calculation with FFT, the PIR defined in (3.4) is $1.08\sqrt{N}$. Therefore, the waveform attenuation for the Chu sequence of length $N = 512$ is about 26.4 dB. In Figure 3.5b, we show the range and Doppler spectrum of a target with a range of 75 m and a velocity of 10 m/s. The automotive radar has two transmit antennas, and two Chu sequences of length $N = 512$ are applied for slow time DDM. It can be seen that the waveform attenuation is about 26 dB.

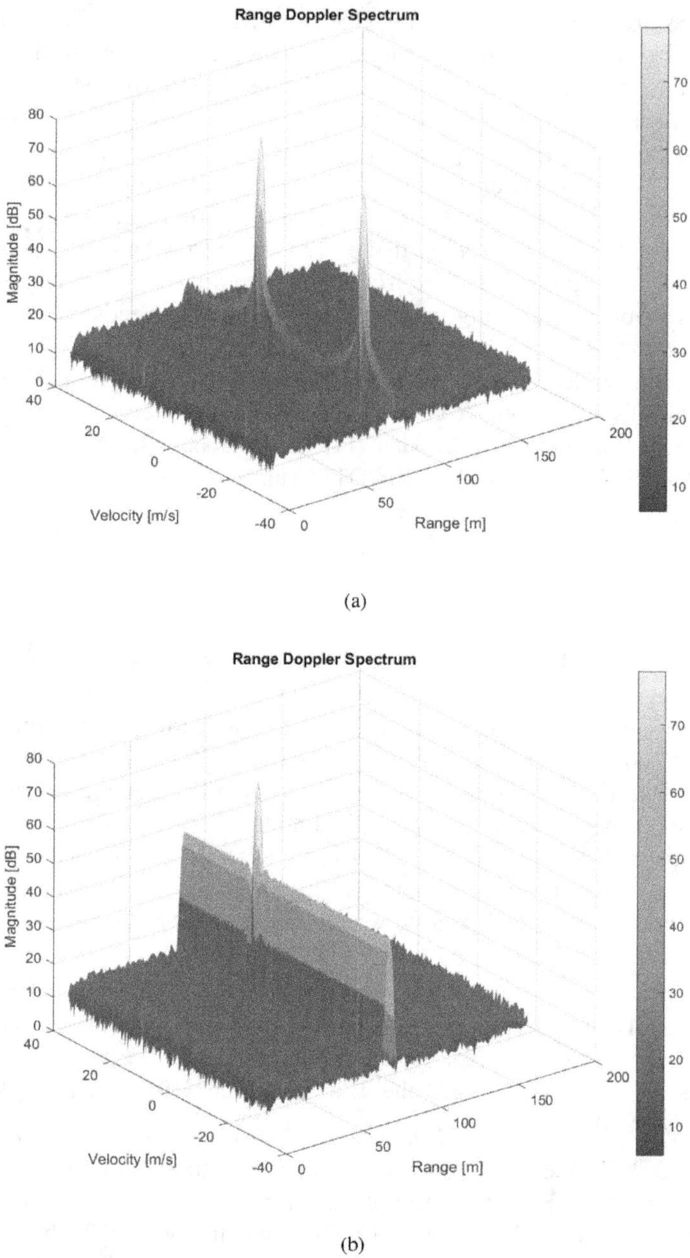

Figure 3.5 *The range and Doppler spectra of a target with a range of 75 m and a velocity of 10 m/s. The automotive MIMO radar has two transmit antennas, and slow-time phase coding of length $N = 512$ is applied for DDM. (a) Phase shift codes defined in (3.3). (b) Two Chu sequences. ©[2020] IEEE. Reprinted with permission from [6].*

In practice, binary phase codes are used due to hardware constraints [20]. The binary phase code sequences are obtained via exhaustive search such that the PIR in (3.4) is low. As the code length increases, the search time will grow exponentially.

The benefit of slow-time phase coding is that the interference from other transmitters does not affect different range bins. The range resolution is only determined by the bandwidth of the FMCW chirp. Therefore, it avoids the range sidelobe issue using fast-time phase coding. However, the Doppler sidelobes would be high due to the residual of the slow-time phase coding. As a result, targets with low radar cross section (RCS), e.g., pedestrians, that are close to the target with strong reflections, e.g., trucks, might be masked by the waveform residual. In other words, the waveform residual reduces the radar dynamic range. Given the code length, the number of phase codes with good correlation properties is limited, or equivalently, the number of antennas that can transmit simultaneously is limited.

3.2.3 *Waveform orthogonality via FDM*

In the FDM scheme, the transmitted signals are modulated by different carrier frequencies. According to Reference [21], the separation of multiple transmit FMCW signals is achieved by shifting the m-th transmit FMCW chirp by an offset frequency $f_{\text{off},m}$. If the differences between all $f_{\text{off},m}$ are larger than twice of the cut-off frequency of the anti-aliasing BPF $f_b^{\max}$, which is determined by the maximum unambiguous detectable range and Doppler, the transmitted signals can be separated at the receive end. Specifically, the received signal at each receiver is first mixed with the same starting carrier frequency f_c. The separation of transmit signals in the mixer output can be implemented by a frequency shift followed by an LPF with cut-off frequency $f_b^{\max}$ [21]. Each receiver needs to carry out such frequency shift and filtering operation M_t times. As a result, high range resolution can be realized using a typical FMCW chirp with large bandwidth. Meanwhile, after FMCW demodulation, frequency shift, and filtering operation, the FDM MIMO scheme can still utilize a low sampling rate determined by the beat signal.

Let us consider the FDM scheme in the context of the example in Section 2, i.e., an FMCW LRR radar with a maximum detectable range of 250 m and a maximum detectable velocity of 120 mph. For bandwidth $B = 150$ MHz and chirp duration $T = 50$ s, the maximum beat frequency is $f_b^{\max} = f_R^{\max} + f_D^{\max} = 5.0274$ MHz. Therefore, the frequency shift for the m-th transmit antenna in the FDM scheme can be chosen as $f_{\text{off},m} = 12\,(m-1)$ MHz. The intermediate frequency should have a bandwidth of $12M_t$ MHz to hold the mixer output.

3.3 Angle finding in automotive MIMO radar

In automotive MIMO radar with M_t transmit and M_r receive antennas, a virtual ULA of M_tM_r elements can be synthesized with inter-element spacing d. The array response can be written as

$$\mathbf{y} = \mathbf{A}(\theta)\,\mathbf{s} + \mathbf{n} \tag{3.5}$$

where $\mathbf{A}(\theta) = [\mathbf{a}(\theta_1), \cdots, \mathbf{a}(\theta_K)]$ is the virtual array steering matrix with

$$\mathbf{a}(\theta_k) = \left[1, e^{j\frac{2\pi}{\lambda}d\sin(\theta_k)}, \cdots, e^{j\frac{2\pi}{\lambda}(M_tM_r-1)d\sin(\theta_k)}\right]^T \tag{3.6}$$

Here, $\mathbf{n}$ is a noise term and $\mathbf{s} = \left[\beta_1, \cdots, \beta_K\right]^T$, where β_k denotes the target reflection coefficient for the k-th target. The array response at a particular time instance consisting of data obtained at all the virtual receivers and corresponding to the same range-Doppler bin is defined as the *array snapshot*. In highly dynamic automotive scenarios, usually, only a small number of array snapshots or even a single snapshot in the worst case is available [22].

In the automotive MIMO radar with virtual ULA, angle finding can be done with digital beamforming (DBF) [23–25] by performing FFTs on snapshots taken across the array elements, , i.e., $\mathbf{y}$ in (3.16) (see Figure 3.6). DBF can be implemented efficiently in an embedded DSP with a single snapshot. However, DBF is not a high-resolution angle-finding method. Higher-resolution angle finding can be achieved with subspace-based methods, such as MUSIC [26] and ESPRIT [27–30], sparse sensing-based methods [31–39], or the iterative adaptive approach (IAA) of [40] and [41]. The performance of subspace-based angle-finding methods relies on accurate estimation of the array covariance matrix with multiple snapshots, which is a challenging task in highly non-stationary automotive radar scenarios. In such a context, spatial smoothing [42] is applied for introducing virtual snapshots for array covariance matrix estimation. While sparse sensing-based methods and IAA have high computational cost, they yield angle estimates based on a single snapshot, which is important for the snapshot-limited automotive radar.

Achieving high angular resolution for L4 and L5 autonomous driving requirements using ULA with $d = \frac{\lambda}{2}$ is very expensive. According to Reference [43], the 3 dB beamwidth of antenna array with aperture size D is $\Delta\theta = 2\arcsin\left(\frac{1.4\lambda}{\pi D}\right)$. To achieve a 3 dB beamwidth of 1 degree, the antenna array aperture should be about $D \approx 51\lambda$. If the antenna array is ULA with inter-element spacing as half wavelength, it should be composed of about 100 array elements. Even with the help of MIMO radar technology, the cost of synthesizing such a large virtual ULA with half-wavelength element spacing is very high. One way to further reduce the cost without sacrificing the high angular resolution is via the use of nonuniform, or sparse linear arrays (SLAs) [44–49], synthesized with MIMO radar technology. In

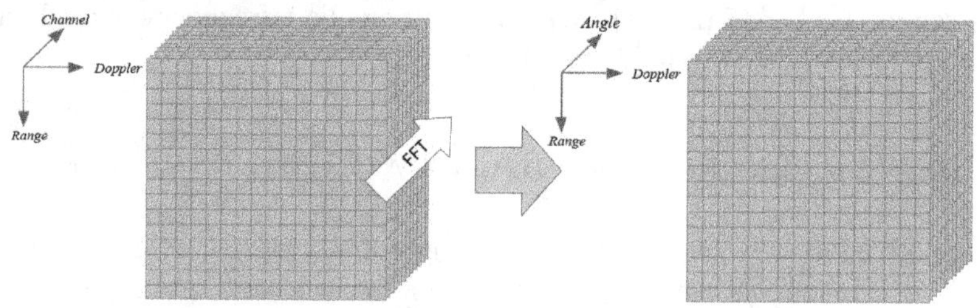

Figure 3.6 Digital beamforming (DBF). © [2020] IEEE. Reprinted with permission from Reference [6].

that context, selecting the locations of array elements and carrying out angle finding with the virtual sparse array are key problems.

3.3.1 High-resolution angle finding with ULA

3.3.1.1 Subspace methods with spatial smoothing

The performance of subspace-based angle-finding methods requires an estimate of the array covariance matrix. Such an estimate is typically obtained based on multiple snapshots. However, in a highly dynamic automotive environment, it is not possible to obtain enough snapshots before the model of (3.16) changes. In such scenarios, spatial smoothing [42] can introduce virtual snapshots for array covariance matrix estimation. In spatial smoothing, the array snapshot, $\mathbf{y}$, is divided into overlapped sub-arrays of length L, and a new sampled array covariance matrix $\mathbf{R} \in \mathbb{C}^{L \times L}$ is obtained based on the sub-array snapshots.

Eigenvalue decomposition of $\mathbf{R}$, along with the Akaike information criteria metric [50], or the minimum description length metric [51], can be used to identify the number of targets. It should be noted, however, that many ideal assumptions in the deduction of these criteria, including additive white Gaussian noise that is uncorrelated with source signal and the availability of enough snapshots for an accurate covariance matrix estimation, might not be satisfied in practice. To this end, forward–backward spatial smoothing techniques [52] have been widely applied to estimate the subarray covariance matrix for subspace methods-based angle estimation. When the subarray length is shorter, more subarrays are available for covariance matrix calculation, and thus a more accurate subspace estimation would be expected. However, when the subarray length is shorter, the DOA estimation resolution is reduced, so a tradeoff between obtaining the number of subarrays and length of subarray should be considered [53].

The target angles can be found by identifying the location of peaks of the MUSIC pseudo-spectrum [26], $P(\theta_i)$, computed at all possible θ_i's, i.e.,

$$P(\theta_i) = \frac{1}{\mathbf{a}_L^H(\theta_i)\,\mathbf{U_n}\mathbf{U_n}^H\mathbf{a}_L(\theta_i)} \tag{3.7}$$

where $\mathbf{U_n}$ is the noise subspace of $\mathbf{R}$ and $\mathbf{a}_L(\theta_i)$ is the array steering vector of length L corresponding to search direction θ_i. The computation cost of the MUSIC algorithm is high due to the angle search process. Alternatively, the ESPRIT algorithm could be used for angle estimation [27]. ESPRIT is also a subspace method, which exploits the array shift invariance properties, and has been widely used in practice. It has lower complexity than MUSIC, which, however, comes at the cost of reduced angular resolution. ESPRIT requires $2L$ sensors with L. To achieve the same angular resolution as MUSIC, ESPRIT needs twice as many sensors as MUSIC. As 2D arrays are needed in automotive radar in order to estimate both azimuth and elevation angles, 2D ESPRIT algorithms [30] can be applied if the array element spacing is uniform rectangular.

3.3.1.2 Compressive sensing

MIMO radars equipped with millimeter-wave technology offer wide bandwidth, thus achieving high range resolution. As a result, there are only a small number of targets that fall in the same range-Doppler bin, and thus, the targets are sparse in the DOA space. This property can be exploited by sparse sensing-based high-resolution methods for target angle estimation. To apply compressive sensing (CS) for DOA estimation, the whole DOA field of view is discretized into a fine grid. Assume that the DOA space is discretized on a fine grid with N points and there are K targets on the grid. The array response in (3.16) can be re-written as

$$\mathbf{y} = \mathbf{A}\mathbf{x} + \mathbf{n} \tag{3.8}$$

where $\mathbf{A} = [\mathbf{a}(\theta_1), \cdots, \mathbf{a}(\theta_N)]$ is the basis matrix with $\mathbf{a}(\theta_i)$ denoting the array steering vector corresponding to the i-th grid point and $\mathbf{x} = [\beta_1, \beta_2, \cdots, \beta_N]^T$ is a sparse vector with K non-zero elements. The value of β_i is non-zero if there is a target at the i-th grid point. The coherence of the basis matrix, defined as

$$\mu \triangleq \max_{i \neq l} \frac{\left|\mathbf{a}^H(\theta_i)\,\mathbf{a}(\theta_l)\right|}{\left\|\mathbf{a}(\theta_i)\right\|_{\ell_2}\left\|\mathbf{a}(\theta_l)\right\|_{\ell_2}} \tag{3.9}$$

needs to be low for obtaining uniform recovery guarantees [54]. When meeting the required coherence conditions, the DOA can be found by solving an ℓ_1 norm optimization problem, such as Dantzig selector [55] defined below:

$$\begin{aligned} &\min \|\mathbf{x}\|_{\ell_1} \\ &\text{s.t} \left\|\mathbf{A}^H(\mathbf{y} - \mathbf{A}\mathbf{x})\right\|_{\ell\infty} < \eta \end{aligned} \tag{3.10}$$

or greedy methods, such as orthogonal matching pursuit (OMP) [56].

In the above formulation, targets are assumed to be on the grid, which is not always possible in practice. While one can make the grid finer in order to capture the targets, the coherence of matrix $\mathbf{A}$ would increase, which would make the ℓ_1 norm solution invalid [57]. Thus, the performance of CS-based methods is sensitive to targets appearing off the grid [58]. Sparse sensing and matrix completion-based methods [38, 39] can avoid grid issues without sacrificing the high-resolution performance.

3.3.1.3 Iterative adaptive approach

The covariance matrix of M array snapshots $\mathbf{y}_l, l = 1, \cdots, M$, can be written as $\mathbf{R} = \mathbf{A}(\theta)\,\mathbf{P}\mathbf{A}^H(\theta)$, where $\mathbf{P}$ is a $K \times K$ diagonal matrix whose diagonal elements contain the power of target reflections. Angle finding in the (IAA) algorithm [40, 41] is carried out by iteratively estimating the reflection coefficient β_k. The estimate is found by minimizing the weighted least square cost function

$$\sum_{l=1}^{M} \left\|\mathbf{y}_l - \beta_k(l)\,\mathbf{a}(\theta_k)\right\|^2_{\mathbf{Q}^{-1}(\theta_k)}, \tag{3.11}$$

where $\|\mathbf{x}\|^2_{\mathbf{Q}^{-1}(\theta_k)} = \mathbf{x}^H\mathbf{Q}^{-1}(\theta_k)\,\mathbf{x}$ and the interference and noise covariance matrix $\mathbf{Q}(\theta_k) = \mathbf{R} - \hat{P}_k\mathbf{a}(\theta_k)\,\mathbf{a}^H(\theta_k)$. The solution is given by [40]

$$\hat{\beta}_k(l) = \frac{\mathbf{a}^H(\theta_k)\,\mathbf{R}^{-1}\mathbf{y}_l}{\mathbf{a}^H(\theta_k)\,\mathbf{R}^{-1}\mathbf{a}(\theta_k)} \tag{3.12}$$

Then matrix $\mathbf{P}$ can be updated as $\hat{P}_k = \frac{1}{M}\sum_{l=1}^{M}\left|\hat{\beta}_k(l)\right|^2$. In IAA algorithm implementation, the DOA space is discretized into a fine grid of N points, and steering matrix $\mathbf{A}$ is constructed in the same way as in CS. In addition, a standard delay-and-sum beamformer is used to initialize $\mathbf{P}$,

$$\hat{P}_k = \frac{\sum_{l=1}^{M}\left|\mathbf{a}^H(\theta_k)\,\mathbf{y}_l\right|^2}{M\left|\mathbf{a}^H(\theta_k)\,\mathbf{a}(\theta_k)\right|^2} \tag{3.13}$$

3.3.2 High-resolution angle finding with SLA

As stated before, the cost of synthesizing a large virtual ULA of D elements with half-wavelength element spacing is very high. One way to further reduce the cost without sacrificing the high angular resolution is via the use of nonuniform or SLAs [44, 45, 59]. With MIMO radar technology, $M_tM_r < D$ virtual array elements can be synthesized. To make the SLA aperture the same as the ULA, two virtual array elements should be deployed at the edge locations of the ULA. For the remaining virtual array elements, there are multiple possibilities to deploy. The main issue with SLA is that the grating lobes may introduce ambiguity in angle finding.

In automotive MIMO radar with virtual SLA, angle finding can still be done with conventional FFT or ESPRIT methods if the holes in the virtual SLA can be filled via interpolation or extrapolation techniques to mitigate the grating lobes [60, 61].

To mitigate the high sidelobes introduced by the sparse arrays, we utilize the matrix completion technology to interpolate/extrapolate the holes in the sparse arrays. Furthermore, matrix completion improves the SNR of array response as there is no loss/holes in the fully recovered arrays.

During one CPI of a typical automotive radar scenario, a dense point cloud with a high volume of targets could be detected in the range-Doppler spectrum [62]. The success of applying matrix completion in irregular one-dimensional (1D) or 2D sparse arrays relies on the following two facts:

1. The number of targets in the same range-Doppler bin that need angle estimation is small since the targets are first separated in the range-Doppler domain. In other words, the targets are sparsely present in the angular domain, and, as a result, the Hankel matrix constructed using the array response is low rank.
2. The SNR in the array snapshot is much higher than that in the echo signal since energy has been accumulated in both range and Doppler domains via the IDFT

and DFT operations. The high SNR in the array snapshot help reduce the matrix completion error and improve the accuracy of angle estimation.

We will illustrate the array interpolation concept through matrix completion in one-dimensional sparse arrays and then extend it to 2D sparse arrays.

3.3.2.1 1-D sparse array interpolation via matrix completion

Figure 3.7 shows an example of the physical array configuration of an automotive radar which is cascaded of two MIMO transceivers, where all transmit and receive antennas are clock synchronized. Let λ denote the wavelength of the carrier frequency. In this example, $M_t = 6$ transmit and $M_r = 8$ receive antennas are deployed on discretized grid points along the azimuth direction with an interval of length equal to 50λ. The interval is discretized uniformly with half-wavelength spacing. The waveform from each transmit antenna is transmitted in a way that, at each receive antenna, the contribution of each transmit antenna can be separated via DDM. Therefore, with MIMO radar technology, a virtual SLA with 48 array elements and aperture of 75λ is synthesized, as shown in Figure 3.7. Compared to a ULA with half-wavelength interelement spacing and the same aperture, a high number of elements at certain locations are "missing" at the rendered virtual SLA (denoted by zeros in the virtual array of Figure 3.7). However, the SLA approach uses only $M_t + M_r = 14$ physical antennas with significantly reduced mutual coupling effects [6].

Suppose an array snapshot contains K targets with DOAs θ_k, $k = 1, \cdots, K$. Without noise, the SLA response can be expressed as

$$\mathbf{y}_S = \mathbf{A}_S \mathbf{s} \tag{3.14}$$

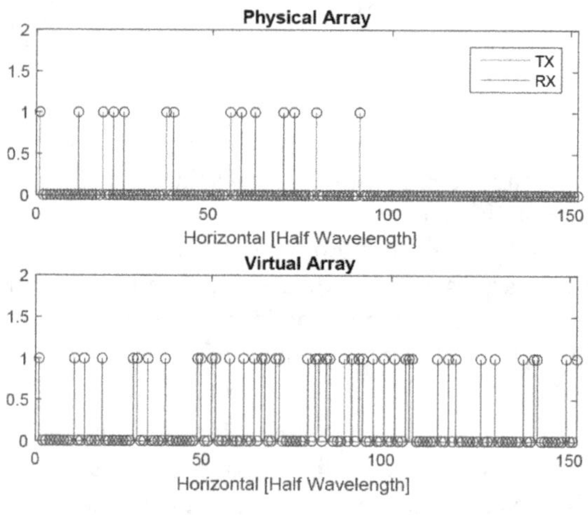

Figure 3.7 Example of an automotive radar cascaded with two transceivers. The virtual array has 48 elements. © [2021] IEEE. Reprinted with permission from Reference [5].

where $\mathbf{A}_S = [\mathbf{a}_S(\theta_1), \cdots, \mathbf{a}_S(\theta_K)]$ is the manifold matrix with

$$\mathbf{a}_S(\theta_k) = \left[1, e^{j\frac{2\pi}{\lambda}d_1 \sin(\theta_k)}, \cdots, e^{j\frac{2\pi}{\lambda}d_{M_tM_r-1}\sin(\theta_k)}\right]^T \tag{3.15}$$

and d_i is the distance between the i-th element of SLA and the reference element. In addition, $\mathbf{s} = [\beta_1, \cdots, \beta_K]^T$, where β_k denotes the amplitude associated with the k–th target.

Consider a virtual ULA that spans the entire array aperture and is filled with antennas spaced by interelement spacing $d = \lambda/2$. The total number of antennas in this virtual ULA is M_o and the noiseless array response is expressed as

$$\mathbf{y}_o = \mathbf{A}_o\mathbf{s} \tag{3.16}$$

where $\mathbf{A}_o = [\mathbf{a}_o(\theta_1), \cdots, \mathbf{a}_o(\theta_K)]$ is the array manifold matrix with

$$\mathbf{a}_o(\theta_k) = \left[1, e^{j\pi \sin(\theta_k)}, \cdots, e^{j\pi(M_o-1)\sin(\theta_k)}\right]^T. \tag{3.17}$$

Let $N_2 = \lfloor M_0/2 \rfloor$ and $N_1 = M_0 - N_2 \geq N_2$. We can formulate $\mathbf{y} \in \mathbb{C}^{M_o \times 1}$ into N_2 overlapped subarrays of length N_1. Based on those subarrays, we formulate a Hankel matrix $\mathbf{Y} \in \mathbb{C}^{N_1 \times N_2}$ with its (i,j)-th element given as $\mathbf{Y}_{ij} = \mathbf{y}_{i+j-1}$ for $i = 1, \cdots, N_1$ and $j = 1, \cdots, N_2$. The Hankel matrix $\mathbf{Y}$ has a Vandermonde factorization [63], expressed as

$$\mathbf{Y} = \mathbf{B}\boldsymbol{\Sigma}\mathbf{B}^T \tag{3.18}$$

where $\mathbf{B} = [\mathbf{b}(\theta_1), \cdots, \mathbf{b}(\theta_K)]$ is the subarray manifold matrix with

$$\mathbf{b}(\theta_k) = \left[1, e^{j\frac{2\pi}{\lambda}d\sin(\theta_k)}, \cdots, e^{j\frac{2\pi}{\lambda}(N-1)d\sin(\theta_k)}\right]^T \tag{3.19}$$

and $\boldsymbol{\Sigma} = \mathrm{diag}(\beta_1, \cdots, \beta_K)$ is a diagonal matrix. Thus, the rank of Hankel matrix $\mathbf{Y}$ is K if $N_2 \geq K$.

We can similarly construct a Hankel matrix $\mathbf{X}$ from the SLA configuration. Unlike matrix $\mathbf{Y}$ constructed from a full ULA, however, matrix $\mathbf{X}$ has many missing entries and thus can be viewed as a subsampled version of $\mathbf{Y}$. Under certain conditions, the missing elements can be fully recovered by solving a relaxed nuclear norm optimization problem conditioned on the observed entries [64]

$$\min \|\mathbf{X}\|_* \quad \text{s.t.} \quad \mathcal{P}_\Omega(\mathbf{X}) = \mathcal{P}_\Omega(\mathbf{Y}) \tag{3.20}$$

where $\|\cdot\|_*$ denotes the nuclear norm of a matrix, and $\mathcal{P}_\Omega(\mathbf{Y})$ is the sampling operator with Ω being the set of indices of observed entries that is determined by the SLA. In practice, the samples are corrupted by noise, i.e., $[\mathbf{X}]_{ij} = [\mathbf{Y}]_{ij} + [\mathbf{E}]_{ij}, (i,j) \in \Omega$, where $[\mathbf{E}]_{ij}$ denotes the noise. In this case, the matrix completion problem is formulated as

$$\min \|\mathbf{X}\|_* \quad \text{s.t.} \quad \|\mathcal{P}_\Omega(\mathbf{X} - \mathbf{Y})\|_F \leq \delta \tag{3.21}$$

where $\|\cdot\|_F$ denotes the Frobenius norm of a matrix and δ is a constant determined by the noise power.

Once the matrix $\mathbf{Y}$ is recovered, the full array response is obtained by averaging its anti-diagonal entries. DOAs can be estimated via standard array processing methods based on the array response corresponding to the completed matrix $\mathbf{Y}$.

To achieve high azimuth angular resolution, multiple automotive radar transceivers are cascaded together to synthesize a large sparse array in azimuth. Here, we consider the same physical array shown in Figure 3.7, where $M_t = 6$ transmit and $M_r = 8$ receive antennas are placed in an interleaved way along the horizontal direction at

$$l_{\text{TX}} = [1, 19, 37, 55, 79, 91]\,\lambda/2,$$
$$l_{\text{RX}} = [12, 22, 25, 39, 58, 62, 70, 73]\,\lambda/2$$

A virtual array with total 48 elements is synthesized. The transmit and receive antennas as well as the virtual array are plotted in Figure 3.7.

Two targets are at the same range $R = 100$ m with a velocity of $v = -10$ m/s. Their respective azimuth angles are $\theta_1 = 0°$ and $\theta_2 = 20°$. The two targets are first separated in range-Doppler. The complex peak values in the range-Doppler spectrum corresponding to every virtual sparse array consist of an array snapshot for azimuth angle finding.

The virtual SLA shown in Figure 3.7 acts as a deterministic sampler of a rank-2 Hankel matrix $\mathbf{Y} \in \mathbb{C}^{N \times N}$ with $N = 76$, which is constructed based on the array response of a ULA with 152 elements. The array response of the SLA is normalized by its first element. Based on the observed SLA response, the Hankel matrix $\mathbf{Y}$ is completed via the singular value thresholding (SVT) algorithm [65]. Let $\hat{\mathbf{Y}}$ denote the completed Hankel matrix. The full ULA response can be reconstructed by taking the average of the anti-diagonal elements of matrix $\hat{\mathbf{Y}}$. The completed full array has an aperture size of 76λ. Intuitively, in this simulation setting, matrix completion contributes around $10 \log 10\,(152/48) \approx 5$ dB SNR improvement for array processing.

In Figure 3.8, we plot the angle spectrum for the two targets. The two azimuth angle spectra are obtained by applying FFT to the original SLA with the holes filled with zeros and to the full array completed via matrix completion, respectively. It is found that the FFT of the SLA generates two peaks corresponding to the correct azimuth directions at a cost of high sidelobes, and thus it is difficult to detect the two targets in azimuth directions under the original SLA. On the contrary, the completed full array shows two clear peaks corresponding to correct azimuth locations in the angle spectrum, and the sidelobes are greatly suppressed in the completed full array.

3.3.2.2 2D sparse array interpolation via matrix completion

To enable driver-over and driver-under functions, automotive radar must measure target's elevation angles accurately. As a result, automotive radar needs to provide point clouds with high angular resolution in both azimuth and elevation directions. Figure 3.9 shows a MIMO radar with 12 transmit antennas and 16 receive antennas

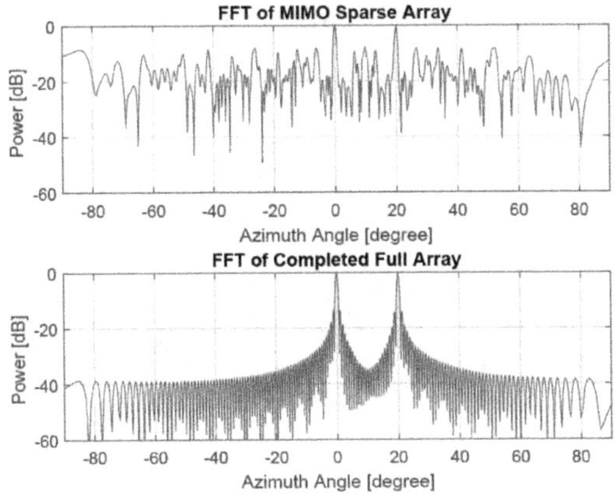

Figure 3.8 The spectrum of two targets with azimuth angles of $\theta_1 = 0°$ and $\theta_2 = 20°$ under MIMO sparse array and fully completed array.

that are obtained by cascading 4 automotive radar transceivers, and the transmit and receive antennas are randomly deployed in an area of $[0, 100]\,(\lambda/2) \times [0, 120]\,(\lambda/2)$ to synthesize a MIMO 2D virtual sparse array of 196 elements. The 2D physical array corresponds to a form factor of about 20×24 cm^2 when the carrier frequency is $f_c = 77$ GHz. It should be noted that a tradeoff between the angular resolution and the radar form factor should be considered in practice so that the radar can be incorporated behind the vehicle bumper. The dimension of the rendered 2D virtual sparse array is $D_y \times D_x = 183(\lambda/2) \times 194(\lambda/2)$, which can be viewed as a spatial sub-Nyquist sampling of a uniform rectangular array (URA) of the same dimension with half-wavelength spacing in both horizontal and vertical directions. The azimuth and elevation angular resolutions are, respectively, expressed as [43]

$$\Delta\theta = 2 \arcsin\left(\frac{1.4\lambda}{\pi D_x}\right) \approx 0.53° \tag{3.22}$$

$$\Delta\phi = 2 \arcsin\left(\frac{1.4\lambda}{\pi D_y}\right) \approx 0.56° \tag{3.23}$$

The angular resolution of imaging radar in this example is comparable to the Velodyne LiDAR HDL-32E whose horizontal resolution is between 0.1° and 0.4° depending on the rotation rate, and the vertical resolution is 1.33° [66].

Consider a general case of an $M_1 \times M_2$ URA with half-wavelength spacing, shown in Figure 3.10, where the URA is on the x-y plane. Assume the k-th point target with azimuth angle θ_k and elevation angle ϕ_k. Let χ_k denote the angle between the k-th target and the x-axis, and φ_k denote the angle between the k-th target and the y-axis. Then, it holds that $\cos(\chi_k) = \sin(\phi_k)\cos(\theta_k)$, $\cos(\varphi_k) = \sin(\phi_k)\sin(\theta_k)$. Therefore,

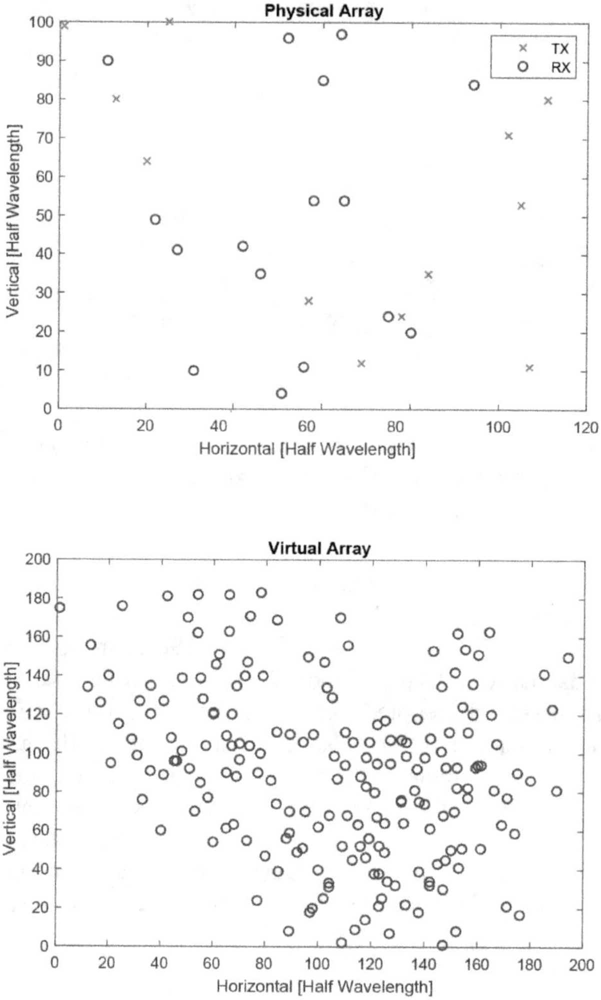

Figure 3.9 *A MIMO radar with* 12 *transmit antennas and* 16 *receive antennas by cascading* 4 *automotive radar transceivers. The transmit and receive antennas are randomly deployed in an area of* $[0, 100]\,(\lambda/2) \times [0, 120]\,(\lambda/2)$ *to synthesize an MIMO 2D virtual array of* 196 *elements. © [2021] IEEE. Reprinted with permission from [5].*

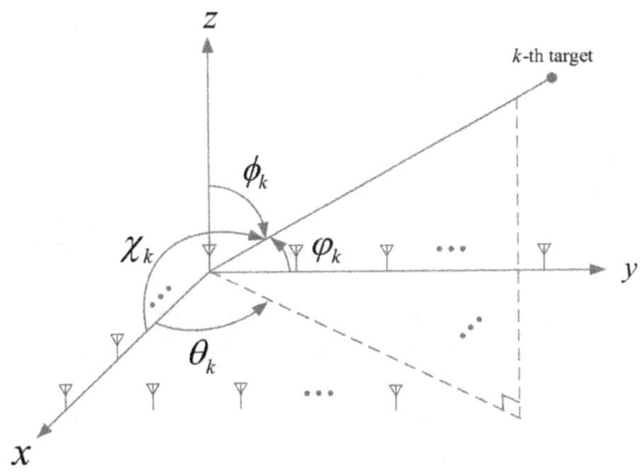

Figure 3.10 5 IEEE. Reprinted with permission from Reference [5].

$$\theta_k = \arctan\left(\frac{\cos(\varphi_k)}{\cos(\chi_k)}\right) \tag{3.24}$$

$$\phi_k = \arcsin\left(\sqrt{\cos^2(\chi_k) + \cos^2(\varphi_k)}\right) \tag{3.25}$$

Once the angles χ_k and φ_k are known, the azimuth angle θ_k and elevation angle ϕ_k can be uniquely determined. To simplify the signal modeling, we use angles χ_k and φ_k for signal modeling in URA.

The (m_1, m_2)-th element of the URA array on the *x-y* plane response with respect to K targets with angle to the x-axis χ_k and angle to the y-axis φ_k, $k = 1, ..., K$, can be written as

$$x_{m_1,m_2} = \sum_{k=1}^{K} \beta_k e^{j\pi((m_1-1)\sin(\chi_k)+(m_2-1)\sin(\varphi_k))} \tag{3.26}$$

for $1 \le m_1 \le M_1$ and $1 \le m_2 \le M_2$. Let $\mathbf{M} = \left[x_{m_1,m_2}\right]_{0\le m_1\le M_1, 0\le m_2\le M_2}$ be the data matrix with entries as the URA array response defined in (3.26). We can construct an $N_1 \times (M_1 - N_1 + 1)$ block Hankel matrix as

$$\mathbf{Y}_E = \begin{bmatrix} \mathbf{Y}_0 & \mathbf{Y}_1 & \cdots & \mathbf{Y}_{M_1-N_1} \\ \mathbf{Y}_1 & \mathbf{Y}_2 & \cdots & \mathbf{Y}_{M_1-N_1+1} \\ \vdots & \vdots & \ddots & \vdots \\ \mathbf{Y}_{N_1-1} & \mathbf{Y}_{N_1} & \cdots & \mathbf{Y}_{M_1-1} \end{bmatrix} \tag{3.27}$$

where

$$\mathbf{Y}_m = \begin{bmatrix} x_{m,0} & x_{m,1} & \cdots & x_{m,M_2-L} \\ x_{m,1} & x_{m,2} & \cdots & x_{m,M_2-L+1} \\ \vdots & \vdots & \ddots & \vdots \\ x_{m,L-1} & x_{m,L} & \cdots & x_{m,M_2-1} \end{bmatrix} \tag{3.28}$$

is an $L \times (M_2 - L + 1)$ Hankel matrix. It can be verified that the rank of matrix $\mathbf{Y}_E$ is K if $N_1 \geq K$ and $L \geq K$ [67].

By choosing the locations of the transmit and receive antennas, we aim to synthesize a sparse 2D array, which can be viewed as spatial subsampling of the URA. The array response of the URA can be obtained via completing the block Hankel matrix $\mathbf{Y}_E$ based on the array response of sparse arrays. The block Hankel matrix completion problem is formulated as

$$\min \|\mathbf{X}_E\|_* \quad \text{s.t.} \quad \mathscr{P}_\Omega(\mathbf{X}) = \mathscr{P}_\Omega(\mathbf{M}) \tag{3.29}$$

where Ω denotes the observation set consisting of the location of 2D sparse virtual array elements, and $\mathbf{X}_E$ is the block Hankel matrix constructed from matrix $\mathbf{X}$ following (3.27) and (3.28). In the noisy observation scenario, $\mathbf{M}$ is replaced by $\mathbf{M}^o = \left[x^o_{m_1,m_2}\right]_{0 \leq m_1 \leq M_1, 0 \leq m_2 \leq M_2}$ with $x^o_{m_1,m_2} = x_{m_1,m_2} + n_{m_1,m_2}$, where $x^o_{m_1,m_2}$ denotes the observed signal and $\mathbf{E} = \left[n_{m_1,m_2}\right]_{0 \leq m_1 \leq M_1, 0 \leq m_2 \leq M_2}$ is the noise term. We assume the noise is bounded, i.e., $\left\|\mathscr{P}_\Omega(\mathbf{E})\right\|_F \leq \delta$. The noisy block Hankel matrix completion problem is formulated as

$$\min \|\mathbf{X}_E\|_* \quad \text{s.t.} \quad \left\|\mathscr{P}_\Omega(\mathbf{X} - \mathbf{M}^o)\right\|_F \leq \delta \tag{3.30}$$

The above optimization problem can be solved in the CVX toolbox [68]. In the simulation, we adopt the SVT algorithm [65] to solve the matrix completion problems, whose computation cost of updating the low-rank matrix in each iteration is of order m with m being the cardinality of the observation set Ω, i.e., $m = |\Omega|$.

We consider the same 2D physical array shown in Figure 3.9 for joint high-resolution azimuth and elevation angle estimation, by cascading 4 automotive radar transceivers. These 12 transmit antennas and 16 receive antennas are randomly deployed in an area of $[0, 100(\lambda/2)] \times [0, 120(\lambda/2)]$ to synthesize a 2D MIMO virtual array of 196 elements. The cascaded automotive radar form factor is around 20×24 cm. In Figure 3.9, the dimension of the 2D sparse array is $D_y \times D_x = 183(\lambda/2) \times 194(\lambda/2)$. A total number of 35, 502 elements are required to construct a URA of the same dimension with half-wavelength interelement spacing. In other words, the virtual sparse array only occupies 0.54% the total elements of the URA.

Two targets with the same range and Doppler bin are considered. Their angles to the x and y directions are $(\chi_1, \varphi_1) = (-20°, 5°)$, $(\chi_2, \varphi_2) = (20°, 10°)$, respectively. The sparse array snapshot is consisted of the complex peak values in the range-Doppler spectrum corresponding to each sparse array element. The input SNR of the array response is set to 20 dB, which is reasonable in automotive radar because the fast-time and slow-time coherent processing provides a high processing gain. We then construct a block Hankel matrix $\mathbf{Y}_E$ of dimension $9,009 \times 8,928$ using one array snapshot of this 2D sparse array with 196 elements. Only 0.78% of the Hankel matrix entries are non-zero. Based on one snapshot of 2D sparse array, the block Hankel matrix is completed via the SVT algorithm and the full URA is then obtained. In this simulation setting, matrix completion contributes around $10 \log 10 \left(35,502/196\right) \approx 22.5$ dB SNR gain for array processing.

Figures 3.11 and 3.12 plot the azimuth-elevation spectra of the two targets under the 2D sparse array and the completed full URA, respectively. It is found that both sparse array and URA generate two peaks corresponding to the correct azimuth and elevation angles of the targets. However, in the azimuth-elevation spectrum of a 2D sparse array, there are high sidelobes over the entire azimuth and elevation FOVs. On the contrary, the high sidelobes are mitigated in the completed URA.

3.3.2.3 SLA optimization

Alternatively, instead of filling the holes, angle finding of the sparse array can be done using spatial CS ideas [69]. In that context, the key problems are how to select the locations of array elements such that the peak sidelobe level (PSL) of the virtual SLA beampattern is low and how to carry out angle finding. There is no analytical solution to determining the antenna locations that achieve a minimum PSL for a given number of antennas [70]. Optimal sparse array design requires global optimization techniques, such as particle swarm optimization [45, 71, 72].

In the SLA scenario, it can be easily verified that the coherence of the basis matrix (see 3.9) is the PSL of the SLA array beampattern [44]. Therefore, the

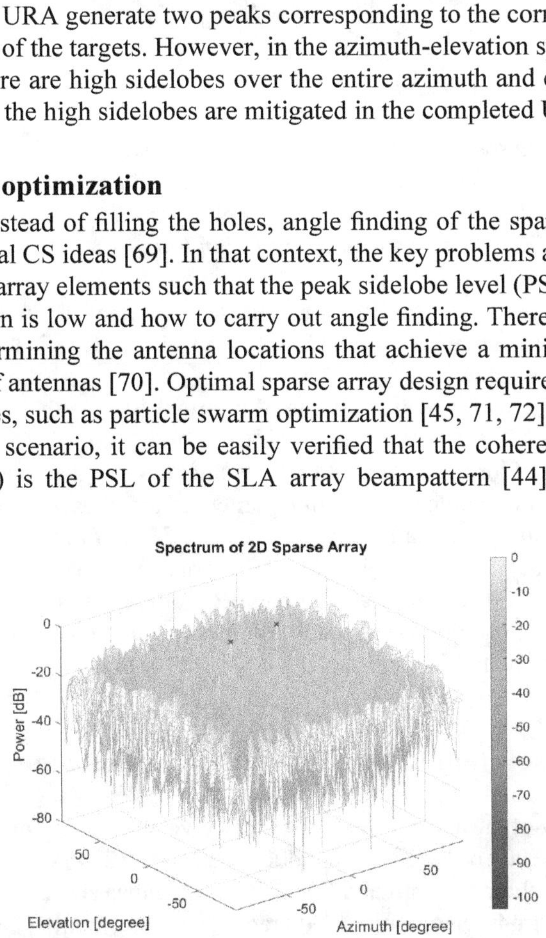

Figure 3.11 *The spectrum of two targets with azimuth and elevation angles of* $(\chi_1, \varphi_1) = (-20°, 5°), (\chi_2, \varphi_2) = (20°, 10°)$ *under the sparse array. The targets' angles are marked with crosses. There are high sidelobes in the spectrum due to the existence of a large number of holes in the sparse array. © [2021] IEEE. Reprinted with permission from [5].*

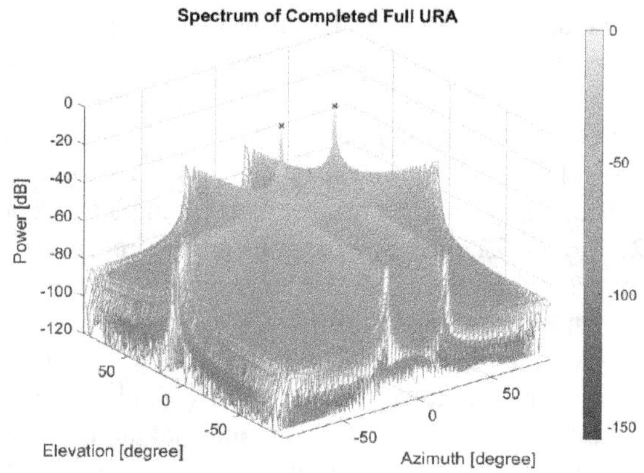

Figure 3.12 *The spectrum of two targets with azimuth and elevation angles of* $(\chi_1, \varphi_1) = (-20°, 5°)$, $(\chi_2, \varphi_2) = (20°, 10°)$ *under the completed full URA. The targets' angles are marked with crosses. © [2021] IEEE. Reprinted with permission from [5].*

coherence, or equivalently, the PSL of a sparse array plays a key role in obtaining uniform recovery guarantees for CS [54]. If the PSL of SLAs is low, angle finding using SLAs can be done via CS or IAA.

In Figure 3.13, we give an example of a virtual SLA with an aperture of 19λ, synthesized with MIMO radar technology using four transmit four receive antennas. The first and fourth of transmit/receive antennas are deployed at the edge of the physical aperture, while the remaining antennas are chosen such that the PSL is −9.1dB. Angle estimation via IAA when using the SLA of Figure 3.13 is illustrated in Figure 3.14. The ground truth involves two targets with azimuth 5° and 10°. The SNR of the received beamvector is set to 30 dB. For comparison, the FFT spectrum is also plotted. One can observe the sharper peaks around the target azimuth angles and the more attenuated sidelobe in the IAA spectrum, as compared to the FFT spectrum.

3.4 High-resolution imaging radar for autonomous driving

Nowadays, many of the automotive radar transceivers designed for ADAS functionality, such as MR3003 of NXP Semiconductor, and AWR2243 of Texas Instruments, can support up to three transmit and four receive antennas. Therefore, using a single automotive radar transceiver with MIMO radar technology, only 12 virtual array elements can be synthesized. In this section, we introduce the high-resolution imaging radar with hundreds of virtual antenna array elements synthesized via MIMO radar technology.

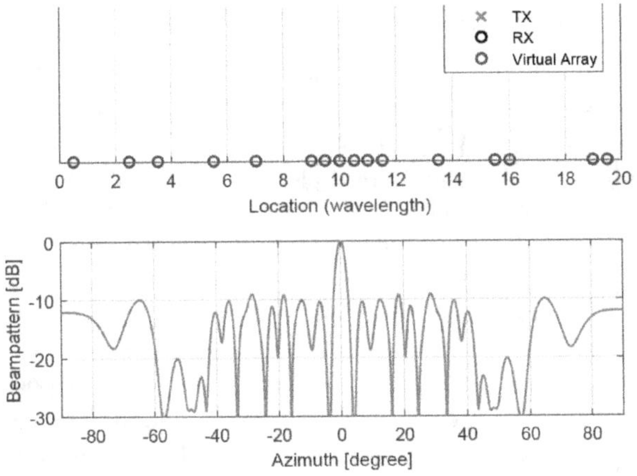

Figure 3.13 *Simulated example of an SLA synthesized with MIMO radar technology using four transmit and four receive antennas. The physical limitation of array aperture is 10λ. We fix the locations of the first and fourth transmit/receive antennas at $0\lambda, 0.5\lambda$ and $10\lambda, 9.5\lambda$, respectively, such that a maximum virtual array aperture of 19λ is achieved. The remaining transmit/receive antennas are chosen such that the PSL of the synthetic virtual array beampattern is −9.1 dB. ©[2020] IEEE. Reprinted with permission from [6].*

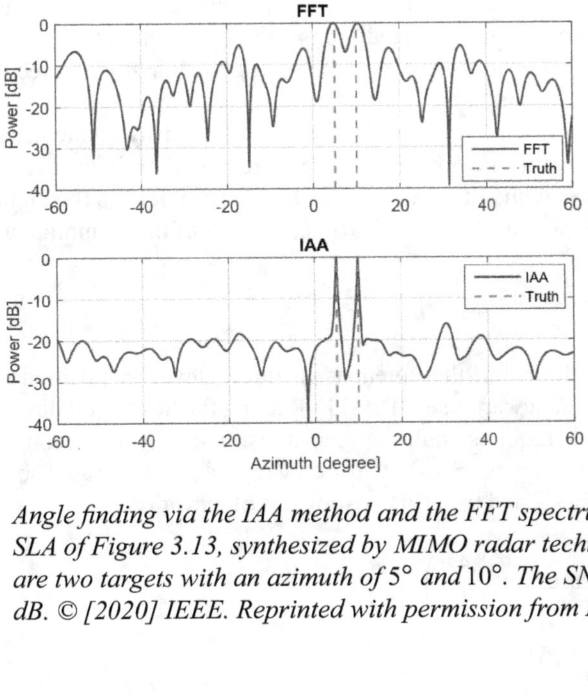

Figure 3.14 *Angle finding via the IAA method and the FFT spectrum using the SLA of Figure 3.13, synthesized by MIMO radar technology. There are two targets with an azimuth of 5° and 10°. The SNR is set to 30 dB. © [2020] IEEE. Reprinted with permission from Reference [6].*

3.4.1 Cascade of multiple radar transceivers

To meet the requirement for Level 4 and Level 5 autonomous driving, multiple automotive radar transceivers would need to be cascaded together, with all transceivers synchronized as a single unit. The received data from all receive antennas would be processed coherently. Cascading provides a cost-effective and scalable solution to achieve high angular resolution. In Reference 73, General Motors and Texas Instruments successfully demonstrated that up to 4 Texas Instruments AWR1243 radar chips can be cascaded together to provide 12 transmit and 16 receive antennas, enabling a synthesis of 192 virtual array elements. In [74], a prototype of cascading 5 Infineon radar chips has been built to synthesize a virtual array of 128 × 4 elements. Such a high number of virtual array elements provide a lot of opportunities in array design. Several azimuth and elevation array configurations can be found in [60]. Several commercial imaging radar products are available with different array configurations, such as forward-looking full-range radar of ZF and ARS540 of Continental [75, 76]. Usually, a trade-off of balancing angular resolution in azimuth and elevation needs to be considered.

3.4.2 Examples of cascaded imaging radars

Figure 3.15 shows an imaging radar design reference board that has 12 transmit and 16 receive antennas, formed by cascading 4 Texas Instruments AWR2243 radar transceivers [77]. The azimuth FOV is $[-70°, 70°]$. One transceiver is selected as master and all the others as slaves for clock distribution. In this way, synchronization can be achieved among 4 transceivers, allowing coherent FMCW transmission from the 12 transmit antennas and joint data processing from the 16 receive antennas. The array configuration of cascaded imaging radar is shown in Figure 3.15. There are three transmit antennas placed along the vertical direction for elevation angle finding and nine transmit antennas placed along the horizontal direction for azimuth angle finding. The virtual array in the horizontal direction is a dense ULA with half-wavelength spacing and consists of 86 virtual array elements (the overlapped virtual array elements are not shown). The array aperture in the azimuth direction is $D_x = 42.5\lambda$. In antenna theory, the 3 dB beamwidth defines the angular resolution. According to Reference [43], the 3 dB beamwidth of the azimuth angle is

$$\Delta\theta_{\text{AZ}} = 2\arcsin\left(\frac{1.4\lambda}{\pi D_x}\right) \approx 1.2°. \tag{3.31}$$

In the vertical direction, the antennas in three elevation positions form multiple minimum redundancy arrays (MRAs) [78] along the horizontal direction. The angle finding in MRA requires multiple snapshots. These MRAs along the horizontal direction can be used as snapshots for elevation angle finding. The elevation array aperture is $D_y = 3\lambda$, and the 3 dB beamwidth of elevation is

$$\Delta\theta_{\text{EL}} = 2\arcsin\left(\frac{1.4\lambda}{\pi D_y}\right) \approx 17°. \tag{3.32}$$

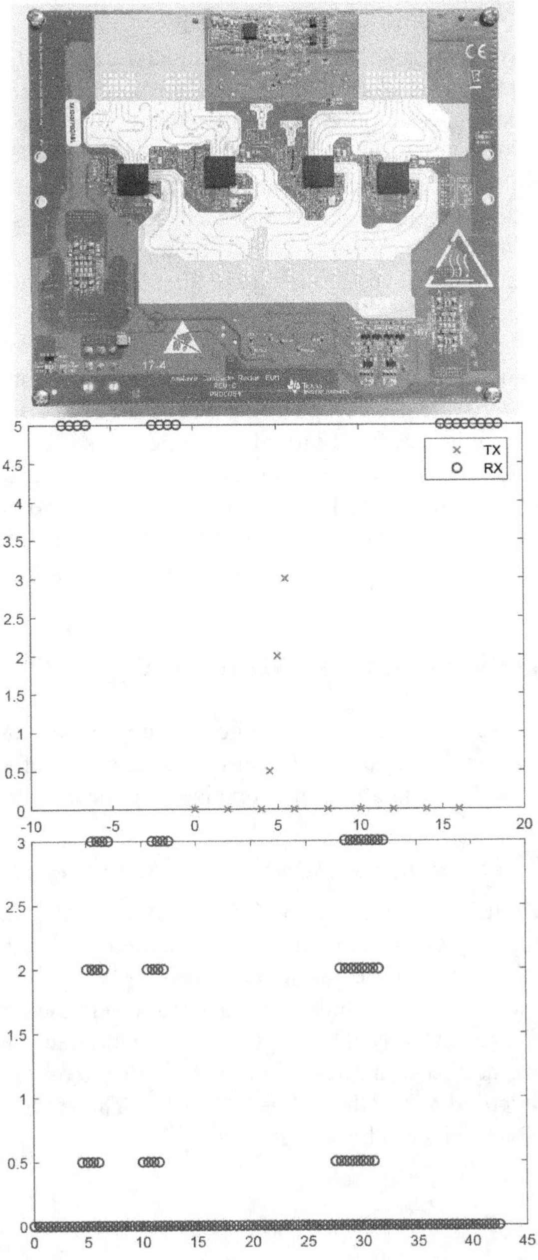

Figure 3.15 *Texas Instruments AWR2243 imaging radar board (top figure) [77]. Four TI AWR2243 radar transceivers are cascaded together, providing* 12 *transmit and* 16 *receive antennas (middle figure), enabling the synthesis of* 192 *virtual array elements (bottom figure).* *© [2020] IEEE. Reprinted with permission from Reference [6].*

3.4.3 Design challenges of imaging radar

Achieving waveform orthogonality in imaging radars using FMCW with a large of number transmit antennas is quite challenging. One strategy could be to divide the transmit antennas into several subgroups. In each subgroup, the transmit antennas would transmit simultaneously with slow-time phase coding (DDM), while antennas of different subgroups would be scheduled to transmit in different time slots (TDM).

Clock distribution among multiple cascaded transceivers is also challenging. For FMCW mixer operation, an LO is shared among master and slaves, and the LO routing from master to all slaves in the circuit should be matched. Also, the additional ADC sampling and data transmission among different transceivers need to be synchronized. It is desirable to develop an automotive radar transceiver that can incorporate a large number of transmit and receive antennas. For example, Uhnder has developed a radar system-on-chip (SoC), which has 12 transmit and 16 receive antennas, enabling synthesis of 192 virtual array elements [79] (discussed in more detail in later chapters). Thus, the four current automotive radar transceivers in the cascaded imaging radar shown in Figure 3.15 can be replaced with a single SoC radar chip. The radar on chip (RoC) developed by Vayyar has 48 transceivers at 76–81 GHz, which can provide synthesis of over 2,000 virtual array elements [80].

3.5 Challenges in automotive MIMO radar

In this section, we discuss the design challenges in automotive MIMO radar, including angle finding in the presence of multipath reflections, waveform orthogonality, and efficient high-resolution angle estimation algorithm development.

3.5.1 Angle finding in the presence of multipath reflections

Automotive radar runs in multipath scenarios [81]. In general, radio propagation in the presence of multipath occurs along four possible routes, i.e., direct/direct, direct/indirect, indirect/direct, and indirect/indirect routes.

Figure 3.16 shows a vertical multipath scenario, where the height of radar and target are h_S and h_T, respectively. The length of the direct/direct path is $d_1 = 2r_1$, the length of the direct/indirect or indirect/direct path is $d_2 = d_3 = r_1 + r_2 + r_3$, and the length of the indirect/indirect path is $d_4 = 2\left(r_2 + r_3\right)$. The received signal, having gone through the four paths can be written as

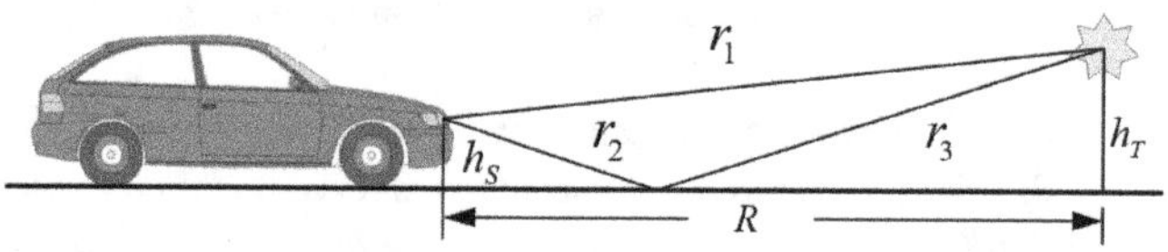

Figure 3.16 Direct path and vertical multipath. © [2020] IEEE. Reprinted with permission from [6].

$$y_r = \sum_{i=1}^{4} \beta_i e^{j\frac{2\pi}{\lambda} d_i} \tag{3.33}$$

where the amplitude term β_i is the function of the antenna gain, the path loss, the road reflection coefficient, and the target RCS. The signal from the indirect paths and that from the direct path would most probably arrive out of phase and thus add up destructively. As a result, the power of the received signal would fluctuate with distance [81], and thus angle finding at SNR nulls would be unstable.

MIMO radar with colocated transmit and receive antennas, also referred to as monostatic MIMO radar, is based on the assumption that DOD and DOA are equal. However, in the presence of multipath, that assumption does not hold and the system becomes bistatic [82, 83], i.e., the transmit and receive antennas view the target from different aspect angles. Figure 3.17 shows a vehicle moving parallel to the guardrail, with an SRR sensor mounted at its front left corner. The length of the direct path of the radar signal is $d_{r_1} = r_1$, corresponding to $\theta_t = \theta_r = \theta_1$. There are also multipath reflections due to the guardrail. The range of the first multipath reflection is $d_{r_2} = (r_1 + r_2 + r_3)/2$, corresponding to $\theta_t = \theta_1, \theta_r = \theta_2$ or $\theta_t = \theta_2, \theta_r = \theta_1$. The range of the second multipath reflection is $d_{r_3} = r_2 + r_3$, corresponding to $\theta_t = \theta_r = \theta_2$. Compared to the direct path, multipath reflections result in longer range and smaller Doppler. For the first type of multipath, the range and Doppler bin are the same as in mirror image target detection. However, as $\theta_t \neq \theta_r$, it turns out that the phase of each virtual array element is corrupted. In other words, the monostatic

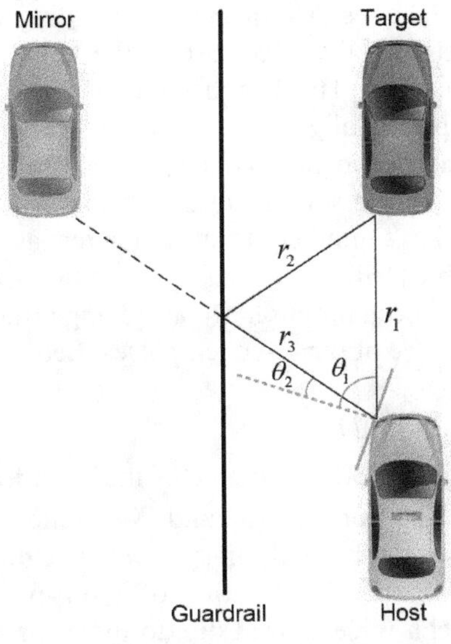

Figure 3.17 A typical multipath reflection scenario along a guardrail for a short-range radar sensor mounted at the front left corner. © [2020] IEEE. Reprinted with permission from [6].

MIMO radar assumption does not hold, which results in a "ghost" target whose direction is different from the mirror target.

To solve this issue, some ideas have been proposed in Reference [82–84]. For example, joint estimation of DOD and DOA is proposed in Reference [82], however, by ignoring the structure of the transmit array, that method cannot enjoy the benefit of the synthesized virtual array. Polarimetric features are exploited in Reference [83] to separate objects in a multipath scenario. However, the approach in Reference [83] can only separate certain real target cases from their mirror targets, for example, when the real target is known and the polarization state change of the multipaths can be recognized. The method in Reference [83] does not work when the ghost target direction is different from that of the mirror target. Doppler information can be exploited to detect moving vehicles in urban areas under multipath [84]. However, Doppler information is not always available when both objects and host vehicles are stationary. In general, there is a need for more research addressing the ghost target issue in MIMO radar due to multipath.

3.5.2 Waveform orthogonality in automotive MIMO radar

Different strategies, such as TDM, DDM, and FDM, can be adopted in automotive FMCW radars to achieve waveform orthogonality. However, several challenges associated with each strategy need to be addressed. For example, in the TDM scheme, the scheduling delay between transmit antennas may introduce phase error for a moving target, which needs to be compensated for; otherwise, the synthesized array beampattern will be distorted. Further, the maximum unambiguous detectable velocity under TDM is reduced by a factor of M_t. In the DDM scheme, the Doppler sidelobes are high due to the residual of the phase coding. As a result, targets with small RCSs, e.g., pedestrians, that are close to the target with strong reflections, e.g., trucks, might be masked. The search time for phase codes using stochastic algorithms increases exponentially as the code length increases. Computationally efficient algorithms are needed to address this problem. In the FDM scheme, though a randomization of frequency shift among transmit antennas could reduce the range–angle coupling, a large number of transmit antennas would be needed for the improvement to be notable [16].

Recently, PMCW has been proposed for achieving orthogonality [85–88]. Each antenna transmits a sequence of phase-coded pulses. Let

$$\mathbf{x}_m = \left[x_m(1), \cdots, x_m(N_p)\right]^T \tag{3.34}$$

be the complex unimodular code sequence of the m-th transmit antenna, where $x_m(n) = e^{j\phi_m(n)}$ is the n-th code of $\mathbf{x}_m$ and N_p is the code length. Here, the phase $\phi_m(n)$ can be chosen arbitrarily in $[-\pi, \pi]$. The duration of a single code sequence is $T_p = N_p T_c$ with T_c the duration of a sub-pulse. In practice, binary code sequences have been widely used due to their simplicity. The bandwidth of PMCW is $B = 1/T_c$. The time–bandwidth product of a code sequence is $BT_p = N_p$. Since the pulses are transmitted continuously, the code sequences should have good periodic auto-correlation and cross-correlation properties

[85]. The periodic cross-correlation of two code sequences $\mathbf{x}_m$ and $\mathbf{x}_l$ at lag k is defined as

$$r_{ml}^{\mathcal{P}}(k) = \sum_{n=1}^{N} x_m(n)\, x_l^*\left(\left(n+k\right) \bmod \left(N_p\right)\right) \tag{3.35}$$

When $m = l$, $r_{ml}^{\mathcal{P}}(k)$ becomes the periodic auto-correlation function of $\mathbf{x}_m$. Good correlation properties require the values of the periodic auto-correlation at non-zero lags, and the values of the cross-correlation at any lag to be low. The Welch lower bound on the cross-correlation between any pair of binary sequences with a period of N_p in a set of M_t sequences equals [89]

$$r_{ml}^{\mathcal{P}}(k) \geq N_p \sqrt{\frac{M_t - 1}{M_t N_p - 1}} \approx \sqrt{N_p} \tag{3.36}$$

Good periodic cross-correlation properties help achieve waveform orthogonality, while good periodic auto-correlation properties make it easier to use matched filters to extract signals reflected from the range bin of interest and suppress signals reflected from other range bins.

As compared to FMCW, PMCW radar has several advantages. PMCW radar is better suited for achieving waveform orthogonality in imaging radars with a large number of transmit antennas. PMCW radar can take advantage of existing sequences with good auto-correlation and cross-correlation properties, previously developed for code-division multiple accessing communications, such as Gold, Kasami, and m-sequences [90–92]. Further, in PMCW radar, each automotive radar sensor can have a unique digital sequence, which may help reduce the automotive radar mutual interference. As a bonus, PMCW radar also provides certain communication capabilities [93] and thus can be explored as a dual functional radar communication system [94].

However, PMCW radar has many implementation challenges. First, the sampling rate of ADC should satisfy the Nyquist rule, i.e., $f_s \geq 2B = 2/T_c$. The high bandwidth required for high range resolution necessitates high-speed ADC and high-speed processing hardware. In practice, it is required to keep the resolution of ADC as low as possible [95]. Second, according to the Welch bound of (3.36), the cross-correlation lower bound of any pair of binary sequences is of the order of $\mathcal{O}\left(\sqrt{N_p}\right)$, which might not provide sufficient separation of transmit waveforms of different antennas. In practice, the auto-correlation and cross-correlation of code sequences are desired to have low sidelobes within a low correlation zone. Furthermore, because there is no mapping relationship between range and beat signal in PMCW, it would be difficult to use high-pass analog filters to reject or attenuate ultra-close-range return signals, including direct path signals from transmit antennas, reflections from Radome, and vehicle bumpers. This escalates the dynamic range challenge, especially when the resolution of ADC must be kept as low as possible [95].

3.5.3 Efficient, high-resolution angle-finding algorithms are needed

A typical duration of a look in automotive radar is around 50 ms, corresponding to the detection update rate of 20 Hz [96]. In such a short duration, the current generation of automotive radar for ADAS can report a maximum of 64–200 detections. With high-resolution imaging radar, the number of cells that can be selected for angle finding from the 2D range and Doppler spectrum is around 10,000 in a single look for a typical road scenario [62]. To achieve point clouds for autonomous driving, angle finding needs to be performed thousands of times in a single look, which is a great challenge for imaging radar with hundreds of virtual array elements. Computationally efficient, high-resolution angle-finding algorithms are highly desirable for real-time implementation in automotive radar.

To reduce the computation complexity, beamspace ESPRIT [28] and unitary ESPRIT [29] algorithms have been proposed. The idea of beamspace ESPRIT is to decompose the original ULA vector into several low-dimensional beamspaces via a transform such as the FFT. Then, if the beamspace transform matrix has the same shift-invariance structure, angle finding can be carried out via ESPRIT on each beamspace in parallel, with reduced computational time [28]. The unitary ESPRIT algorithm takes advantage of the unit magnitude property of the phase factors representing the phase delays between the two subarrays and is formulated in terms of real-valued computations. As a result, it achieves a great reduction in computational complexity [29].

The computation cost of each IAA iteration is $2NM^2 + NM + M^3$, where M is the number of array snapshots and N is the number of discretized grids. Fast and superfast IAA algorithms have been proposed in Reference [97–99], respectively. The fast IAA algorithm exploits the FFT operation as well as Gohberg–Semencul representation of matrix $\mathbf{R}^{-1}$. As a result, the computation cost of each fast IAA iteration is $M^2 + 12\zeta(2M) + 3\zeta(N)$, where $\zeta(N)$ stands for the computation cost of performing FFT of size N, i.e., $O(N\log N)$ [98]. The superfast IAA uses a conjugate gradient algorithm to approximate the matrix $\mathbf{R}^{-1}$, which further reduces the computation cost.

The strengths and limitations of each DOA estimation algorithm discussed in Section 3.3 when applied to the automotive radar scenario are summarized in Table 3.1. For subspace-based high-resolution DOA estimation methods, such as MUSIC and ESPRIT, the automotive radar array needs to be ULA, and multiple snapshots are required to estimate the array covariance matrix accurately. However, automotive radars operating in a highly dynamic environment typically rely on a single snapshot. While multiple snapshots can be generated via spatial smoothing or by dividing a chirp into sub-chirps, the associated cost is respectively reduced array aperture or reduced SNR. SLAs have been widely used in automotive radar to further reduce the hardware cost. However, it is not straightforward to apply MUSIC or ESPRIT to SLA-based automotive radar. On the other hand, DBF and sparsity-based high-resolution methods, such as OMP and IAA, apply to SLA- as well as ULA-based automotive radar and work with a single snapshot. In the DBF method, the number of targets can be estimated by counting the number of peaks

Table 3.1 Summary of different DOA estimation algorithms in automotive radar scenario

Algorithms	**Resolution**	**Snapshot**	**Arrays**	**Grid free**	**Rank estimation**	**Robustness**	**Complexity**
DBF	Low	Single	ULA/SLA	No	No	Strong	Low
MUSIC	High	Multiple	ULA	No	Yes	Medium	High
ESPRIT	High	Multiple	ULA	Yes	Yes	Medium	Medium
OMP	High	Single	ULA/SLA	No	No	Medium	High
IAA	High	Single	ULA/SLA	No	No	Strong	High

in the DOA spectrum. DBF is not sensitive to coherent or correlated signals which, in subspace-based methods [100], need special preprocessing via spatial smoothing. It has been shown that DBF is robust to array element position errors and has low computational cost [100]. However, DBF is not a high-resolution method. Also, as OMP and IAA are iterative schemes, they involve high computation cost, which limits their applicability in low-cost embedded DSPs typically used in current generation automotive radar. Further, the methods of DBF, MUSIC, OMP, and IAA assume targets are on the grid and suffer from errors when the targets arise between grid points. In summary, more research is needed on developing computationally efficient, high-resolution DOA estimation algorithms that are robust to noise and are applicable to automotive radar using SLAs with low PSL under a single snapshot.

References

[1] C.R Q., Su H., Mo K., Guibas L.J. 'PointNet: deep learning on point sets for 3D classification and segmentation'. Presented at Proceedings the IEEE Conference on Computer Vision and Pattern Recognition (CVPR); Honolulu, HI.

[2] Qi C.R., Yi L., Su H., Guibas L.J. 'PointNet++: deep hierarchical feature learning on point sets in a metric space'. Presented at Proceedings 31st Conference on Neural Information Processing Systems (NIPS); Long Beach, CA.

[3] Brisken S., Ruf F., Höhne F. 'Recent evolution of automotive imaging radar and its information content'. *IET Radar, Sonar & Navigation*. 2018, vol. 12(10), pp. 1078–81.

[4] Danzer A., Griebel T., Bach M., Dietmayer K. '2D CAR detection in radar data with pointnets'. *ArXiv*. 2019, arXiv:1904.08414.

[5] Sun S., Zhang Y.D. '4D automotive radar sensing for autonomous vehicles: A sparsity-oriented approach'. *IEEE Journal of Selected Topics Signal Process*. 2021, vol. 15(4), pp. 879–91.

[6] Sun S., Petropulu A.P., Poor H.V. 'MIMO radar for advanced driver-assistance systems and autonomous driving: advantages and challenges'. *IEEE Signal Processing Magazine*. 2020, vol. 37(4), pp. 98–117.

[7] Bergin J., Guerci J.R. *MIMO radar: theory and application*. Boston, MA: Artech House; 2018.

[8] Duly A.J., Love D.J., Krogmeier J.V. 'Time-division beamforming for MIMO radar waveform design'. *IEEE Transactions on Aerospace and Electronic Systems*. 2013, vol. 49(2), pp. 1210–23.

[9] Rambach K., Yang B. 'Colocated MIMO radar: cramer-rao bound and optimal time division multiplexing for DOA estimation of moving targets'. Proceedings of IEEE 38th on Acoustics, Speech, and Signal Processing (ICASSP); Vancouver, Canada, 2013.

[10] Zoeke D., Ziroff A. 'Phase migration effects in moving target localization using switched MIMO arrays'. Presented at Proceedings of European Radar Conference (EuRAD); Paris, France.

[11] Bleh D., Rosch M., Kuri M., *et al.* 'W-band time-domain multiplexing FMCW MIMO radar for far-field 3-D imaging,'. *IEEE Transactions on Microwave Theory and Techniques*. 2017, vol. 65(9), pp. 3474–84.

[12] Bechter J., Roos F., Waldschmidt C. 'Compensation of motion-induced phase errors in TDM MIMO radars'. *IEEE Microwave and Wireless Components Letters*. 2017, vol. 27(12), pp. 1164–6.

[13] Jansen F., Filippi A., Zivkoic Z. 'MIMO radar system'. *European Patent EP3021132A1*. 2016.

[14] Roos F., Bechter J., Appenrodt N., Dickmann J., Waldschmidt C. 'Enhancement of doppler unambiguity for chirp-sequence modulated TDM-MIMO radars'. Presented at Proceedings of Int Conference on Microwaves for Intelligent Mobility (ICMIM); Apr 2018.

[15] Mecca V.F., Ramakrishnan D., Krolik J.L. 'MIMO radar space-time adaptive processing for multipath clutter mitigation'. *Proceedings of IEEE Workshop on Sensor Array and Multichannel Processing (SAM)*. 2006.

[16] Sun H., Brigui F., Lesturgie M. 'Analysis and comparison of MIMO radar waveforms'. Proceedings of Int Radar Conference; 2014.

[17] Madsen N., Cao S. 'Slow-time waveform design for MIMO GMTI radar using CAZAC sequences'. *Proceedings of IEEE Radar Conference*; 2018.

[18] Benedetto J.J., Konstantinidis I., Rangaswamy M. 'Phase-coded waveforms and their design'. *IEEE Signal Processing Magazine*. 2009, vol. 26(1), pp. 22–31.

[19] Chu D. 'Polyphase codes with good periodic correlation properties (corresp)'. *IEEE Transactions on Information Theory*. 1972, vol. 18(4), pp. 531–32.

[20] Feger R., Haderer H., Stelzer A. 'Optimization of codes and weighting functions for binary phase-coded FMCW MIMO radars'. Proceedings of International Conference on Microwaves for Intelligent Mobility (ICMIM); San Diego, CA.

[21] Feger R., Pfeffer C., Stelzer A. 'A frequency-division MIMO FMCW radar system based on delta–sigma modulated transmitters'. *IEEE Transactions on Microwave Theory and Techniques*. 2014, vol. 62(12), pp. 3572–81.

[22] Häcker P., Yang B. 'Single snapshot DOA estimation'. *Advances in Radio Science*. 2010, vol. 8, pp. 251–6.

[23] Barton P. 'Digital beam forming for radar'. *IEEE Proceedings of Communications, Radar and Signal Processing*. 1980, vol. 127(4), p. 266.

[24] Alland S., Searcy J. 'Radar system and method of digital beamforming'. *U.S. Patent 2009/0085800*. 2009.

[25] Tokoro S., Kuroda K., Kawakubo A., Fujita K., Fujinami H. 'Electronically scanned millimeter-wave radar for precrash safety and adaptive cruise control system'. *Proceedings of IEEE Intelligent Vehicles Symposium*; Columbus, OH, 2003.

[26] Schmidt R. 'Multiple emitter location and signal parameter estimation'. *IEEE Transactions on Antennas and Propagation*. 1986, vol. 34(3), pp. 276–80.
[27] Roy R., Kailath T. 'ESPRIT - estimation of signal parameters via rotation invariance techniques'. *IEEE Transactions on Acoustics, Speech, Signal Process*. 1989, vol. 17(7), pp. 984–95.
[28] Xu G., Silverstein S.D., Roy R.H., Kailath T. 'Beamspace ESPRIT'. *IEEE Transactions on Signal Processing: A Publication of the IEEE Signal Processing Society*. 1994, vol. 42(2), pp. 349–56.
[29] Haardt M., Nossek J.A. 'Unitary ESPRIT: how to obtain increased estimation accuracy with a reduced computational burden'. *IEEE Transactions on Signal Processing*. 1995, vol. 43(5), pp. 1232–42.
[30] Zoltowski M.D., Haardt M., Mathews C.P. 'Closed-form 2-D angle estimation with rectangular arrays in element space or beamspace via unitary ESPRIT'. *IEEE Transactions on Signal Processing*. 1996, vol. 44(2), pp. 316–28.
[31] Strohmer T., Friedlander B. 'Compressed sensing for MIMO radar - algorithms and performance'. Proceedings of 43th Annual Asilomar Conference on Signals Systems, and Computers; Pacific Grove, CA, 2009.
[32] Yu Y., Petropulu A.P., Poor H.V. 'MIMO radar using compressive sampling'. *IEEE Journal of Selected Topics in Signal Processing*. 2010, vol. 4(1), pp. 146–63.
[33] Yu Y., Petropulu A.P., Poor H.V. 'Measurement matrix design for compressive sensing–based MIMO radar'. *IEEE Transactions on Signal Processing*. 2011, vol. 59(11), pp. 5338–52.
[34] Yu Y., Petropulu A.P., Poor H.V. 'CSSF MIMO radar: compressive-sensing and step-frequency based MIMO radar'. *IEEE Transactions on Aerospace and Electronic Systems*. 2012, vol. 48(2), pp. 1490–504.
[35] Yu Y., Sun S., Madan R.N., Petropulu A. 'Power allocation and waveform design for the compressive sensing based MIMO radar'. *IEEE Transactions on Aerospace and Electronic Systems*. 2014, vol. 50(2), pp. 898–909.
[36] Strohmer T., Friedlander B. 'Analysis of sparse MIMO radar'. *Applied and Computational Harmonic Analysis*. 2014, vol. 37(3), pp. 361–88.
[37] Sun S. *Mimo radars with sparse sensing*. Ph.D. dissertation. New Brunswick, NJ: Rutgers University; 2016.
[38] Sun S., Bajwa W.U., Petropulu A.P. 'MIMO-MC radar: a MIMO radar approach based on matrix completion'. *IEEE Transactions on Aerospace and Electronic Systems*. 2015, vol. 51(3), pp. 1839–52.
[39] Sun S., Petropulu A.P. 'Waveform design for MIMO radars with matrix completion'. *IEEE Journal of Selected Topics in Signal Processing*. 2015, vol. 9(8), pp. 1400–14.
[40] Yardibi T., Li J., Stoica P., Xue M., Baggeroer A.B. 'Source localization and sensing: a nonparametric iterative adaptive approach based on weighted least squares'. *IEEE Transactions on Aerospace and Electronic Systems*. 2010, vol. 46(1), pp. 425–43.

[41] Roberts W., Stoica P., Li J., Yardibi T., Sadjadi F.A. 'Iterative adaptive approaches to MIMO radar imaging'. *IEEE Journal of Selected Topics in Signal Processing*. 2010, vol. 4(1), pp. 5–20.

[42] Shan T., Wax M., Kailath T. 'On spatial smoothing for direction-of-arrival estimation of coherent signals'. *IEEE Trans. Acoust., Speech, Signal Process.* 1985, vol. 33(4), pp. 506–811.

[43] Richards M.A. *Fundamentals of radar signal processing*. 2nd Ed. New York: McGraw-Hill; 2014.

[44] Carin L., Liu D., Guo B. 'Coherence, compressive sensing, and random sensor arrays'. *IEEE Antennas and Propagation Magazine*. 2011, vol. 53(4), pp. 28–39.

[45] Schmid C., Feger R., Wagner C., Stelzer A. 'Design of a linear non-uniform antenna array for a 77-ghz MIMO FMCW radar'. Proceedings of IEEE MTT-S Intl. Microwave Workshop on Wireless Sensing, Local Positioning, and RFID; Cavtat, Croatia, 2009.

[46] Alcalde C.A. 'Radar system and method for virtual antenna signals'. *U.S. Patent 9 664 775 B2*. 2017.

[47] Li Z., Alcalde C.A. 'Angle finding for a detector having a paired staggered array'. *U.S. Patent 2018/0267555 A1*. Sept. 20, 2018.

[48] Alland S. 'MIMO antenna with improved grating lobe characteristics'. *U.S. Patent 2015/0253419 A1*. Sept. 10, 2015.

[49] Searcy J., Alland S. 'MIMO antenna with elevation detection'. *U.S. Patent 9 541 639 B2*. 2017.

[50] Akaike H. 'A new look at the statistical model identification'. *IEEE Transactions on Acoustics, Speech, Signal Process*. 1989, vol. 37(1), pp. 8–15.

[51] Barron A., Rissanen J., Bin Yu, Yu B. 'The minimum description length principle in coding and modeling'. *IEEE Transactions on Information Theory*. 1998, vol. 44(6), pp. 2743–60.

[52] Pillai S.U., Kwon B.H. 'Forward/backward spatial smoothing techniques for coherent signal identification'. *IEEE Transactions on Acoustics, Speech, and Signal Processing*. 1989, vol. 37(1), pp. 8–15.

[53] Ertan A.E., Ali M. 'Spatial and temporal smoothing for covariance estimation in super-resolution angle estimation in automotive radars'. Proceedings of International Conference on Acoustics, Speech, Signal Process. *(ICASSP)*; 2020.

[54] Candès E., Romberg J. 'Sparsity and incoherence in compressive sampling'. *Inverse Problems*. 2007, vol. 23(3), pp. 969–85.

[55] Candès E.J., Tao T. 'The Dantzig selector: Statistical estimation when p is much larger than n'. *The Annals of Statistics*. 2007, vol. 35(6), pp. 2313–51.

[56] Cai T.T., Wang L. 'Orthogonal matching pursuit for sparse signal recovery with noise'. *IEEE Transactions on Information Theory*. 2011, vol. 57(7), pp. 4680–8.

[57] Tang G., Bhaskar B.N., Shah P., Recht B. 'Compressed sensing off the grid'. *IEEE Transactions on Information Theory*. 2013, vol. 59(11), pp. 7465–90.

[58] Chi Y., Scharf L.L., Pezeshki A., Calderbank A.R. 'Sensitivity to basis mismatch in compressed sensing'. *IEEE Transactions on Signal Processing*. 2011, vol. 59(5), pp. 2182–95.

[59] Lo Y.T. 'A mathematical theory of antenna arrays with randomly spaced elements'. *IEEE Transactions on Antennas and Propagation*. 1964, vol. 12(3), pp. 257–68.

[60] Alland S., Bordes J.P., Davis C., Ali M. 'Virtual radar configuration for 2D array'. (U.S. patent 9869762). 2018, vol. 9, pp. 869–762.

[61] Spreng T., Ziegler V., Siart U., Schumacher H., Valenta V., Yuan S. 'Wideband 120 ghz to 140 ghz MIMO radar: system design and imaging results'. Proceedings of European Microwave Conference; Paris, France, 2015.

[62] Meinl F., Stolz M., Kunert M., Blume H. 'An experimental high performance radar system for highly automated driving'. Proceedings of International Conference on Microwaves for Intelligent Mobility (ICMIM); Nagoya, Japan, 2017.

[63] Ying J., Cai J.-F., Guo D., Tang G., Chen Z., Qu X. 'Vandermonde factorization of Hankel matrix for complex exponential signal recovery-application in fast NMR spectroscopy'. *IEEE Transactions on Signal Processing*. 2018, vol. 66(21), pp. 5520–33.

[64] Candès E.J., Recht B. 'Exact matrix completion via convex optimization'. *Foundations of Computational Mathematics*. 2009, vol. 9(6), pp. 717–72.

[65] Cai J.-F., Candès E.J., Shen Z. 'A singular value thresholding algorithm for matrix completion'. *SIAM Journal on Optimization*. 2010, vol. 20(2), pp. 1956–82.

[66] Velodyne *HDL-32E high definition real-time lidar, datasheet [online]*. 2021. Available from https://velodynelidar.com/products/hdl-32e/

[67] Hua Y. 'Estimating two-dimensional frequencies by matrix enhancement and matrix pencil'. *IEEE Transactions on Signal Processing*. 1992, vol. 40(9), pp. 2267–80.

[68] Grant M., Boyd S. CVX: Matlab software for disciplined convex programming, version 2.1. 2014. Available from http://cvxr.com/cvx.

[69] Rossi M., Haimovich A.M., Eldar Y.C. 'Spatial compressive sensing for MIMO radar'. *IEEE Transactions on Signal Processing*. 2014, vol. 62(2), pp. 419–30.

[70] Feger R., Wagner C., Schuster S., Scheiblhofer S., Jager H., Stelzer A. 'A 77-ghz FMCW MIMO radar based on an sige single-chip transceiver'. *IEEE Transactions on Microwave Theory and Techniques*. 2009, vol. 57(5), pp. 1020–35.

[71] Jin N., Rahmat-Samii Y. 'Advances in particle swarm optimization for antenna designs: real-number, binary, single-objective and multiobjective implementations'. *IEEE Transactions on Antennas and Propagation*. 2007, vol. 55(3), pp. 556–67.

[72] Gonalez-Huici M., Mateos-Nunez D., Greiff C., Simoni R. 'Constrained optimal design of automotive radar arrays using the weiss-weinstein bound'.

Proceedings of International Conference on Microwaves for Intelligent Mobility; Munich, Germany, 2018.

[73] Bilik I., Villeval S., Brodeski D, *et al.* 'Automotive multi-mode cascaded radar data processing embedded system'. Proceedings of IEEE Radar Conference; Oklahoma City, OK, 2018.

[74] Och A., Pfeffer C., Schrattenecker J., Schuster S., Weigel R. 'A scalable 77 ghz massive MIMO FMCW radar by cascading fully-integrated transceivers'. Proceedings of Asia-Pacific Microwave Conference (APMC); Kyoto, Japan, 2018.

[75] *Product for cars [online]*. 2021 May. Available from https://www.zf.com/products/en/cars/products_58368.html

[76] *Continental automotive world [online]*. 2021 May. Available from https://www.continental-automotive.com

[77] *Texas instruments inc.,design guide:TIDEP-01012 imaging radar using cascaded mmwave sensor reference design (REV. A). [online]*. 2020. Available from https://www.ti.com/lit/ug/tiduen5a/tiduen5a.pdf

[78] Moffet A. 'Minimum-redundancy linear arrays'. *IEEE Transactions on Antennas and Propagation*. 1968, vol. 16(2), pp. 172–5.

[79] Giannini V., Welland D., Lobo R, *et al.* 'A 192-virtual-receiver 77/79ghz GMSK code-domain MIMO radar system-on-chip'. Proceedings of IEEE International Solid-State Circuits Conference (ISSCC); San Francisco, CA, 2019.

[80] *Vayyar*. Available from https://vayyar.com/

[81] Ioffe A., Doerr W., Yan H., Vu D.H., Arage A.H. 'RCS characteristics of street curbs and the applications in automotive radar classification'. Proceedings of European Radar Conference (EuRAD); London, 2016.

[82] Engels F., Wintermantel M., Heidenreich P. 'Automotive MIMO radar angle estimation in the presence of multipath'. Proceedings of European Radar Conference (EuRAD); Nuremberg, Germany, 2017.

[83] Visentin T., Hasch J., Zwick T. 'Analysis of multipath and DOA detection using a fully polarimetric automotive radar'. Proceedings of European Radar Conference (EuRAD); Nuremberg, Germany, 2017.

[84] Fertig L.B., Baden J.M., Guerci J.R. 'Knowledge-aided processing for multipath exploitation radar MER)'. *IEEE Aerospace and Electronic Systems Magazine*. 2017, vol. 32(10), pp. 24–36.

[85] Thillo W.V., Gioffré P., Giannini V., Guermandi D., Brebels S., Bourdoux A. 'Almost perfect auto-correlation sequences for binary phase-modulated continuous wave radar'. Proceedings of European Radar Conference (EuRAD); Nuremberg, Germany, 2013.

[86] Giannini V., Guermandi D., Shi Q., *et al.* 'A 79 GHz phase-modulated 4 GHz-BW cw radar transmitter in 28 nm CMOs'. *IEEE Journal of Solid-State Circuits*. 2014, vol. 49(12), pp. 2925–37.

[87] Bourdoux A., Ahmad U., Guermandi D., Brebels S., Dewilde A., Thillo W.V. 'PMCW waveform and MIMO technique for a 79 ghz CMOS automotive radar'. Proceedings of IEEE Radar Conference; Philadelphia, PA, 2016.

[88] Guermandi D., Shi Q., Dewilde A, *et al.* 'A 79-ghz 2x2 MIMO PMCW radar soc in 28-nm CMOS'. *IEEE Journal of Solid-State Circuits*. 2017, vol. 52(10), pp. 2613–26.

[89] Welch L. 'Lower bounds on the maximum cross correlation of signals (Corresp)'. *IEEE Transactions on Information Theory*. 1974, vol. 20(3), pp. 397–9.

[90] Gold R. 'Optimal binary sequences for spread spectrum multiplexing (Corresp)'. *IEEE Transactions on Information Theory*. 1967, vol. 13(4), pp. 619–21.

[91] Kasami T. 'Weight distribution formula for some class of cyclic codes'. *Report*. 1966, pp. 1–32.

[92] Dinan E.H., Jabbari B. 'Spreading codes for direct sequence CDMA and wideband CDMA cellular networks'. *IEEE Communications Magazine*. 1998, vol. 36(9), pp. 48–54.

[93] Hueber G., Niknejad A.M. *Millimeter-Wave Circuits for 5G and Radar*. Cambridge, UK: Cambridge University Press; 2019.

[94] Liu F., Masouros C., Petropulu A.P., Griffiths H., Hanzo L. 'Joint radar and communication design: applications, state-of-the-art, and the road ahead'. *IEEE Trans. Commun., to appear*.

[95] Thillo W.V., Giannini V., Guermandi D., Brebels S., Bourdoux A. 'Impact of ADC clipping and quantization on phase-modulated 79 ghz CMOS radar'. Proceedings of European Radar Conference (EuRAD); Rome, Italy, 2014.

[96] Rohling H., Meinecke M.-M. 'Waveform design principles for automotive radar systems'. Proceedings of the CIE International Conference on Radar; Beijing, China, 2001.

[97] Xue M., Xu L., Li J. 'IAA spectral estimation: fast implementation using the gohberg-semencul factorization'. *IEEE Transactions on Signal Processing: A Publication of the IEEE Signal Processing Society*. 2011, vol. 59(7), pp. 3251–61.

[98] Glentis G.-O., Jakobsson A. 'Efficient implementation of iterative adaptive approach spectral estimation techniques'. *IEEE Transactions on Signal Processing*. 2011, vol. 59(9), pp. 4154–67.

[99] Glentis G.O., Jakobsson A. 'Superfast approximative implementation of the IAA spectral estimate'. *IEEE Transactions on Signal Processing*. 2012, vol. 60(1), pp. 472–8.

[100] Godara L.C. 'Application of antenna arrays to mobile communications. II. Beam-forming and direction-of-arrival considerations'. *Proceedings of the IEEE*. 1997, vol. 85(8), pp. 1195–245.

Chapter 4

Interference and interference mitigation

Faruk Uysal[1]

Recent improvements in self-driving technology emphasize the importance of sensors that are being used in vehicles. Achievements in integrated circuits and the semiconductor industry made the low-cost mass production of single-chip automotive radars possible. While other sensors exist, automotive radar acts as the digital eyes of self-driving vehicles due to its proven all-weather, day, and night capabilities, which makes the automotive radar one of the key elements for self-driving technology. Most of today's vehicles are already equipped with radar systems to improve situational awareness and road safety. Current mmWave automotive radar sensors share a spectrum space from 76 to 81 GHz [1, 2]. The increasing number of radar-equipped vehicles on the roads has already been an issue for the coexistence of multiple automotive radars in congested traffic because the unwanted radar signals generated by other radar—also known as interference—negatively affect the functionality of the radar systems by decreasing their sensing capability. Since a lot of equal or similar waveforms and transmission strategies are presently used in automotive radar applications, interference occurs between multiple radar units. This kind of interference may raise the noise floor and reduce the signal-to-noise ratio which degrades the probability of target detection. On the other hand, interference generated by other radar systems may cause (ghost) false targets that reduce the target tracking ability of the radar.

Recent automotive radar systems are taking advantage of multiple-input multiple-output (MIMO) antenna arrays to provide the azimuth information of targets. Depending on the MIMO antenna configuration, it is also possible to exploit the azimuth and elevation information of the targets. While interference is mostly generated from other radars, there might be self-interference from the strong return signal reflected by the radome* or induced by the mutual-coupling (spill-over) effect between transmitters and receivers. It is crucial to mitigate interference or reduce its

[1]Radar Technology Department, Netherlands Organisation for Applied Scientific Research (TNO), Netherlands

*A radome is a protective cover between radar antenna and the environment with minimal impact to the radar performance. Radome is usually constructed of dielectric material that effectively transparent to radio waves. It can be different shape and size (such as a bumper or brand logo in modern cars).

effect because the objects with low radar cross sections (RCSs), such as pedestrians or cyclists, may not be detected or be completely lost during tracking. Therefore, interference leads to dangerous situations and becomes a bottleneck for driving assistance and autonomous vehicles. Especially in fully autonomous vehicles, where any human intervention will no longer be present, the dependability on the sensors is extremely high and there is no tolerance for sensing failures due to interference.

In this chapter, we examine different types of automotive radar interference, their characteristics, and their effects on radar system performance, as well as provide a review of the current state of the art for interference mitigation techniques.

4.1 Automotive radar interference

The interference mitigation problem has been an emerging topic in automotive radar signal processing and has received remarkable attention during the last decade to improve the coexistence of multiple radar systems. Joint radar and communication (which eventually addresses the interference issue) also bring a new era that aims to combine the sensing and vehicle-to-vehicle (V2V) communication functionalities in a single sensor.

While there are some system characteristics and recommendations [2], as well as standards[†] for testing short-range automotive radar equipment, there are currently no regulations on waveform structure and modulation schemes for automotive radars, and thus, many different approaches are studied to mitigate interference or compensate its effects. However, since the various criteria should be taken into account for real-time automotive radar such as the realization of the hardware and processing cost, it is still not clear which approach is the best for the interference countermeasure technique. When a large number of sensors are operating at the same time, one single interference mitigation approach will not be sufficient [3]. Hence, applicable and efficient interference countermeasure methods, as well as waveforms are still a focus of interest.

4.1.1 Signal model for radar and the interference

There are different kinds of waveforms that have been used in automotive radar applications. While in the early development years, continuous-wave (CW) signal was often used; in the last decade, frequency-modulated continuous-waveform (FMCW) radar systems have been preferred. Phase-modulated (PM) radar waveforms[‡] such as phase-coded continuous-wave (PC-CW) are an emerging waveform in the automotive radar field.

The transmitted signal s_t for a continuous radar waveform (CW) is written as[§]

[†]ETSI EN 303 396 (Short Range Devices – Meas. Techniques for Automotive Radar).

[‡]Phase modulated waveform is also known as phase coded (PC) waveform in literature.

[§]Note that in reality, the radar signal is a real signal if an IQ receiver is not used. For simplicity in the derivation and easy representation of different kinds of signals, we prefer to use the complex representation of the signal through this chapter.

$$s_{t_{CW}}(t) = rect\left(\frac{t - T_p/2}{T_p}\right) e^{j2\pi f_c t} \tag{4.1}$$

where $rect\,()$ is the rectangle function, T_p is the duration of the transmit signal, and f_c is the carrier frequency. Similarly, the transmitted signal for FMCW is given as,

$$s_{t_{FMCW}}(t) = rect\left(\frac{t - T_p/2}{T_p}\right) e^{j2\pi\left(f_c t + \frac{1}{2}kt^2\right)} \tag{4.2}$$

where $k = B/T_p$ is the chirp rate (also known as the slope of the linear frequency modulated) defined as the ratio of the bandwidth of the waveform B and pulse duration T_p. The spectrum of the FMCW signal in (4.2) spans the frequency band $[f_c - B/2,\ f_c + B/2]$. As seen in (4.1) and (4.2), the FMCW signal is equivalent to CW when slope rate $k = 0$.

The transmitted signals propagate in free space and reflect from a target and are collected by the receiver antenna, which can be written as

$$s_{r_{target}}(t) = rect\left(\frac{t - T_p/2 - \tau}{T_p}\right) \sqrt{P_{target}}\, e^{j2\pi\left(f_c(t-\tau) + \frac{1}{2}k(t-\tau)^2\right)} \tag{4.3}$$

where τ is the time delay between the radar and the target. In general, the radar return signal includes the returns from multiple targets as well as the return signal from, jammers¶, interference and clutter at the scene. For simplicity, we narrow down our derivation on a single target and a single interference case.

The received interference, denoted by subscripts $()_{int}$, can be written as

$$s_{r_{int}}(t) = rect\left(\frac{t - T_p/2 - \tau_{int}}{T_p}\right) \sqrt{P_{int}}\, e^{j2\pi\left(f_{c_{int}}(t-\tau_{int}) + \frac{1}{2}k_{int}(t-\tau_{int})^2\right)} \tag{4.4}$$

where τ_{int} is the time delay between the victim and interfering radars, $f_{c_{int}}$ is the carrier frequency, k_{int} is the chirp rate, and T_{int} is the pulse duration of the interfering radar, respectively.

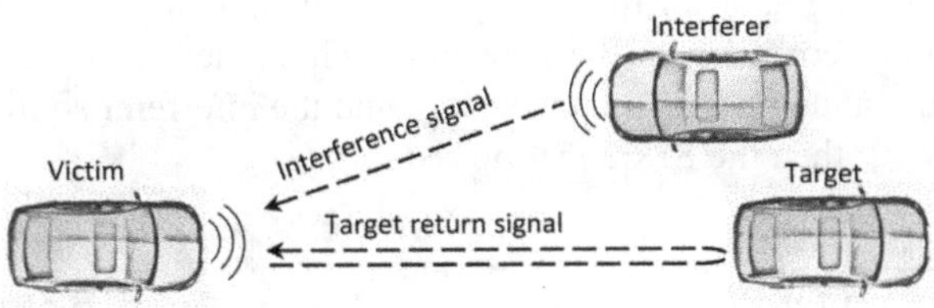

Figure 4.1 *The target return signal power is subject to two-way round trip delay while the received interference signal has one-way propagation delay*

¶Both jammer and interference signals distort the system performance of victim radar. While 'jammer signal' is an intentional disruption that aims to prevent listening transmitted waveform, the returns from 'interference source' are unintentional forms of disruption.

The received power of the target return signal P_{target} and the received interference signal P_{int} are subject to different round trip propagation delays, as illustrated in Figure 4.1.

While the target return signal is subject to two-way round trip delay,

$$P_{target} = \frac{P_t G_t G_r \lambda^2 \sigma}{(4\pi)^3 R^4{}_{target}} \tag{4.5}$$

received interference signal power computed based on one-way propagation delay

$$P_{int} = \frac{P_{t_{int}} G_{t_{int}} G_r \lambda^2{}_{int}}{(4\pi)^2 R^2{}_{int}} \tag{4.6}$$

Here λ is the wavelength, G_t is the gain of the transmitting antenna, and G_r is the gain of the receiving antenna for the victim radar. Similarly, λ_{int} is the wavelength, $P_{t_{int}}$ is the transmitting power, and $G_{t_{int}}$ is the gain of the transmitting antenna for the interferer radar. The σ is the RCS of the target. As seen in (4.5) and (4.6), the received power of the interference signal is inversely proportional to the interference range R_{int}, as $P_{int} \propto R_{int}^{-2}$, while the target return signal power is inversely proportional to target range R_{target}, described as $P_{target} \propto R_{target}^{-4}$, which may lead to the reception of a strong interferer signal that may mask the weak target return signal which is known as masking. In such cases, the signal-to-interference ratio (*SIR*) at the victim radar's receiver can be measured as

$$SIR_{receiver} = \frac{P_{target}}{P_{int}} = \frac{\sigma}{(4\pi)} \frac{P_t}{P_{t_{int}}} \frac{G_t}{G_{t_{int}}} \frac{\lambda^2}{\lambda_{int}^2} \frac{R_{int}^2}{R_{target}^4} \tag{4.7}$$

After the processing SIR is given as

$$SIR_{proc} = \frac{G_p}{G_{p_{int}}} SIR_{receiver} \tag{4.8}$$

where G_p is the processing gain in the victim radar for its transmit signal, whereas $G_{p_{int}}$ is the processing gain for the interference signal. Moreover, if we assume the victim and the interfere radar are the same kind (which means the wavelength, transmitter power, and antenna gain are the same), and the interferer source on the target ($R = R_{target} = R_{int}$), then the SIR equation reduces to

$$SIR \approx \frac{G_p}{G_{p_{int}}} \frac{\sigma}{(4\pi) R^2} \tag{4.9}$$

For a reliable target detection, bigger *SIR* is preferred; thus, in the processing and waveform selection, $G_p >_{p_{int}}$ should be considered.

4.1.2 Characteristics of the automotive radar interference

The characteristics of the automotive radar interference might be divided into subgroups based on different criteria. The automotive radars interference can be divided into three subgroups based on interference source (induced by the self, cross, and other sources). And based on synchronization, it can be divided into two subgroups (synchronous and asynchronous interference) [4].

The synchronous (self- and/or cross-) interference is usually caused by the radar system itself, and most of these kinds of interference are mitigated in radar system design since the interference's characteristics are well known. An example of this kind of interference might be a spillover from transmitters to receivers that can be mitigated by an adaptive cancellation system by an analog subtraction of the interference signal based on knowledge of the transmitted waveforms [5, 6].

When multiple radar systems operate at the same time and same frequency, they cause mutual interference. Unlike the self- and cross-interference, interference caused by other radar systems is usually asynchronous and dynamic, because the interferer radar may change operation mode or system parameters. Moreover, the characteristics of the interference change based on victim and interferer radar specifications.

Assume that two FMCW automotive radars are operating in the same vicinity as illustrated in (4.1). Based on their chirp rate, there might be narrowband or wideband interference, as illustrated in Figure 4.2. When the victim and interferer radars have different chirp rates, wideband interference occurs and causes a noise floor increase at the victim radar. On the other hand, when the victim radar has the same (or similar) chirp rate as the victim radar, a narrowband interference occurs which leads to ghost targets at the victim radar [7].

The narrowband interference generates a single beat frequency (if it has the same slope as transmit signal) or a narrow spread in the beat frequency spectrum (if it has a similar chirp rate). While the wideband interference has a short duration and usually generates multiple beat frequencies during the duration of the interference (see Figure 4.2a) and spreads out a large frequency band, thus wideband interference increases noise floor. The duration of the interference is bounded by the upper and lower bound of the low-pass filter (LPF) in the radar system.

To better understand the FMCW interference on an FMCW victim radar, we can look at the instantaneous frequency of the received interference signal. For simplicity of the derivation, we assume first the received interference signal (4.4) is mixed with the complex conjugate of the reference signal (4.2) in analog domain** to compute the base-band signal as

$$y_{int} = s^{*}_{t_{FMCW}} s_{r_{int}} \tag{4.10}$$

which can be written explicitly as

$$y_{int} = rect\left(\frac{t - T_p/2 - \tau_{int}}{T_p}\right)\sqrt{P_{int}}\, e^{-j2\pi\left(f_{c\,int}\tau_{int} + t(f_c - f_{c\,int} + k_{int}\tau_{int}) + t^2\left(\frac{1}{2}k - \frac{1}{2}k_{int}\right) - \frac{1}{2}k_{int}\tau_{int}^2\right)} \tag{4.11}$$

**This process is known as mixing, dechirping, and stretch processing.

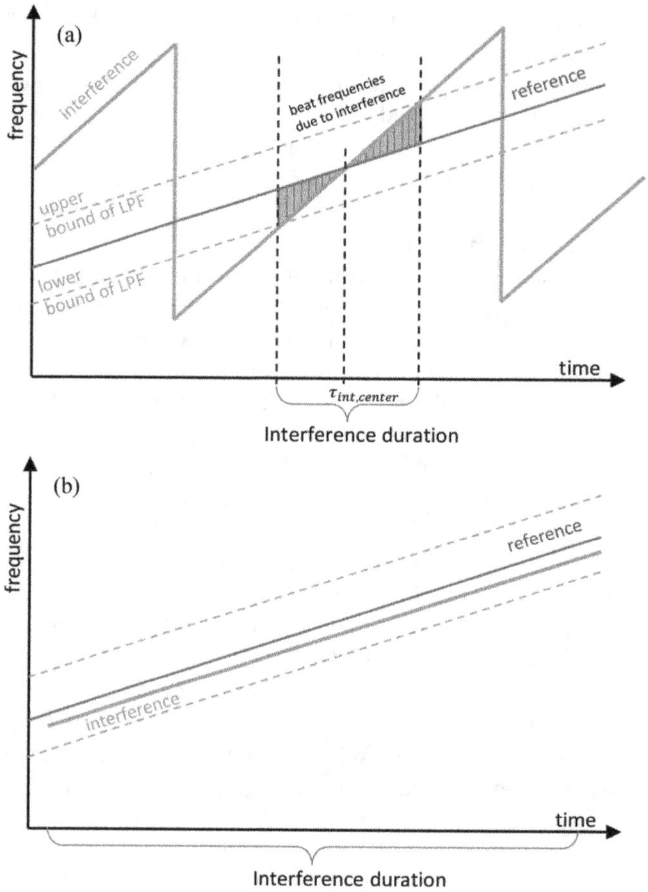

Figure 4.2 Illustration of (a) wideband interferences and (b) narrowband interferences

The beat frequencies resulting from dechirping are usually much smaller than the bandwidth; thus, an analog LPF is applied. If the slope (chirp rate) of the interference is different from the reference signal, the beat frequency signal due to interference removed by LPF is illustrated in Figure 4.2a. To identify the limits of the interference and for simplicity in derivations, we assume a perfect Brick wall filter with a cut-off frequency of f_{LFP} is used as

$$H(f) = \text{rect}\left(\frac{f}{f_{LPF}}\right) \tag{4.12}$$

The phase of the received interference signal (4.11) can be written as

$$\phi_{int}(t) = -2\pi\left(f_{c\,int}\tau_{int} + t\left(f_c - f_{c\,int} + k_{int}\tau_{int}\right) + t^2\left(\frac{1}{2}k - \frac{1}{2}k_{int}\right) - \frac{1}{2}k_{int}\tau_{int}^2\right)$$

The received interference signal's instantaneous frequency $f_{int}(t)$ can be computed by taking the derivative of its phase as

$$f_{int}(t) = \frac{1}{2\pi}\frac{d}{dt}\phi_{int}(t) \tag{4.13}$$

$$= t(k_{int} - k) + (f_{c_{int}} - f_c) - k_{int}\tau_{int} \tag{4.14}$$

Based on (4.14), the instantaneous frequency of the interference after dechirping can be described with three terms as follows: the difference between chirp slopes for a given time $t(k_{int} - k)$; the difference between carrier frequencies $(f_{c_{int}} - f_c)$; and the corresponding beat frequencies of the interference as $(-k_{int}\tau_{int})$.

The difference between chirp slopes for a given time $t(k_{int} - k)$ defines the spread of interference in the range domain. This interference signal is usually removed or clipped by LPF and only some part of it is digitized by ADC because it is bounded by the cut-off frequency of the LPF. This bound is defined as [8, 9]

$$-f_{LPF} \leq \left(t(k_{int} - k) + (f_{c_{int}} - f_c) - k_{int}\tau_{int}\right) \leq f_{LPF} \tag{4.15}$$

The lower and upper bounds of the beat frequencies are illustrated in Figure 4.2a. Considering the fact that the interference is symmetric around the reference signal due to these bounds, the duration of the interference T_{int} over the beat signal can be defined as [8, 9]

$$T_{int} \leq \left|\frac{2f_{LPF}}{k_{int} - k}\right| \tag{4.16}$$

Note that the interference is time-limited and has an oscillatory behavior (quadratic-phase signal), as shown in (4.1.2) [10]. By using the possible duration of the interference, the number of samples that are being interfered can be computed as

$$N_{int} \leq T_{int}f_s. \tag{4.17}$$

As mentioned before, the FMCW interference is symmetric and is centered around the time instance $t_{int,center}$, as illustrated in Figure 4.2. By taking into account that the interference has a duration of T_{int}, the final interference signal after LPF can be written as

$$y_{int,LPF}(t) = rect\left(\frac{t - \tau_{int,center}}{T_{int}}\right) y_{int}(t) \tag{4.18}$$

The center of the interference can be defined as the time instance when the frequencies of the victim and the interferer radar are the same, which is

$$\tau_{int,center} = \frac{-(f_{c_{int}} - f_c) + k_{int}\tau_{int}}{(k_{int} - k)} \tag{4.19}$$

By considering all (4.11), (4.15), and (4.18), a post-LPF received signal consisting of target and interference returns can be written as

$$y_L = \begin{cases} y_{target} + y_{int} & t \leq \left|\frac{f_{LPF} - (f_{c_{int}} - f_c) + k_{int}\tau_{int}}{(k_{int} - k)}\right| \\ y_{target} & \text{otherwise,} \end{cases} \tag{4.20}$$

where the received target signal after LPF is

$$y_{target} = s^*_{tFMCW} s_{r_{target}} \tag{4.21}$$

$$= rect\left(\frac{t - T_p/2 - \tau}{T_p}\right) \sqrt{P_{target}} e^{j\pi k(\tau^2 - 2t\tau)} e^{-j2\pi f_c \tau} \tag{4.22}$$

Here, the second term $e^{-j2\pi f_c \tau}$ is due to the Doppler shift and is exploited in Doppler processing to estimate the velocity of the target. The first term $e^{j\pi k(\tau^2 - 2t\tau)}$ consists of a residual phase $k\tau^2$, and the beat frequency $f_{beat} = k\tau$. Thus, it is called usually the beat signal.

4.1.2.1 Duration of the interference

To better understand and illustrate the interference in some example data, the FMCW victim and FMCW interferer case is considered. In this controlled experiment, two MIMO radar systems†† were used as a victim and the interferer radars. The center frequency of the victim radar was set to 78.8 GHz, while the bandwidth is 1.0 GHz. A controlled moving target (0 dBm^2 at 77 GHz) is located 5 m away from the victim radar unit. The interferer radar that was located 5 m away from the victim radar unit with a 30° degree offset was set to operate with 1 GHz bandwidth at the 78.6 GHz center frequency. Both automotive radar units were set to operate in MIMO mode. The victim radar was set to transmit 256 chirps with a duration of 36.66 μs per processing frame‡‡, whereas interferer radar's chirp duration is 40.05 μs to create wideband asynchronous interference. A total of 512 ADC samples (per pulse) were collected for victim radar and investigated for different pulses. The specifications of both radar units are summarized in Table 4.1.

The real part of the collected beat signal that consists of 512 fast-time samples for pulse number 58 is shown in Figure 4.3. As a comparison, Figure 4.4 shows one pulse of a collected beat signal for pulse number 59, where the interference

Table 4.1 Some of the radar system parameters for the victim and the interferer radars

	Victim radar	**Interferer radar**
Bandwidth	~1 GHz	~1 GHz
Carrier frequency	78.8 GHz	78.6 GHz
Samples per chirp	512	256
Number of chirps	256	256
Chirp time	36.66 μs	40.05 μs
Mode	MIMO 3 TX - 4 RX	MIMO 3 TX - 4 RX

††Such as Texas Instruments IWR1443 Single-Chip 76- to 81-GHz mmWave Sensor.

‡‡A processing frame is also known as coherent processing interval (CPI) in the radar field which is the time frame where usually a range-Doppler response is computed.

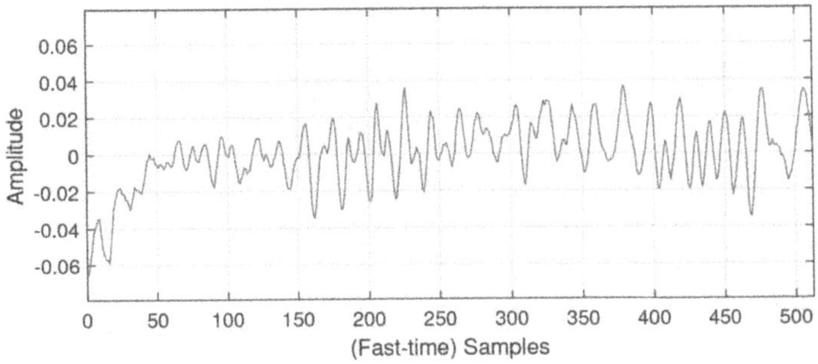

Figure 4.3 *Beat signal of the collected signal by victim radar. Illustration of the real part of pulse number 58.*

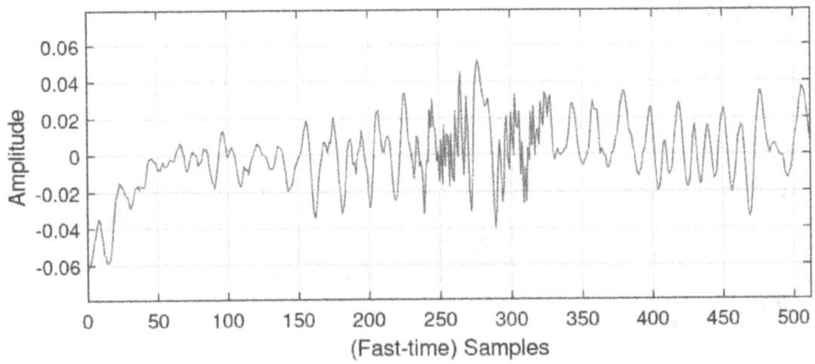

Figure 4.4 *Beat signal of the collected signal by victim radar in the presence of interference. Illustration of the real part of pulse number 59.*

is present. In these two figures, it is hard to observe neither the discussed features of the interference nor the beat signal that belongs to the target, due to the clutter around the radar system. Though, it is still possible to observe interference visually due to its high power which appeared between samples roughly 230 through 330.

A ground clutter filter that is nothing more than a high-pass filter (HPF) can be used to further investigate the collected radar data. It should be noted that clutter is an important target for automotive radar; moreover, the ground clutter does not span at zero Doppler bin when the automotive radar platform is in motion. In this aforementioned controlled experiment, since the victim radar is stationary, a perfect HPF was applied to eliminate stationary clutter so that the beat signal belongs to the moving target as well as the interference signals' features can be observed clearly.

The collected beat signal after ground clutter filtering is shown in Figure 4.5. As seen from the figure, the clutter is eliminated and the beat signal that belongs to the moving target is observable as a single-tone signal that is of the form of (4.22). Likewise, Figure 4.5 shows the collected beat signal that is under the influence of interference after ground clutter filtering. While the beat signal due to the moving target is still observable (with a different starting phase due to being a different

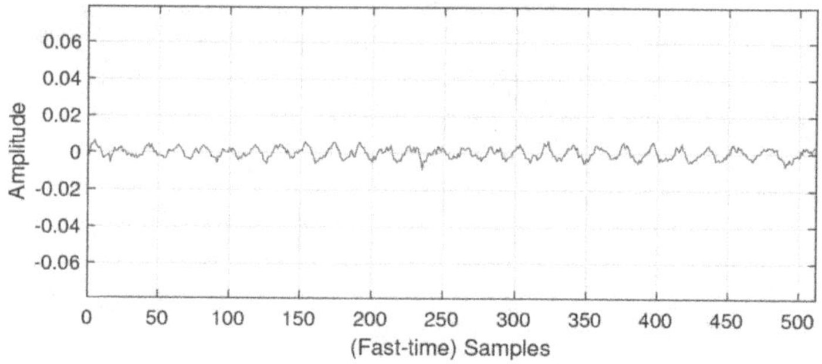

Figure 4.5 Beat signal after ground clutter filtering from the collected signal by victim radar. Illustration of the real part of pulse number 58 where a single target response is observable, which is in the form of a single tone signal.

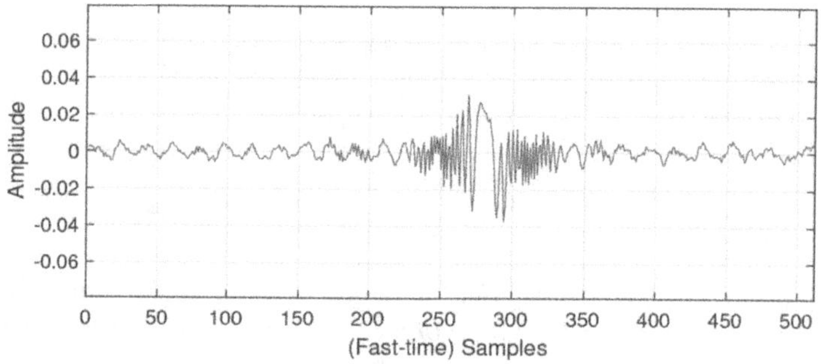

Figure 4.6 Illustration of the interference in the beat signal after ground clutter filtering from the collected signal by victim radar. The real part of pulse number 59 where a single target response and interference are observable.

pulse), the wide-band interference is also observable. In this case, the interference is time-limited due to the internal LPF. If we consider a perfect LPF with a cutoff frequency of 5MHz, the duration of interference can be computed by (4.16) as

$$T_{int} \approx \left| \frac{2f_{LPF}}{k_{int} - k} \right| \approx \left| \frac{10\text{MHz}}{\left(\frac{1\text{GHz}}{40.05\mu\text{s}}\right) - \left(\frac{1\text{GHz}}{36.66\mu\text{s}}\right)} \right| \approx 4.33\mu\text{s} \tag{4.23}$$

For a sampling frequency of 20 MHz, the number of fast-time samples affected by interference can be estimated as $N_{int} \approx T_{int}f_s \approx 87$ samples. In Figure 4.5, the interference roughly spans between samples 230 and 330, which is slightly more than the theoretical estimate value, 87 samples.

The duration of the interference defined (4.16) is a function of the cut-off frequency of the LPF that is assumed to be perfect (ideal filter)[§§] in the sense that it removes all frequency components above a given cut-off frequency, without affecting lower frequencies, and has linear phase response.

The transition region present in practical filters does not exist in an ideal filter. An ideal LPF can be realized mathematically (theoretically) by multiplying a signal by the rectangular function in the frequency domain or, equivalently, convolution with its impulse response, a sinc function, in the time domain. This ideal filter is impossible to realize without having signals of an infinite duration of time, and thus needs to be approximated in real-time applications.

The analog LPF used in automotive radar approximates the ideal filter by truncating an ideal filter due to the limited time. This results in ringing artifacts due to the Gibbs phenomenon. Moreover, this non-ideal behavior creates a transition region that does not exist in an ideal filter. The ringing artifacts, as well as transition region, are based on the design (such as cut-off frequency, the rate of frequency roll-off, filter order, etc.) and choice (such as the type) of the analog filter. This (the use of non-ideal analog LPF) clearly explains why observed interference in Figure 4.6 spans over a larger duration than theoretical computation.

In the literature, it was shown that the duration of interference is reduced by increasing the filter order [11, 12]. This is expected because increasing the filter order results in a steeper frequency roll-off (which is a better approximation of an ideal filter) and thus reduces the interference. On the other hand, it was also shown that a blind increase in filter order does not necessarily lead to a reduction in the duration of interference [12] because the duration of interference (4.16) is not only a function of the cut-off frequency of LPF but also the difference between victim and interference chirp rates. Due to this dependency, in higher-order filters, the transient response time (ringing) may become more dominant, which leads to the increased duration of the observed interference pulses [12].

4.1.2.2 Probability of interference and vulnerability

While the type and duration of the interference are based on both victim and the interferer radar system parameters, the probability of interference being present in a given pulse depends on only victim radar parameters.

Let us define a vulnerability interval called V, in which the interference signal from an interfering radar may be received by the victim radar. Obviously, the duration of the vulnerability region is defined by ADC sampling time $T_s = 1/f_s$ and the chirp duration, while the frequency spread of the vulnerability region is defined by the bandwidth of the transmit signal, as illustrated in Figure 4.7. The received interference in victim radar is occurred at time

$$\tau_{int} = \tau_0 + \tau_P + \tau_D \tag{4.24}$$

[§§]The ideal filter frequency response is of the from of a rectangular function thus also known as a brick-wall filter.

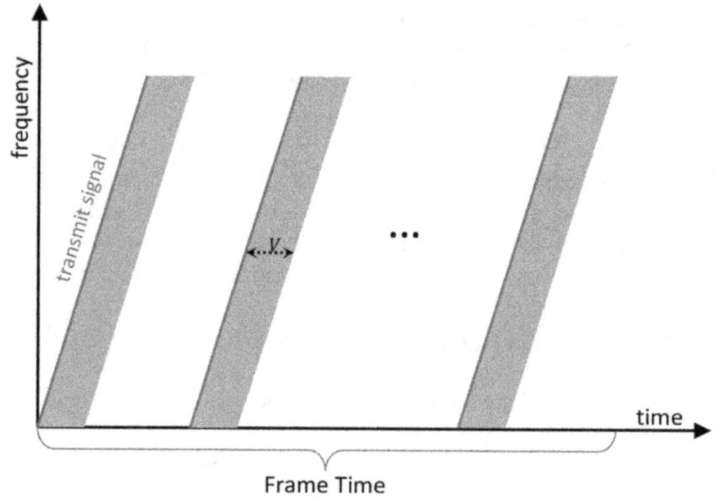

Figure 4.7 Illustration of vulnerability regions with in a frame time T_f

where τ_0 is the interference transmission time at interferer radar, $\tau_P = R_0/c$ is the propagation delay, and τ_D is the time delay due to the Doppler shift of moving interferer. If we recall the beat frequency $f_{beat} = k\tau$ and the maximum beat frequency $f_{beat,max} = f_s/2 = k\tau_{max}$, then the maximum possible time delay can be written as

$$\tau_{max} = \frac{T_p}{2BT_s} \tag{4.25}$$

When the received interference at the victim radar starts at τ_{int}, the interference occurs at $\tau_{int} \in \left[0, \frac{T_P}{2BT_S}\right]$. If the imperfect LPF at the hardware is taken into account [13], which causes mutual interference especially when the interference power is high, then interference also occurs when $\tau_{int} \in \left[-\frac{2T_P}{2BT_S}, -\frac{T_P}{2BT_S}\right]$.

The Doppler shift due to moving interferer is perceived by victim radar as a time delay $\tau_D = f_D/k$. If we recall the Doppler shift of the interference signal $f_D = 2v_{int}/\lambda$ and assume the maximum velocity of the interferer is equal to the maximum detectable velocity of the victim radar $v_{max} = \lambda/4T_p$, then the maximum time delay due to the Doppler shift can be computed as

$$\tau_{D,max} = \frac{f_{D,max}}{k} = \frac{1}{2B} \tag{4.26}$$

which is nothing more than $\tau_{D,max} = \pm\frac{1}{4B}$ for approaching or receding interference.

By putting all together, the interference interval, maximum possible delays (4.25), and the maximum time delay due to Doppler shift (4.26), the vulnerability interval in which the interference signal may be received by the victim radar is defined as

$$V = \left[-\frac{3T_p}{2BT_s} - \frac{1}{4B}, \frac{T_p}{2BT_s} + \frac{1}{4B} \right] \approx \left[-\frac{3T_p}{2BT_s}, \frac{T_p}{2BT_s} \right] \tag{4.27}$$

which can be further approximated due to T_D <p and T_s <p as [13]

$$|V| \approx \frac{2T_p}{BT_s} \tag{4.28}$$

In case the radar has access to in-phase and quadrature (IQ) samples, similar to our signal model, the receiver uses the full spectrum unambiguously such that the maximum beat frequency becomes $f_{beat,max} = f_s$ instead of $f_s/2$. Thus, the vulnerable period for the IQ receiver can be reduced to $|V| \approx T_p/(BT_s)$.

It is possible to estimate the probability of interference for a single chirp based on vulnerability interval as [13],

$$P_{int,chirp} \approx \frac{|V|}{T_p} \approx \frac{2}{BT_s} \tag{4.29}$$

The interference probability on a single chirp can easily be extended to frame level. If the victim radar transmits N chirps within a frame time T_f, it only uses $T_f/(NT_p)$ seconds for transmission. Any interfering radar signal sequence starting $(N-1)T_p$ up to the end of the radar transmission fall in the victim radar transmit time and results in mutual interference. Thus, the probability of interference for a frame is reduced to [13]

$$P_{int,frame} \approx \frac{(2N-1)\,T_p\,P_{int,chirp}}{T_f} \tag{4.30}$$

Since the number of chirps used for range-Doppler processing is usually large in automotive radar applications, it can be further approximated to

$$P_{int,frame} \approx \frac{2\,N\,T_p\,P_{int,chirp}}{T_f} \tag{4.31}$$

from which it is possible to compute the number of corrupted frames in a certain period.

While (4.31) is correct from a theoretical point of view, there are some differences in transmission and reception schemes for the real automotive radar system. Figure 4.8 illustrates a chirp with different parameters. Based on the transceiver, there might be an idle time T_{idle} before (may also be called Dwell time) and after each chirp generation. DAC may need time to generate a reliable chirp signal; this time gap is also known as settling time T_{settle}. The ADC sampling occurs only for the reliable part of the chirp signal that is illustrated with the blue region. There might be a small gap after ADC sampling ends; this part and the settling part are usually neglected to avoid imperfection of the chirp ramp. Finally, ADC needs time T_{reset} to reset the instantaneous frequency. As seen from the figure, the effective bandwidth B_{eff} is smaller than the intended bandwidth B as well as the sampled chirp time. For more precise determination in all analysis, the chirp time T_p should be considered as T_{ADC}, and the bandwidth B should be considered as B_{eff}.

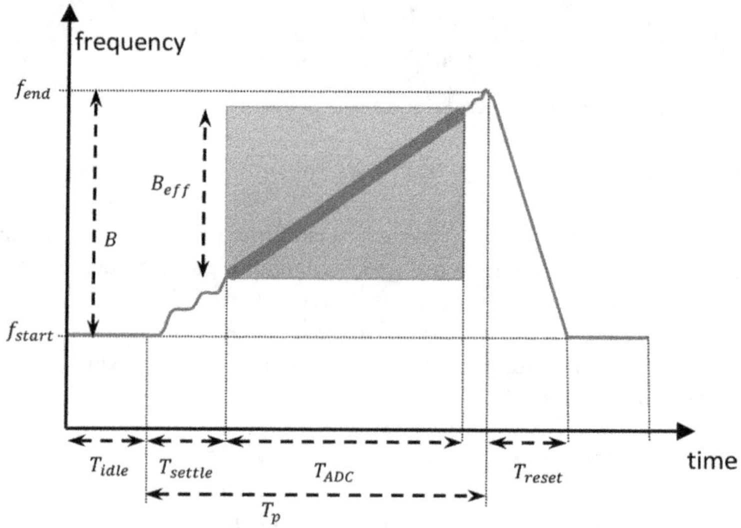

Figure 4.8 Illustration of chirp with different parameters. Note that each of these regions does not necessarily appear in every system, their appearance and duration are subject to different manufacturing designs.

In our controlled experiment, the radar uses $N = 256$ chirps to compute one range-Doppler frame in MIMO mode by using three transmitters sequentially and four synchronous receivers. The pulse time is $T_P = 30.6\mu s$ and the frame time is $T_f = 28.1$ ms. In this case, the interference probability for a single pulse is

$$P_{int,chirp} \approx \frac{2}{1\ \text{GHz}\ .\ 50\ \rho\text{sec}} \approx 0.0398\% \tag{4.32}$$

and the probability of interference for a frame is

$$P_{int,frame} \approx \frac{2\ .\ 256\ .\ 30.6\ \mu\text{sec}\ .\ 0.0398}{28.1\ \text{msec}} \approx 0.0221\% \tag{4.33}$$

which means $\sim$ 6 pulses out of 256 pulses should be affected by interference. In our aforementioned experiment, we observed an average of 7 out of 256 pulses in a frame were affected by interference which is close enough to the theoretically computed probability.

4.1.2.3 Interference effects on range profiles

The range profile can be achieved by applying spectral estimation methods (such as applying Fourier Transform to collected beat signals. Note that the frequency domain is directly proportional to the range domain. If we recall the beat frequency $f_{beat} = k\tau$ where $\tau = 2R/c$, then the range bins can be estimated from spectral analyses as

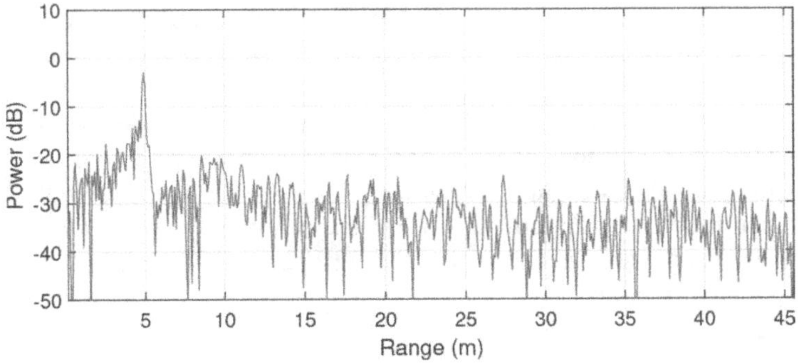

Figure 4.9 *Range profile of a single pulse of the collected signal by victim radar (No interference). Controlled moving target is observable at 5 m. Pulse number 58.*

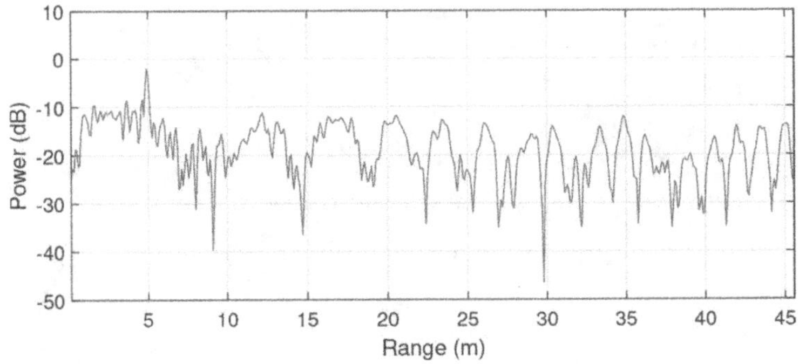

Figure 4.10 *Range profile of a single pulse of the collected signal by victim radar in the presence of interference. Pulse number 59.*

$$r = \frac{cf}{2k} = \frac{cfT_p}{2B} \tag{4.34}$$

Since we are only interested in positive range bins, only positive frequencies are used $0_s/2$ after spectral analysis; thus, range spans 0_{max}. The effect of the interference in the range domain shows differences based on the type, duration, and power of the interference.

If the interference signal has the same chirp rate as the victim radar $k_{int} = k$, the interference energy appears at an arbitrary range bin, and a ghost target appears. It is usually hard to discriminate ghost targets from real targets in the range domain. When the interferer's chirp rate starts to deviate from the victim radar chirp rate $k_{int} \neq k$, the interference energy starts to spread multiple range bins and become wider and wider; thus, it creates a wideband interference. This kind of interference is usually observed as a noise-floor increase in the range domain.

Figure 4.9 illustrates the range profile for the aforementioned controlled experiment. As seen from the figure, the target appears at 5 m away from the radar. Likewise, the range profile that is affected by the interference is illustrated in Figure 4.10. While the target is almost in the same range bin (note that the target is also moving), the noise floor between two pulses is increased. In both Figures 4.9 and 4.10, the ground clutter filter is applied to illustrate the noise floor increase, and thus only controlled moving target is observable at 5 m in the range plot. The noise floor increase is based on how wide is the interference and how much interference energy is passed from the LPF.

4.1.2.4 Time–frequency analysis and V-shaped interference

Another popular method for investigating interference is to examine how the frequency content of the signal change over time. Time–frequency analysis methods,

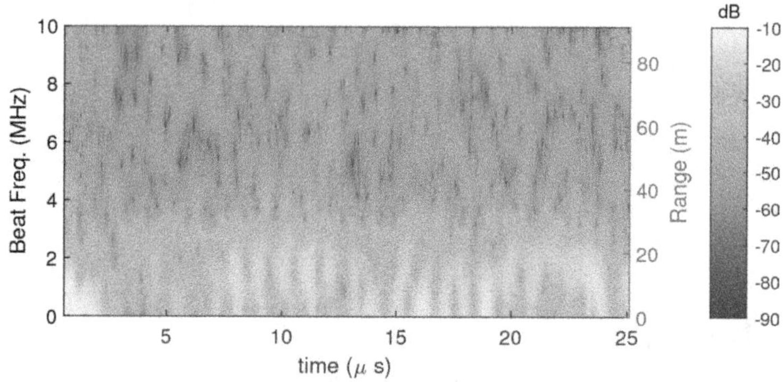

Figure 4.11 *Spectrogram (STFT) of single pulse (no interference). Target, as well as the ground clutter can be observable up to 20 m. Pulse number 58.*

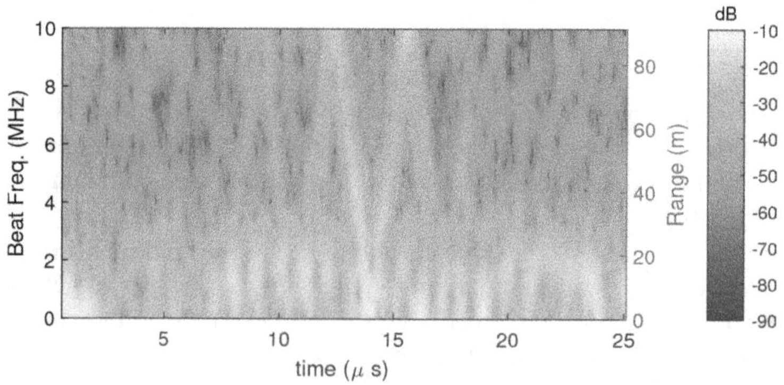

Figure 4.12 *Spectrogram of the collected signal by victim radar in presence of interference. V-shaped interference observable. Pulse number 59.*

such as short-time Fourier transform (STFT), can be used to further investigate the interference. Since the frequency domain is directly proportional to the range, it also shows how the target's location changes within a pulse duration. It is expected for automotive radar applications that the location of targets or their beat frequencies do not change dramatically within a pulse duration. Figure 4.11 illustrates the beat frequency (range) change of target and clutter within ADC sampling time (in this real case T_{ADCP}). As seen from the figure, beat frequencies do not change during a pulse. Figure 4.12 illustrates the same scene but one pulse later, which is distorted due to interference. Since it is a wide band interference, it creates a V-shape in time–frequency analysis. The V-shape signature of the interference widens, when the interferer radar's chirp rate k_{int} gets closer to the victim radar's chirp rate k. If the chirp rates are equal $k_{int} = k$, it forms a horizontal line signature in the time–frequency domain and appears as a ghost target.

It is also possible to observe other important features from time–frequency analysis such as the effects of LPS and its transition region. In our controlled experiment, the analog LPF cut-off frequency was selected as 10 MHz, while the sampling frequency was selected as 20 MHz for illustration purposes. Due to the transition region of analog LPF, strong interference was not filtered out perfectly and creates aliased interference signature. These signal components are observed in Figure 4.12 like a continuation of V-shaped interference. After the arms of the V-shape touch the f_s (or $f_{beat,max}$), the interference signal signature folds back over the first Nyquist sampling zone and creates an upside-down W-shape, while the new arms (of W-shape) lose their energies when the interference signal frequency reaches the stop-band of the LPF. As we have discussed before, the duration of the interference is observed slightly larger than the theoretical value due to the imperfect of the LPF.

4.1.2.5 Interference effects on range-Doppler map

Range-Doppler map that is also known as a radar frame can be computed by exploiting the Doppler shift induced by the target motion after range processing. Spectral estimation methods (such as Fourier transform) can be applied to slow-time (pulse) dimension to achieve Doppler frequency, which is proportional to the target velocity $f_D = 2v_{int}/\lambda$. It is also possible to directly get a range-Doppler map from the beat signal by applying 2D Fourier Transform.

The effect of the interference on range-Doppler map also shows differences based on interference type (synchronous or asynchronous), and it is period (periodic or aperiodic). Figure 4.13 shows a typical range-Doppler frame for a stationary radar with a single moving target (aforementioned experiment). It is wise to mention that for a radar platform in motion, the ground clutter shifts from the zero-Doppler region and does not perform a straight line, as illustrated in this figure.

Nevertheless, as an example, the same scene is illustrated in Figure 4.14 on one frame time later when interference is present. In this frame, there is a total of seven pulses affected by the interference, which form diagonal streaks in the range-Doppler map. This is typical behavior for asynchronous and aperiodic interference and can be classified as a noise floor increase.

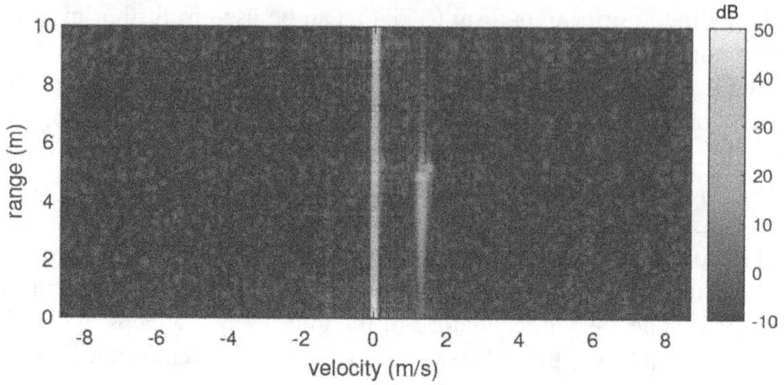

Figure 4.13 *Range-Doppler map of the collected signal by victim radar. Frame number 20.*

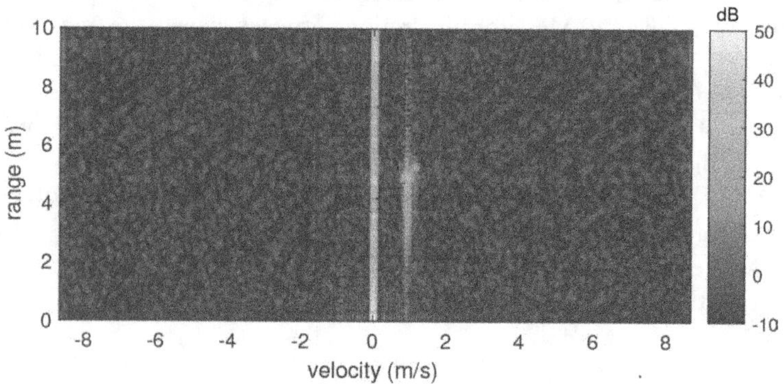

Figure 4.14 *Range-Doppler map of the collected signal by victim radar in the presence of interference. The interference signature is expected for an aperiodic and asynchronous interference that creates some diagonal streaks (lines) on the range-Doppler map. Frame Number 21.*

When the chirp rates of the victim and interferer radars are the same, the ghost target is observed at arbitrary range and interferer radar's velocity. Or vertical lines may be observed when the interferer radar has a narrowband signal like CW interference.

4.1.2.6 Generalized radar interference equation

In previous sections, we have illustrated interference for a selected case, in which the victim and the interferer radar have LFM (chirp) waveform. In that case, the interference is aperiodic and synchronous. While this is the most usual case, there are automotive radars in service which use different kinds of waveforms such as

CW or PMCW. Thus, it is important to have a generalized interference equation to mathematically describe as many cases as possible.

Kumbul *et al.* [14] use the PC-FMCW [15, 16] to generalize the radar-to-radar interference mitigation for continuous waveform that are mostly used in current automotive radars. The transmitted signal for a phase-coded FMCW radar is defined as

$$s_{t_{PC-FMCW}}(t) = \sqrt{P_t} e^{\left(j2\pi(f_c t + \frac{1}{2}kt^2) + j\phi(t)\right)} \tag{4.35}$$

where $\phi(t)$ represents the phase coding (or extra modulation). Since we would like to generalize the equation, we prefer to keep the phase coding term as a separate signal component as

$$s_{t_{PC-FMCW}}(t) = \sqrt{P_t} c(t) e^{j2\pi(f_c t + \frac{1}{2}kt^2)} \tag{4.36}$$

where $c_t(t) = e^{j\phi(t)}$ is a single period of a transmitted phase code. For simplicity, the coding signal is defined by using a phase term, as follows:

$$c_t(t) = \sum_{n=1}^{La} e^{j\phi_n} \mathrm{rect}\left(\frac{t - (n - 1/2)T_a}{T_a}\right) \tag{4.37}$$

where L_a is the number of chips within one sweep, T_a is the single-chip duration, calculated as $T_a = T/L_a$, and ϕ_n denotes the phase corresponding to the nth bit of the L_a bit sequence. Since this is a very general case, we widen our scope to the current technology in most automotive radars and concentrate on binary phase-shift keying where the phase term spans $\phi_n \in \{0, \pi\}$¶¶.

The PC-FMCW signal (4.36) is general enough for interference investigation since it converges to CW signal when $k = 0$ and $L = 0$; to FMCW signal when $k \neq 0$ and $L = 0$; to PMCW signal when $k = 0$ and $\neq 0$; and finally PC-FMCW signal when $k \neq 0$ and $L \neq 0$.

Setting the victim radar's waveform as PC-FMCW, the received radar signal can be written for the target (by omitting the constant phase term due to the scattering coefficient from the target) as [14]

$$s_{r_{target}}(t) = \sqrt{P_{target}} \sum_{n=1}^{La} e^{j\phi_n} \mathrm{rect}\left(\frac{t-\tau-(n-1/2)T_a}{T_a}\right) e^{j2\pi(f_c(t-\tau)+\frac{1}{2}k(t-\tau)^2)} \tag{4.38}$$

Similarly, if we use the PC-FMCW waveform for interferer radar, the received interference signal is written as:

¶¶For different coding techniques (such as different phase behavior over time) a detailed analysis and re-derivation of formulas are needed.

$$s_{r_{int}}(t) = \sqrt{P_{int}} \sum_{n=1}^{L_b} e^{j\phi_n} \mathrm{rect}\left(\frac{t-\tau_{int}-(n-1/2)T_b}{T_b}\right) e^{j2\pi(f_{c\,int}(t-\tau_{int})+\frac{1}{2}k_{int}(t-\tau_{int})^2)} \tag{4.39}$$

where subscripts $()_a$ and $()_b$ are used for victim and interferer radar accordingly such as L_a and L_b are the number of chips within one sweep, $T_a = T_P/L_a$ and $T_b = T_{p,int}/L_b$ are the chip durations of the victim and the interfering radar, respectively, whereas the pulse duration of the victim and interfering radar are defined as T_P and $T_{P,int}$, respectively.

The beat signal including the target and the interference signals is given as [14]

$$y_{beat}(t) = \sqrt{P_{target}} e^{j\phi_{target}(t)} + \sqrt{P_{m_{int}}} e^{j\phi_{int}(t)} \tag{4.40}$$

The phase of the beat signal related to the target echo is given as

$$\phi_{target}(t) = -2\pi\left(f_c\tau + k\tau t - \tfrac{1}{2}k\tau^2\right) + \phi_{tar_{code}}(t) \tag{4.41}$$

where the victim radar phase due to coding (modulation) is

$$\phi_{tar_{code}}(t) = \sum_{n=1}^{L_a} \phi_n \mathrm{rect}\left(\frac{t-\tau-(n-1/2)T_a}{T_a}\right) \tag{4.42}$$

Similarly, the phase of the beat signal related to the received interference can be written as

$$\begin{aligned}\phi_{int}(t) = -2\pi(&t(f_c - f_{c_{int}} + k_{int}\tau_{int}) + f_{c_{int}}\tau_{int} \\ &+ t^2(\tfrac{1}{2}k - \tfrac{1}{2}k_{int}) - \tfrac{1}{2}k_{int}\tau_{int}^2) \\ &+ \phi_{int_{code}}(t)\end{aligned} \tag{4.43}$$

where the interference phase due to coding (modulation) is

$$\phi_{int_{code}}(t) = \sum_{n=1}^{L_b} \phi_n \mathrm{rect}\left(\frac{t-\tau_{int}-(n-1/2)T_b}{T_b}\right) \tag{4.44}$$

These equations together with Table 4.2 can be used to represent interference (as well as the target response) for any given victim radar and interferer radar waveform pairs. For instance, for FMCW (victim) to FMCW (interferer) case, we set $L_a = 0$ and $L_b = 0$, and thus we achieve the phase of beat signal related to interference that converges to the same as (4.1.2).

Moreover, it is sometimes useful to know the instantaneous frequency of each signal component because it provides additional information about the signal features. Such as we can estimate what kind of interference shape to expect in time–frequency analysis. Instantaneous frequency for the generalized radar-to-radar interference equation can be obtained by taking the derivative of the phase. The instantaneous frequency of the beat signal related to the target signal is given as

$$\begin{aligned}f_{i_{target}}(t) &= \tfrac{1}{2\pi}\tfrac{d}{dt}\phi_{target}(t) \\ &= \tfrac{1}{2\pi}(-2\pi(k\tau)) + \tfrac{1}{2\pi}\tfrac{d}{dt}\phi_{tar_{code}}(t) \\ &= -k\tau + \tfrac{1}{2\pi}\sum_{n=1}^{L_a}(\phi_{n+1} - \phi_n)\delta(t - \tau - nT_a)\end{aligned} \tag{4.45}$$

Table 4.2 Different interference scenarios [14]

Victim	Interference	L_a	k	L_b	k_{int}
CW	CW	$=0$	$=0$	$=0$	$=0$
FMCW	CW	$=0$	$\neq 0$	$=0$	$=0$
CW	FMCW	$=0$	$=0$	$=0$	$\neq 0$
FMCW	FMCW	$=0$	$\neq 0$	$=0$	$\neq 0$
PMCW	CW	$\neq 0$	$=0$	$=0$	$=0$
CW	PMCW	$=0$	$=0$	$\neq 0$	$=0$
PMCW	PMCW	$\neq 0$	$=0$	$\neq 0$	$=0$
FMCW	PMCW	$=0$	$\neq 0$	$\neq 0$	$=0$
PMCW	FMCW	$\neq 0$	$=0$	$=0$	$\neq 0$
PC-FMCW	CW	$\neq 0$	$\neq 0$	$=0$	$=0$
CW	PC-FMCW	$=0$	$=0$	$\neq 0$	$\neq 0$
FMCW	PC-FMCW	$=0$	$\neq 0$	$\neq 0$	$\neq 0$
PC-FMCW	FMCW	$\neq 0$	$\neq 0$	$=0$	$\neq 0$
PMCW	PC-FMCW	$\neq 0$	$=0$	$\neq 0$	$\neq 0$
PC-FMCW	PMCW	$\neq 0$	$\neq 0$	$\neq 0$	$=0$
PC-FMCW	PC-FMCW	$\neq 0$	$\neq 0$	$\neq 0$	$\neq 0$

Similarly, the instantaneous frequency of the beat signal related to the interference is given as,

$$\begin{aligned} f_{i_{int}}(t) &= \frac{1}{2\pi}\frac{d}{dt}\phi_{m_{int}}(t) \\ &= t(k_{int} - k) + (f_{c\,int} - f_c - k_{int}\tau_{int}) \\ &\quad + \frac{1}{2\pi}\frac{d}{dt}\phi_{int_{code}}(t) \\ &= t(k_{int} - k) + (f_{c\,int} - f_c - k_{int}\tau_{int}) \\ &\quad + \frac{1}{2\pi}\sum_{n=1}^{L_b}(\phi_{n+1} - \phi_n)\delta(t - \tau_{int} - nT_b) \end{aligned} \tag{4.46}$$

4.1.3 Final remarks

It is not feasible to present or demonstrate all possible interference cases since the interference signal has many degrees of freedom. Thus, in this section, only a single case, FMCW-to-FMCW interference, from real radar data is presented, and the effect of this kind of interference is discussed with details to give an intuition to the readers about the automotive radar interference. Moreover, a generalized radar equation is presented for interested readers to cover interferences generated by different waveforms. The phenomenology of the interference is investigated for different cases and validated by experiments in the literature. Interested readers are referred to References [12, 14, 17] for different kinds of interference.

4.2 Interference mitigation

The interference problem in automotive radar application is getting more challenging since the number of the automotive radar system on the roads has been increasing, while the allocated spectrum for automotive radar application stays mostly the same***. The automotive radar interference issue was pointed out by the project MOSARIM (MOre Safety for All by Radar Interference Mitigation), which started in 2010 intending to investigate possible automotive radar interference mechanisms [18, 19]. Although many appropriate countermeasure and mitigation techniques have been investigated and developed since then, there is still no single interference mitigation technique that is effective for all interference cases. Generally, mitigation and avoidance techniques rely on nearly orthogonal transmission and reception principles in some domains (e.g., polarization, time, and frequency, or their joint domains) [20]. Improvement in automotive radar hardware technology allows the use of different waveforms and modulation schemes to enhance orthogonality, hence making possible implementation of advanced approaches for interference mitigation.

One may consider classifying the interference mitigation techniques into two groups. While the techniques in the first group aim to mitigate the adverse effects of interference in hardware or digital domain after it happens, the second group tries to reduce the vulnerability region and avoids the interference by adapting the radar system parameters. Nonetheless, approaches and techniques in both groups are implemented individually for each sensor and may not provide robust interference mitigation in congested areas. In that case, more advanced methods, such as channel access control similar to communication protocols, are needed. While many techniques are available in the literature, we have discussed some of them throughout this section.

4.2.1 Detection of interference

It is important to detect the presence of interference in the received signal since precise interference detection may help the selection of mitigation and/or avoidance techniques.

The detection of interference is usually easy when the interference power is high. Though, this case is not preferred since the effect of the interference is worse. When the interference power is higher than the allowable power by the front end, it saturates the receiver, and thus the received signal is clipped (reach out to maximum value) at ADC. The clipped samples are identified as distorted by interference. These samples may not be recovered if the number of affected samples is also large relative to the number of total samples.

When the interference power is low enough not to saturate the receiver but higher than the target signal power, a detector with a fixed threshold can be used to

***Automotive radar sensors migrated from 24 GHz band to 79 GHz band (76–81 GHz). The 24 GHz ultra-wideband (UWB) band is not available after January 1, 2022.

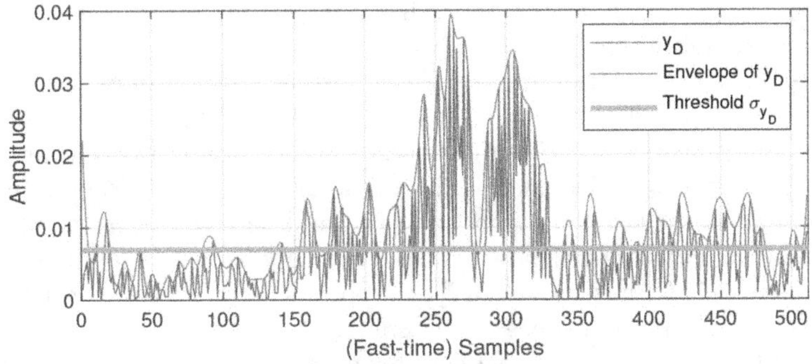

Figure 4.15 *Interference detection: Detector applied to the received signal (which is illustrated in Figure 4.4). The detection performance is not satisfactory.*

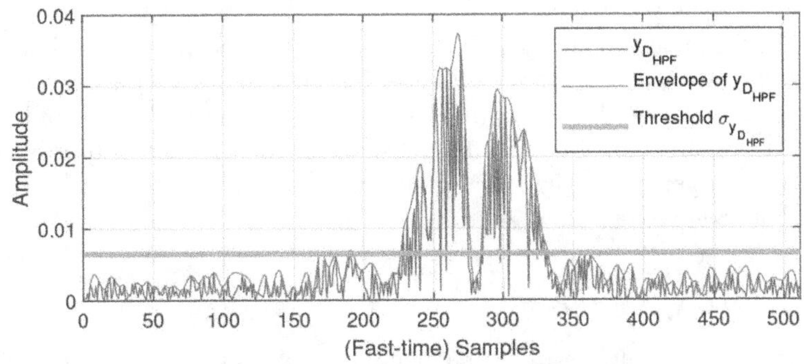

Figure 4.16 *Interference detection after HPF: detector applied to received signal after HPF (which is illustrated in Figure 4.6)*

detect the change in the signal power, thus the interference. The output signal for the detector that aims to identify the change of signal by time can be defined as [8, 21]

$$y_D(t) = \left| \frac{d}{dt} y_L(t) \right| \tag{4.47}$$

and its threshold can be computed based on its standard deviation as σ_{y_D}. In addition to this, a smoothing filter (like a moving average filter) or envelope estimation may be applied to the detector output to help cluster the output signal. Note that different kinds of detectors, such as CFAR-based methods [22] or an energy detector [23], can also be used for interference detection.

Although this detector is simple and effective for time-limited interference and can be applied easily to real-time applications, the success of this detector is based on interference, target, and clutter power. Figure 4.15 illustrates the output of the detector together with its envelope signal and the determined threshold value. In this example, clutter and the target were present, and thus the detection performance of a fixed threshold value is not very satisfactory. It is hard to accurately identify the samples affected by interference because there are multiple sample clusters that passed the threshold value.

One approach to address this issue is to use a dynamic threshold computed by one of the CFAR algorithms. Another approach is to suppress the contribution of targets and clutter so that detection can be done mostly on leftover interference signal. It is known from the radar equation that, after a certain distance, the target power does not exceed the noise floor at the victim radar receiver. If a power level is greater than the receiver noise power, then it can only be related to the presence of the interference [22]. Thus, Laghezza *et al.* proposed removing the power relative to the close target by HPF to increase the interference detection.

The performance of the simple detector on high-pass filtered received signal y_{DHPF} is illustrated in Figure 4.16, where a perfect ground clutter filter is used to eliminate the power from the clutter returns. The samples between 221 and 331 are detected as affected by interference. There are also two outliers (samples 191 and 359) that also passed the threshold, which can easily be neglected by clustering algorithms. In this example, the perfect ground clutter filter eliminates the returns from clutter since the radar is stationary. Thus, the detection performance is dramatically increased. In reality, to achieve the same performance, it is important to identify the cut-off frequency of the HPF for optimal detection since the application of HPF with a correct cut-off frequency allows the detection of interference with lower power than the close targets [22]. The HPF cut-off frequency f_H at 3 dB can be calculated using the following expression [22]

$$f_H > \frac{2B_{eff}}{cT_{ADC}} \sqrt[4]{\frac{P_t G_t G_r \sigma_{max} \lambda^2}{(4\pi)^3 P_n}} \tag{4.48}$$

where P_n is the noise power of the receiver and σ_{max} is the maximum RCS. After this HPF, the threshold value may be computed by using CFAR methods for the detection of the samples affected by the interference.

4.2.2 *Interference mitigation and avoidance*

The detection of the interference leads to different signal processing steps that aim to mitigate or avoid the interference. The radar system may take precautions and (if possible) change the radar system parameters to avoid interference in future radar frames, while a broad range of signal-processing techniques are used to mitigate the adversary effects of the interference in affected frames. The current approach for real-time interference mitigation in the industry focuses on various techniques that include but are not limited to removing and repairing distorted pulses, identifying and avoiding used frequencies, using waveform and frequency diversity, using beamforming and null steering.

4.2.2.1 Nulling and signal restoration

One simple and traditional approach is removing (in literature also referred to as deleting, zeroing, or nulling) the samples affected by the interference. Although removing the detected interference samples from the received signal (and replacing

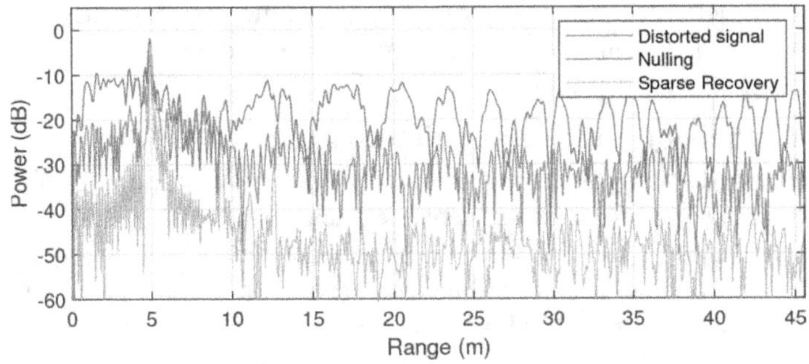

Figure 4.17 Illustration of the interference (same as Figure 4.10) and its mitigation by nulling and sparsity-based method

them with zeros) before range processing reduces the noise floor generated by interference, the discontinuity in the received signal (thus, in the instantaneous bandwidth of the beat signal) rises the sidelobes around the strong target peaks after range processing, which may hinder the detection of the weak targets.

The nulling may also be applied in different domains, such as time, frequency, and their joint domains[†††]. The sidelobes generated by nulling in time and/or frequency domain can be avoided by repairing the missing samples. There are different approaches for preparing the missing samples including but not limited to interpolation, autoregressive modeling [24], and sparse recovery [9, 25].

While nulling is an effective method when the number of interfered samples is relatively small to the number of ADC samples, it does not offer a reliable solution for wideband interference in which the number of samples affected by interference is relatively large compared to the total number of ADC samples in the pulse. In that case, sparsity-based signal separation methods can be used to blindly separate the beat signals from the interference by exploiting the known features of interference [4, 9, 26]; thus, they do not need any interference detection prior to applying the algorithm.

While there are many methods available in the literature, two of these mitigation methods are applied to the beat signal and their results are illustrated in Figure 4.17. In the first mitigation method (namely, nulling), an interference detection algorithm was applied to the beat signal, and then the detected samples were taken out. For comparison, a sparsity-based blind signal separation algorithm [4] was also applied to the same signal without any interference detection. Following FFT processing was performed on both signals to achieve the range profile. Nulling approach mitigates the interference, but the range profile still suffers from the high sidelobes due to the gap in the nulled signal spectrum. Sparsity-based signal separation algorithm results in better interference mitigation. It should be noted that, even though the

[†††]It is also possible to apply nulling in angle domain which is executed during the beamforming as null steering. See section 4.2.2.4.

sparsity-based methods have superior performance, they are currently not preferred in automotive radar systems due to their computational complexities.

4.2.2.2 Frequency altering and timing jitter

In the frequency-hopping approach, first interference detection is performed over the received beat signal for each pulse of each detection frame. If the interference is detected within a pulse, the interferer's center frequency $f_{c_{int}}$ is estimated as above or below the victim radar's center frequency f_c. Based on this estimate, the center frequency of the victim radar shifted up or down to the available frequency band [27]. Another approach might be to change the carrier frequency of the victim radar randomly (or in a defined order) after every sweep or every frame [28].

These frequency-based interference avoidance methods are very effective when there is an available spectrum to use. For instance, the adaptive cruise control (ACC) radar may use 1 GHz bandwidth (usually between 76 and 77 GHz) [2]; hence, four ACC radars with each having 1 GHz bandwidth can share the 76–81 GHz automotive radar spectrum without interfering with each other by using frequency hopping. On the other hand, different automotive radar applications, such as short-range corner radar, may need a higher range resolution. Therefore, radar may use 2–4 GHz bandwidth (usually between 77 and 81 GHz). In such a case, it is not possible to hop another frequency; instead, each pulse can be divided into subpulses (short chirps) and transmitted in randomly hopped center frequencies. In the receiver, the phase alignment algorithm is applied to remove the phase discontinuities between received subpulse signals [29], so the traditional range estimation can be applied.

Similar to frequency altering or hopping, a timing jitter can also be introduced in the time domain to avoid interference, which can be done by changing the temporal position of transmit signal or deviating from its true periodicity. While timing jittering and frequency hopping in victim radar can be well controlled, the transmission strategy of the interfere radar cannot. Thus, it does not always guarantee full interference avoidance by these methods.

4.2.2.3 Cooperative time and frequency scheduling

Current radar systems were designed to coexist. The increasing number of operational units makes coexisting challenging due to limited time and spectrum possibilities. To address the issue of uncontrolled time and frequency altering, a cooperative time and frequency scheduling approach was proposed [13, 30]. In this approach, it is possible to avoid interference by coordinating automotive radars through a communications channel. By using the communication channel, such as the 802.11 p standard, vehicle-to-vehicle (V2V), or vehicle-to-everything (V2X) communication, each radar can decide and assign orthogonal time, frequency, and space resources to avoid interfering with other radars.

While it is possible to use an external communication channel, it is desired to have communication capabilities on the radar system to reduce system complexity. Thus, integrating the sensing and communication on a single (chip) radar system has become an emerging research field since these approaches may provide potential

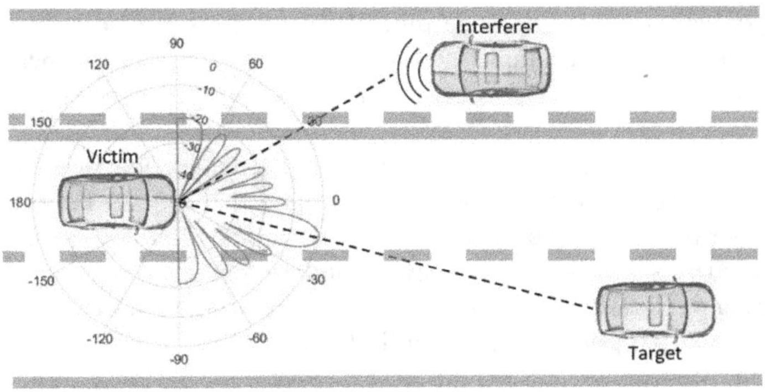

Figure 4.18 *Example scenario for beamforming and null steering. Array response (in Figure 4.19) is superimposed on the scenario and illustrated in the polar coordinate system.*

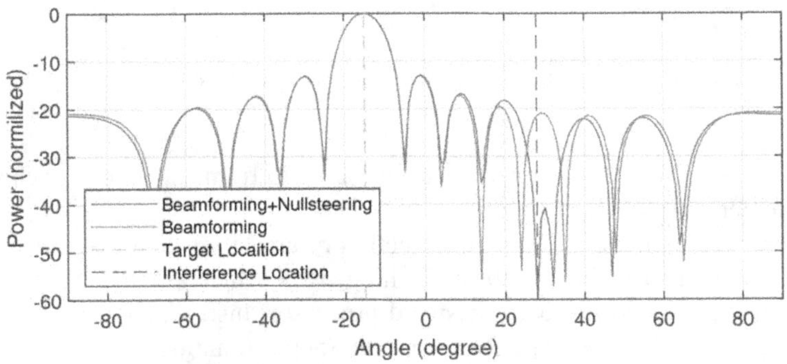

Figure 4.19 *Array response for beamforming and null steering. Target and interference are located at −15° and 28°, respectively.*

gains in performance, size, cost, power consumption, and robustness [30, 31]. This can be accomplished completely in the digital domain, without changing the available hardware, by using waveform for suitable both sensing and communication [31, 32]. This approach needs synchronization of radar units [33], which is another challenge because current automotive radar hardware has been designed for neither synchronization nor communication. Another approach proposes to exploit the similarity of the radar circuitry to standard communication hardware and upgrade automotive radars to joint radar communication units by implementing data link and physical layer functionalities that use the same hardware [30].

4.2.2.4 Digital beamforming and null steering

Modern automotive radar uses MIMO technology to exploit the angle of arrival of return signals. While the main objective of beamforming is to maximize the power at a point of interest (such as target), it is also possible to reduce the energy in a specific direction for interference mitigation, which is known as null steering. Figure 4.18 illustrates an example scenario where the beam steering (beamforming) focuses on the target while null steering mitigates the interference (which is illustrated in Figure 4.19).

Simple null steering can be computed by calculating the steering vector at the desired (interfere) direction and removing its response from the beamforming response. As explained in the MIMO section, let virtual array steering vector be of the form of $\mathbf{A}(\theta) = \mathbf{a}_t(\theta) \otimes \mathbf{a}_r(\theta)$, where $\mathbf{a}_t$ and $\mathbf{a}_r$ are the transmitter and receiver array steering vectors, respectively. Then, the virtual array steering vector at the target direction θ_{tar} is given by $\mathbf{A}_{tar}$. Similarly, for the interference at θ_{int}, the virtual array steering vector is $\mathbf{A}_{int}$. Then the steering vector for the target direction with a null at the interference direction can be written as [34]

$$\mathbf{A} = \mathbf{A}_{tar} - \mathbf{A}_{int}\,\mathbf{r}_{int} \tag{4.49}$$

where the array response at the interference direction is given by

$$\mathbf{r}_{int} = \frac{(\mathbf{A}_{int})^{\mathrm{T}}\mathbf{A}_{tar}}{(\mathbf{A}_{int})^{\mathrm{T}}\mathbf{A}_{int}} \tag{4.50}$$

The final beamforming output of the spatial filtering is given by

$$\mathbf{y}_{BF} = \mathbf{A}^{\mathrm{T}}\mathbf{y} \tag{4.51}$$

where $\mathbf{y} = [y_0, y_1, \ldots, y_{(M-1)}]$ consists of the samples from each receiver channel of the M element virtual ULA.

To use such a method, the exact direction of arrival of the interferer has to be known or estimated before the beamforming, which increases the computational complexity and the workload at the signal processor. Instead, an adaptive interference mitigation technique that does not require any estimation can be used to compute steering vectors in an adaptive manner [21].

It should be noted that, however, the beamforming and null-steering methods are effective interference mitigation methods in the spatial domain, it causes loss of target signal at the interference direction; thus, it may not possible to detect (especially weak) targets at or around the interference direction.

4.2.2.5 Orthogonality and waveform

Orthogonal transmission and reception are some of the key factors for mutual interference mitigation in the MIMO radar system. As discussed in the MIMO section, orthogonality can be achieved in different ways (and in different domains) such as using the time-division multiple access (TDMA), Doppler-division multiple access, and frequency-division multiple access (FDMA). Unfortunately, each of these methods has some shortcomings, for instance, TDM reduces the maximum unambiguous detectable velocity, as well as increases the scan time by the number of transmitters. Moreover, these methods may not be fully effective for interference generated by other radar systems.

Waveform design for interference mitigation is one of the most emerging fields in automotive radar. The waveform design approach aims to have (quasi)orthogonal

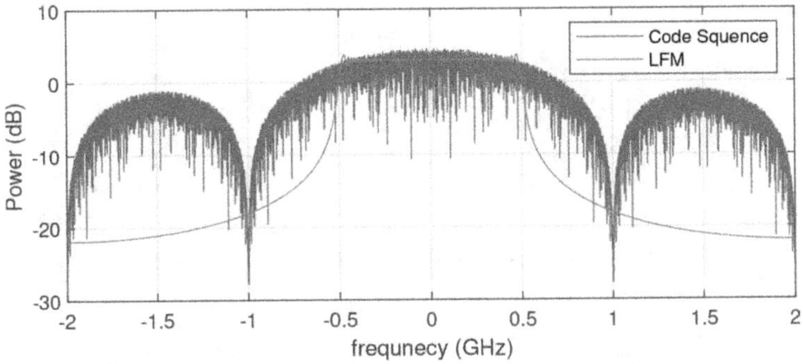

Figure 4.20 Comparison of PMCW and FMCW radar spectrum

waveform in transmit since the interference mitigation comes due to the cross-correlation nature of the orthogonal waveform in the receiver processing. Some of the promising approaches may use spread spectrum techniques such as PMCW or a specialized form of multiplexing such as OFDM [35].

For PMCW radar, each transmitter on the victim radar and/or on the interferer radars in the vicinity can have a unique code sequence to reduce the mutual interference between these units. In the current automobile radar system, binary codes as in (4.37) are usually preferred for simplicity. If a random code sequence is not selected, then one of the existing code sequences with good autocorrelation and cross-correlation properties such as Gold, almost perfect autocorrelation sequences, and zero-correlation zone (ZCZ) sequences, may be preferred [36, 37][‡‡‡]. The cross-correlation indicates the amount of interference, while the auto-correlation indicates the performance of the code like peak-to-side-lobe ratio.

Figure 4.20 illustrates the spectrum of a PMCW and FMCW waveform at the baseband frequency. A 1,024-bit ZCZ code sequence and an LFM waveform are used as PMCW and FMCW, respectively. In code sequence, the chip duration (duration of each bit) is inversely proportional to the bandwidth of the waveform and controls the range resolution; since the bandwidth of a code sequence is inversely proportional to the chip time $T_c = 1/B$, both waveforms have the same range resolution $\delta R = \frac{cT_C}{2} = \frac{c}{2B}$ and same bandwidth B. Unlike the FMCW waveform, the range estimation in PMCW radar can be done by using either matched filtering or circular auto-correlation; thus, a full-band ADC is required. For velocity estimation, pulse-to-pulse phase compensation must be applied before Doppler processing. While the phase-coded waveforms perform well for slow-moving targets, in the presence of high-speed targets, their performance degrades due to the frequency shifts caused by Doppler effects [37].

[‡‡‡]All code sequences that are provided as an example are binary sequences. There are many (binary and nonbinary) code sequences are available in the literature with good auto- and cross-correlation properties.

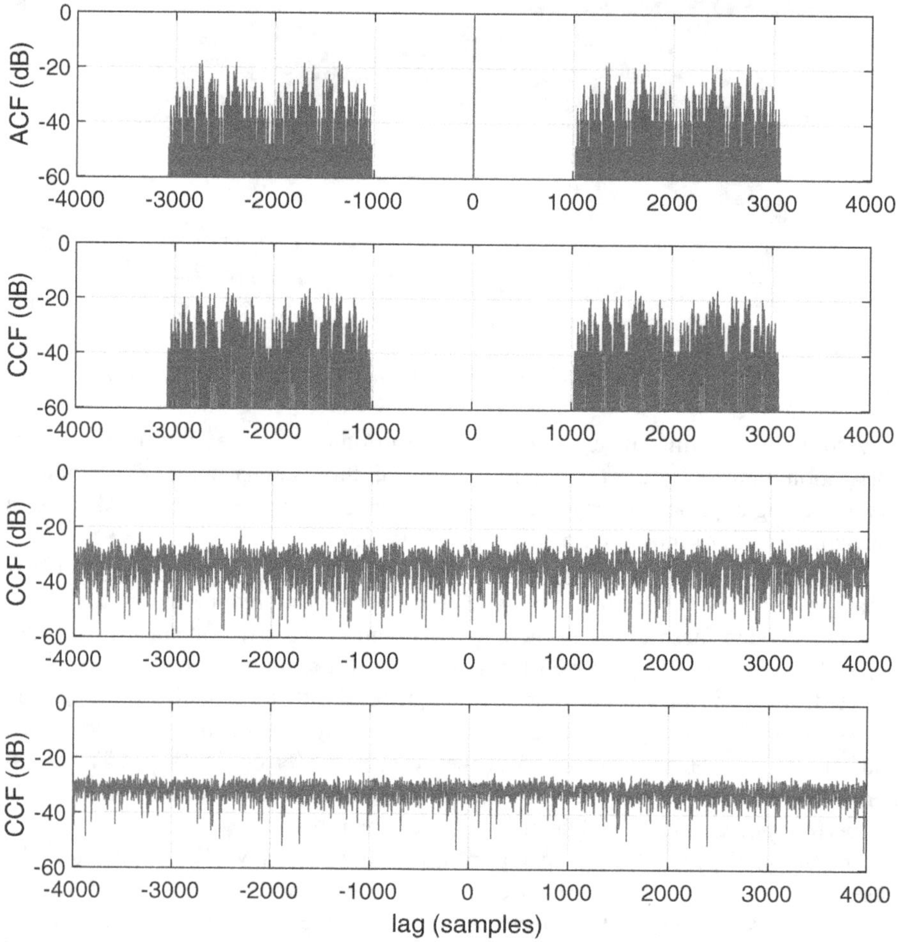

Figure 4.21 *Illustration of ZCZ sequences with length 1,024 (4 times over-sampled). Top to bottom: auto-correlation function; cross-correlation function (CCF) with another ZCZ code; CCF function with LFM waveform; and CCF function with random sequence.*

As an example, correlation properties of a 1,024-bit ZCZ code sequence are illustrated in Figure 4.21 ZCZ code sequence generates a zero correlation zone between lags one and $L_c/4$. The code sequences are usually up-sampled before transmission to satisfy the range resolution requirement. In our example, the code sequence is four times over-sampled (as illustrated in Figure 4.20), and zero correlation zone spans from lag 1 to 1,024, which might be enough for many short-range applications. The cross-correlation of this sequence with another ZCZ code with the same properties is also illustrated in the figure. Such a code set is useful for MIMO radar since the cross-correlation is zero for a given interval; thus, the mutual interference between antenna elements does not exist. In the same figure, its cross-correlation with LFM (with the same center frequency and same bandwidth) and with a random sequence PMCW signal is also illustrated. As a general trend, phase coding mitigates any interference by spreading its energy through the spectrum thus signal to interference ratio is related to code length as $SIR = 10\, log_{10}(Lc)$ dB. The longer codes should be preferred for not only better interference mitigation but also bigger unambiguous range, since the unambiguous range increases by increasing the period of the code for a given chip duration.

To address the full-band sampling issue of the spread spectrum approach, the PC-FMCW waveform is introduced [16, 32]. The PC-FMCW waveform, which is of the form of (4.36), does not spread the RF spectrum of the FMCW radar but spreads the beat frequency spectrum; thus, the bandwidth of RF hardware stays the same as a classical FMCW radar. Similar to the PMCW radar, longer code sequence is preferred for PM-FMCW radar for better interference mitigation, though the number of code sequences bounded by the number of samples of LFM waveform. In the PC-FMCW approach, standard FMCW radar receiver is used. Following a group delay filter is applied to the received signal to compensate for the different delays of each beat frequency. After the delay is compensated, coding signal is applied for decoding. Since the spectrum of beat frequency is proportional to the range, any interfering signal without the proper code sequence spreads over the range domain and loses its energy by $10\,\log_{10}(\mathrm{Lc})$; thus, the interference signal is mitigated.

4.2.3 Final remarks

This section makes an introduction to the interference detection problem and discussed some of the possible approaches for interference mitigation and avoidance. As pointed out before, there is not a single method to mitigate automotive radar interference since the interference signal has many degrees of freedom and has many different forms. The lack of standardization of automotive radar systems, as well as the emerging hardware technology makes the interference control challenging. In the near future, it is expected to have multi-functional single-chip sensors for joint sensing and communication. Multi-functional radar systems are expected to operate in cooperation instead of coexistence, which eventually will address the interference problem.

Acknowledgements

The presented work was performed during the author's employment at the Technical University of Delft (TUD), the Netherlands.

References

[1] Hasch J., Topak E., Schnabel R., Zwick T., Weigel R., Waldschmidt C. Millimeter-wave technology for automotive radar sensors in the 77 GHz

frequency band. *IEEE Transactions on Microwave Theory and Techniques*. 2012, vol. 60(3), pp. 845–60.

[2] Series M. 'Systems characteristics of automotive radars operating in the frequency band 76–81 GHz for intelligent transport systems applications'. *Recommendation ITU-R, M*; 2014. pp. 2057–1.

[3] Roos F., Bechter J., Knill C., Schweizer B., Waldschmidt C. 'Radar sensors for autonomous driving: modulation schemes and interference mitigation'. *IEEE Microwave Magazine*. 2019, vol. 20(20), pp. 58–72.

[4] Uysal F. 'Synchronous and asynchronous radar interference mitigation'. *IEEE Access: Practical Innovations, Open Solutions*. 2019, vol. 7, pp. 5846–52.

[5] Eshraghi A., Bordes J.P., Davis C, *et al.* 'Vehicular radar system with self-interference cancellation'. *US Patent*. 2017, vol. 9(791), p. 551.

[6] Bauduin M., Bourdoux A. *Spillover Cancellation in Radar Systems. European Patent Application EP3588127A1*. 2020.

[7] Goppelt M., Blocher H.L., Menzel W. 'Analytical investigation of mutual interference between automotive FMCW radar sensors'. *German Microwave Conference*; 2011. pp. 1–4.

[8] Schipper T., Mahler T., Harter M. 'An estimation of the operating range for frequency modulated radars in the presence of interference'. *European Radar Conference*; 2013. pp. 227–30.

[9] Uysal F., Sanka S. 'Mitigation of automotive radar interference'. *IEEE Radar Conference (RadarConf18)*; Oklahoma City, OK, 2018. pp. 0405–10.

[10] Bechter J., Waldschmidt C. 'Automotive radar interference mitigation by reconstruction and cancellation of interference component'. *IEEE MTT-S International Conference on Microwaves for Intelligent Mobility (ICMIM)*; Heidelberg, Germany, 2015. pp. 1–4.

[11] Sanka S. 2017. "Radar to radar interference for 77GHz automotive radar', M.S. Thesis'.Delft University of Technology

[12] Norouzian F., Pirkani A., Hoare E., Cherniakov M., Gashinova M. 'Phenomenology of automotive radar interference'. *IET Radar, Sonar & Navigation*. 2021, vol. 15(9), pp. 1045–60.

[13] Aydogdu C., Garcia N., Hammarstrand L., *et al.* 'Radar communications for combating mutual interference of FMCW radars'. *IEEE Radar Conference (RadarConf)*; 2019. pp. 1–6.

[14] Kumbul U., Uysal F., Vaucher C.S., Yarovoy A. 'Automotive radar interference study for different radar waveform types'. *IET Radar, Sonar & Navigation*. 2022, vol. 16(3), pp. 564–77.

[15] Reneau J., Adhami R.R. 'Phase-coded LFMCW waveform analysis for short range measurement applications'. *IEEE Aerospace Conference*; 2014. pp. 1–6.

[16] Uysal F. 'Phase-coded FMCW automotive radar: system design and interference mitigation'. *IEEE Transactions on Vehicular Technology*. 2020, vol. 69(1), pp. 270–81.

[17] Bourdoux A., Parashar K., Bauduin M. 'Phenomenology of mutual interference of FMCW and PMCW automotive radars'. *IEEE Radar Conference (RadarConf17)*; Seattle, WA, 2017. pp. 1709–14.

[18] Schipper T., Harter M., Zwirello L., *et al.* 'Systematic approach to investigate and counteract interference-effects in automotive radars'. *9th European Radar Conference*; 2012. pp. 190–3.

[19] Kunert M. Final report, *European Commission: MOre Safety for All by Radar Interference Mitigation (MOSARIM)*. Luxembourg; 2012.

[20] Alland S., Stark W., Ali M., Hegde M. 'Interference in automotive radar systems: characteristics, mitigation techniques, and current and future research'. *IEEE Signal Processing Magazine*. 2019, vol. 36(5), pp. 45–59.

[21] Rameez M., Dahl M., Pettersson M.I. 'Adaptive digital beamforming for interference suppression in automotive FMCW radars'. *IEEE Radar Conference (RadarConf18)*; 2018. pp. 0252–56.

[22] Laghezza F., Jansen F., Overdevest J. 'Enhanced interference detection method in automotive FMCW radar systems'. *20th International Radar Symposium (IRS)*; 2019. pp. 1–7.

[23] Pirkani A., Norouzian F., Hoare E., Cherniakov M., Gashinova M. 'Automotive interference statistics and their effect on radar detector'. *IET Radar, Sonar & Navigation*; vol.16 (1), 2022. pp. 9–21.

[24] Rameez M. *Interference Mitigation Techniques in Fmcw Automotive Radars. [Phd. Thesis]*. Blekinge Tekniska Högskola; 2020.

[25] Bechter J., Roos F., Rahman M, *et al.* "Automotive radar interference mitigation using a sparse sampling approach'. *European Radar Conference*; 2017. pp. 90–93.

[26] Correas-Serrano A., Gonzalez-Huici M.A. "Sparse reconstruction of chirplets for automotive FMCW radar interference mitigation'. *IEEE MTT-S International Conference on Microwaves for Intelligent Mobility*; 2019. pp. 1–4.

[27] Bechter J., Sippel C., Waldschmidt C. 'Bats-inspired frequency hopping for mitigation of interference between automotive radars' in ICMIM'. *2016 IEEE MTT-S International Conference on Microwaves for Intelligent Mobility*; 2016. pp. 1–4.

[28] Luo T.-N., Wu C.-H.E., Chen Y.-J.E. 'A 77-GHz CMOs automotive radar transceiver with anti-interference function'. *IEEE Transactions on Circuits and Systems I: Regular Papers*. 2013, vol. 60(12), pp. 3247–55.

[29] Moon T., Park J., Kim S. 'BlueFMCW: random frequency hopping radar for mitigation of interference and spoofing'. *EURASIP Journal on Advances in Signal Processing*. 2021, vol. 7.

[30] Aydogdu C., Keskin M.F., Carvajal G.K., *et al.* 'Radar interference mitigation for automated driving: exploring proactive strategies'. *IEEE Signal Processing Magazine*. 2020, vol. 37(4), pp. 72–84.

[31] Ma D., Shlezinger N., Huang T., Liu Y., Eldar Y.C. 'Joint radar-communication strategies for autonomous vehicles: combining two key automotive technologies'. *IEEE Signal Processing Magazine*. 2020, vol. 37(4), pp. 85–97.

[32] Uysal F., Orru S. 'Phase-coded FMCW automotive RADAR: application and challenges'. *2020 IEEE International RADAR Conference (RADAR)*; Washington, DC, USA, 2020. pp. 478–82.

[33] Lampel F., Uysal F., Tigrek F., *et al.* 'System level synchronization of phase-coded FMCW automotive radars for radcom'. *2020 14th European Conference on Antennas and Propagation (EuCAP)*; Copenhagen, Denmark, 2020. pp. 1–5.

[34] Harry L., Trees V. *Optimum array processing*. New York: John Wiley & Sons, Inc. Publication; 2002.

[35] Carvajal G.K., Keskin M.F., Aydogdu C, *et al.* 'Comparison of automotive FMCW and OFDM radar under interference'. *2020 IEEE Radar Conference (RadarConf20)*; Florence, Italy, 2020. pp. 1–6.

[36] Overdevest J., Jansen F., Laghezza F., Uysal F, *et al.* 'Uncorrelated interference in 79 GHz FMCW and PMCW automotive radar'. *2019 20th International Radar Symposium (IRS)*; Ulm, Germany, 2019. pp. 1–8.

[37] Overdevest J., Jansen F., Uysal F, *et al.* 'Doppler influence on waveform orthogonality in 79 GHz MIMO phase-coded automotive radar'. *IEEE Transactions on Vehicular Technology*. 2020, vol. 69(1), pp. 16–25.

Chapter 5
mmWave radar tracking and sensor fusion with camera

Renyuan Zhang[1] *and Siyang Cao*[1]

5.1 Introduction

W-band millimeter wave (mmWave) radar is becoming an essential sensor in advanced driver assistance system (ADAS) and autonomous driving fields [1–5], including adaptive cruise control, pedestrian detection, collision avoidance, lane changing monitoring, and emergency braking. ADAS systems comprise of multiple sensors, such as radar, camera, and LiDAR. Each type of sensor offers its advantage, e.g., radar has a reliably moving target detection and is robust to adverse weather conditions. But radar has a relatively poor angular resolution comparing with its optical counterparts, which can achieve 0.1–1 degree resolution in both azimuth and elevation directions. The shortcoming of radar will result in uncertainty, inaccuracy, and missing or ghost detections at certain distances. One way to reduce the uncertainties and failures is to fuse radar outputs with heterogeneous sensors [4, 6].

mmWave radar research has gained immense popularity with the recent increasing demand of automotive radars in ADAS and autonomous driving industry. Among all the researches on mmWave radar applications, tracking [7–9], target recognition [10–12], and sensor fusion [13–15] are the most popular topics. In this chapter, we will start with a reliable radar tracking method using extended Kalman filter (EKF) and discuss the nonlinearities of radar tracker. Then, we will further develop a radar–camera sensor fusion by constructing a fusion-extended Kalman filter (fusion-EKF) in the remaining of the chapter.

Sensor fusion has been developing rapidly in recent years. However, there are limited studies discussing fusion of mmWave radars with other sensors since radars provide a limited number of detection points representing targets-of-interest [4], which make it difficult to recognize from a snapshot of radar detection. But if fusion can be achieved before the target classification of radar, a fusion system may fully take the advantage of radar, i.e., increasing the reliability of detecting moving

[1]Department of Electrical and Computer Engineering, University of Arizona, Tucson, Arizona, United States

targets, avoiding blockage, and tracking dramatically. In this chapter, we aim to increase mmWave radar's informative capability about targets and further its versatility by fusion with monocameras as an example.

Specifically, we introduce a fusion-EKF, which is designed to fuse data from heterogeneous sensors such as mmWave radar and monocamera with real-time fusion algorithm running for tracking. Sensor fusion and association are done within the fusion-EKF using a homography estimation method (HEM) [16], timeline alignment, and region search. Reliable detection and cross-validated target tracking are also realized. As we do not use any machine learning-based approaches to realize the fusion, the introduced fusion method achieves a low computational complexity, which is ideal for implementing in real-time systems. The experimental results also show that the proposed system can provide a reliable tracking and detecting result with low calculation costs. An embedded system like Arduino or Raspberry Pi can be utilized to process the data for real-time applications.

For the new fusion-EKF, a new concept is introduced, i.e., error bounds (EBs), which is defined as the sensor's region of approximation. EBs are not from the uncertainty of sensors [17] but the sensors' resolutions from their respective perspectives. An HEM is applied in associating heterogeneous sensors via their EBs. The fusion-EKF is designed to take both radar and camera as inputs and associate the data inside the filter to obtain ideal target tracking outputs. Data association of the fusion-EKF is capable to support tracking of multiple targets.

The structure of this chapter is as follows. In section 5.2, related work on mmWave radar tracking and sensor fusion are studied and presented. In section 5.3, radar-EKF tracking methodology is shown. In section 5.4, the sensor fusion fusion-EKF is presented with preprocessing, data association and sensor synchronization. In section 5.5, results of radar–camera fusion-EKF and root mean square errors (RMSEs) of improved EBs are shown.

5.2 Related work

mmWave radar tracking has been raised interest over years. In Reference [18], the unscented Kalman filter (UKF) with the joint probabilistic data association algorithm is developed and tested. It shows a superior performance. However, UKF is not an optimal solution compared to EKF. Because radar's nonlinearity comes from Doppler, which is the first-order approximation of range. Thus, UKF can be unstable and its results will be biased when estimating radar's Doppler velocities. The radar-EKF is designed based on the Kalman filter (KF). KF typically uses a series of measurements observed over time, containing statistical noise and other inaccuracies, and produces estimates of unknown variables that tend to be more accurate than those based on a single measurement alone [19]. The difference between EKF [20] and traditional KF is that EKF can handle linear equations as well as can adapt to nonlinearities. EKF uses the Jacobian matrix to linearize nonlinear functions, for instance, radar's Doppler velocities (range rates) in our case. Fusion-EKF extends EKF's ability in handling data from heterogeneous sensors. Thus, the tracked targets can be utilized to implement classification algorithms like micro-Doppler signatures from radar [10, 21] to classify targets.

mmWave radars have been implemented to fuse with LiDARs [22, 23] in recent years. Hajri and Rahal [22] introduced radar and LiDAR real-time sensor fusion that used the mean square error with radar, LiDAR, and the fusion variance in the experiment, to yield convincing results. By default, radar and LiDAR, both can provide detections in the bird's-eye view (BEV) plane, which is the top view looking onto the ground [24]. However, LiDAR works on the infrared band, which is prone to interference, cost-intensive, and bulky in size. Not many studies have been conducted with associating heterogeneous sensors like cameras or infrared sensors with radars [25, 26], primarily due to the fact that the camera's and radar's detections are unrelated and difficult to align without certain assumptions. In this case, we are implementing HEM and creating a track-oriented association and fusion algorithm for calibration and tracking.

Alternate approaches to sensor fusion have been studied using machine learning methods. The sensor fusion is typically performed via concatenating tensors inside the neural network [27]. However, it requires large neural networks for high-resolution radar and cameras, which makes it difficult to implement for real-time applications. Similarly, Zhong *et al* [28] introduced the Texas Instruments' (TI) commercial TDA3x/TDA2x boards for camera radar fusion. It is so far the most reliable fusion system for the mmWave radar's detection compounded with the camera's vision to the best of our knowledge. In their paper, a complete fusion system provides effective tracking and detections over time. The shortcomings of this research are that, the object detections and classifications rely on machine learning techniques with online computer vision libraries. Therefore, the complete pipeline is difficult to implement for real-time applications. In the meantime, radar noise handling is not ideal for various scenarios. Using fusion-EKF, we can strongly overcome these shortcomings.

Muresan and Nedevschi [29] present a two-step data association method for LiDAR. It is based on spreading parameter weights from KF. In this chapter, the weights are further studied to support both the heterogeneous sensors. Specifically, the EBs are introduced in the data association phase as radar provides raw Doppler velocities and camera vision generates the bounding boxes (BBoxes) of targets. EB supports the two sensors to form a clear gating for the association of the fusion-EKF. Thus, comparing to other weight-based or probability-based association methods, the proposed fusion-EKF with EBs will provide an extra dimension in association for heterogeneous sensors. Motion model selection for multi-target tracking can also be an important part of KF tracking. In Reference [30], the survey on different motion models is introduced and compared. In our setup, the constant acceleration (CA) model is used because we assumed targets can have the linearity of the state transition equation, which allows an optimal propagation of the state probability distribution of humans and cars. Yaw angle and turn rate are less essential in these models. Thus, it allows tracks to have more flexibility in association from heterogeneous sensors.

Some papers [25, 26, 31, 32] illustrated the alignment between radars and other heterogeneous sensors. They assumed radars and other sensors are working at identical principle. But in fact, they are heterogeneous in dimensions, processing chain and sensing wavelength. In Reference [31], vision lateral position improvement of radar detections is introduced that worked for radar alignment onto the vision well using cross-correlation. However, the errors are not corrected during detections. Folster and Rohling [32] introduced a lateral velocity estimation for automotive radar sensors for ADAS. The method simply refers to the location of the target on the actual detections. The disadvantages are obvious: there are not many cross-correlations between sensors, and the approximation relies only on locations. To overcome limitations in these cases, EBs proposed in this chapter are essential to apply to radar-vision fusion-EKF to improve region search and tracking. The HEM further enhances the transformation and track-oriented association between heterogeneous sensors, which makes the proposed fusion system to takes both sensors' advantages, and be robust even one sensor is temporarily out of work.

5.3 Radar-EKF: radar tracking methodology

In this section, we will go through current state-of-art radar tracking using EKF and further expand to fusion-EKF in section 5.4.

5.3.1 FMCW radar

For FMCW radars preprocessing, the radar sensor configuration and antenna determine its range resolution ($\Delta\rho$), bearing resolution ($\Delta\theta$) and Doppler resolution (Δv_d).

Range resolution $\Delta\rho$ can be calculated from radar sweeping chirp bandwidth:

$$\Delta\rho = \frac{c_0}{2B} \tag{5.1}$$

where c_0: the speed of electromagnetic waves; B: the bandwidth of the radar chirp signal.

The bearing resolution physically depends on the number of virtual antennas multiple-input and multiple-output (MIMO) radar emulated. The angular resolution is given as

$$\Delta\theta = \frac{c_0}{f_c d N_{RX} N_{TX} \cos\theta_i} \tag{5.2}$$

where

f_c: center frequency of linear frequency modulated signal;
d: receiver element spacing, typically $\lambda/2$;
λ: carrier signal wavelength;
N_{RX}: number of receivers (RX);
N_{TX}: number of transmitters (TX);
θ_i the angle of interest.

Doppler resolution provides the target range changing rate and is defined as

$$\Delta v_d = \frac{c_0}{2f_c T N_d} \tag{5.3}$$

where

T: the pulse repetition interval;
N_d: number of chirps per radar frame.

T is commonly the total chirp time of a single frequency modulated signal. Thus, Δv_d is typically restricted by number of chirps and chirp repetition interval.

5.3.2 FMCW radar noise

An important part for frequency modulated continuous wave (FMCW) radar sensing is that FMCW radar measures targets' scatters with compound noises. These noises come from the following effects:

1. Multipath effect [33], that results in detecting targets to undergo ghost or fading where a phase-shifted reflection is produced by walls or ground.
2. mmWave radar has its inherent measurement noise ($\mathbf{R_{radar}}$). In this case, we compute the noise model by calculating the variance of the point cloud and signal-to-noise ratio (SNR) of actual measurement based on the set radar configuration.
3. Measurement processing noise from stretch processing, and fast Fourier transforms (FFTs) produce noisy point cloud points. The noise is typically dense when multiple targets are being detected. At the meantime, a higher processing rate causes the processing noise to lift further.
4. Antenna gain contributes to a varied SNR at different angles. This may cause the SNR at the center (line of sight) to be higher than SNR at the sides.

For the first case, which is the multipath effect in 77 GHz automotive radar, a typical symptom is the presence of ghost targets that are produced with random phase but similar shape, Doppler, and amplitude compared to the original target. A method to reduce when applied to tracking is to use a multilayer perceptron [34]. In this case, the ghost is largely alleviated and then the tracking becoming more reliable.

For the second case, $\mathbf{R_{radar}}$ obeys the Gaussian distribution. $\mathbf{R_{radar}}$ is then:

$$\mathbf{R_{radar}} \sim \mathcal{N}(\begin{pmatrix} \mu_\rho \\ \mu_\theta \\ \mu_{v_d} \end{pmatrix}, \begin{pmatrix} \sigma_\rho^2 \\ \sigma_\theta^2 \\ \sigma_{v_d}^2 \end{pmatrix}) \tag{5.4}$$

The $\begin{pmatrix} \mu_\rho & \mu_\theta & \mu_{v_d} \end{pmatrix}^{\mathrm{T}}$ vector corresponds to the targets' detection matrix. It specifies target's range, bearing, and Doppler velocity. On the other hand, the

$\left(\sigma_\rho^2 \quad \sigma_\theta^2 \quad \sigma_{v_d}^2\right)^\mathsf{T}$ vector is the variances of the targets, which, in turn, correspond to the uncertainty of the target's range, bearing, and Doppler velocity. Therefore, the $\left(\sigma_\rho^2 \quad \sigma_\theta^2 \quad \sigma_{v_d}^2\right)^\mathsf{T}$ vector corresponds to the EBs of mmWave radars. With the resolutions discussed in section 5.4.1, the relations are below:

$$\begin{pmatrix} \Delta\rho^2 \\ \Delta\theta^2 \\ \Delta{v_d}^2 \end{pmatrix} = \begin{bmatrix} \frac{1}{4} & 0 & 0 \\ 0 & \frac{1}{4} & 0 \\ 0 & 0 & \frac{1}{4} \end{bmatrix} \begin{pmatrix} \sigma_\rho^2 \\ \sigma_\theta^2 \\ \sigma_{v_d}^2 \end{pmatrix} \tag{5.5}$$

Here, we have the assumptions that the discrete Fourier transform (DFT) size is the same as input data size. However, as direction of arrival (DOA) estimation can be estimated by conventional DFT/FFT method and several high-resolution algorithms, e.g., MUSIC algorithm [35] and ESPIRIT algorithm [36], the Δ_θ^2 term is modified to $\Delta_{\theta_{\mathrm{DOA}}}^2$ according to number of angle bins. The 0s in the matrix represents the uncorrelated from range to angle, angle to Doppler, and range to Doppler, respectively. The 1/4 term is the minimum requirement to avoid aliasing. This way, the EB of mmWave radar is defined.

For the third case, we assume that the mmWave radar is a linear system with an output range ρ_o, which is the response of an input signal $\mathbf{s}$, the measurement noise $\mathbf{e}$ can then be obtained by

$$\rho_o = \rho + \mathbf{e} = \mathbf{hs} + \mathbf{e} \tag{5.6}$$

where ρ is the raw measurement from sensor and $\mathbf{h}$ is the transform from input signal to output signal. Using least-squares estimation from orthogonal principles [37], and choosing appropriate $\mathbf{h}$ to minimize the Euclidean distance, the minimized projection error $\mathbf{e}$ is orthogonal to every column of $\mathbf{s}$ if and only if [38]:

$$\mathbf{s}^\mathsf{T}\mathbf{e} = 0 \tag{5.7}$$

Therefore, by orthogonality:

$$\mathbf{h} = (\mathbf{s}^\mathsf{T}\mathbf{s})^{-1}\mathbf{s}^\mathsf{T}\rho_o \tag{5.8}$$

The measurement processing noise is minimized with the least sum of $\|\mathbf{e}\|^2$ to reduce effects onto tracking. In the later fusion-EKF stages, this measurement noise only depends on the time variance of the receiving signal from equations. Therefore, the inherent measurement noise $\mathbf{R}_{\mathrm{radar}}$ is used to produce the EKF estimations.

In this application, the processing noise is then the variance from the processing time from the radar board. The processing covariance noise from radar, $\mathbf{Q}_{\mathrm{radar}}$, is related to the motion variance of acceleration of the object and the processing time variance Δt from radar. Similarly, in section 5.4 fusion-EKF part, we have camera's $\mathbf{Q}_{\mathrm{cam}}$ that is associated with the camera's processing time variance.

In the final scenario, the antenna gained results in SNR difference, which results in the center detections to carry less noise than the surroundings. When it comes to sensor fusion in section 5.4, the radial detections around corners are noisy and could lead to missed detections. This potential failure should be compensated by other sensors like cameras. Note that the antenna gain in this application only reflects the SNR at different detecting angles. Measurements in the anechoic chamber are conducted and the antenna gain at 0 degrees is 10 dB compared to 0 dB at ±60 degrees for the TI mmWave radar that we used in this study. The reference can be seen in TI's document [39].

5.3.3 *EKF prediction*

Model selection for radar tracking is CA model as stated before because targets can have the linearity of the state transition and radar's Doppler information can be a better help to determine the velocity of CA model. If we have a state vector of a target $\mathbf{x}$ with position p and velocity v, we can define the state vector as a four-element vector with position and velocity projections onto x-axis and y-axis. That is

$$\mathbf{x} = (p, v)^{\mathsf{T}} = (p_x, p_y, v_x, v_y)^{\mathsf{T}} \tag{5.9}$$

The linear motion model from position and velocity can be simply expressed as

$$p_k = p_{k-1} + v_k \Delta t \tag{5.10}$$

With a more complexity form, also considering the velocity change, the new linear model is

$$\begin{pmatrix} p_k \\ v_k \end{pmatrix} = \begin{bmatrix} 1 & \Delta t \\ 0 & 1 \end{bmatrix} \begin{pmatrix} p_{k-1} \\ v_{k-1} \end{pmatrix} \tag{5.11}$$

The state transition matrix is

$$\mathbf{x}_k = \mathbf{F}_k \mathbf{x}_{k-1} + \mathbf{B}_k \mathbf{u}_k + \mathbf{w} \tag{5.12}$$

where

$\mathbf{F}_k$: state transition matrix;
$\mathbf{x}_{k-1}$: previous state; $k-1$: state vector;
$\mathbf{B}_k$: control input matrix;
$\mathbf{u}_k$: control input vector;
$\mathbf{w}$: compound noise from system (motion noise and processing noise).

If we have no motion on sensors $\mathbf{u}_k = (0, 0)^{\mathsf{T}}$, left the 5.12 only the state and noise component:

$$\mathbf{x}_k = \mathbf{F}_k \mathbf{x}_{k-1} + \mathbf{w} \tag{5.13}$$

The compound noise $\mathbf{w}$ obeys zero mean Gaussian distribution:

$$\mathbf{w} \sim \mathcal{N}(0, \mathbf{Q}) \tag{5.14}$$

where $\mathbf{Q}$ is the processing noise introduced before as $\mathbf{Q}_{\mathbf{radar}}$ or $\mathbf{Q}_{\mathrm{cam}}$. As time stamps are changing, the time stamp difference Δt should also be updated throughout the process.

From the normal distribution of noise, an uncertainty covariance can be obtained for linearity. The uncertainty covariance matrix, also often referred to priori error covariance, for the prediction step is

$$\mathbf{P}_k = \mathbf{F}_k\mathbf{P}_{k-1}\mathbf{F}_k^{\mathsf{T}} + \mathbf{Q}_{\mathbf{k}} \tag{5.15}$$

So for the state transition matrix $\mathbf{F}$ from 5.13, the updated with time stamp differences is

$$\mathbf{F} = \begin{bmatrix} 1 & 0 & \Delta t & 0 \\ 0 & 1 & 0 & \Delta t \\ 0 & 0 & 1 & 0 \\ 0 & 0 & 0 & 1 \end{bmatrix} \tag{5.16}$$

At the meantime, the compound noise can be expressed as

$$\mathbf{w} = \left(w_{px}, w_{py}, w_{vx}, w_{vy}\right)^{\mathsf{T}} \tag{5.17}$$

When a CA model is applied, we set the model with a random acceleration vector $\mathbf{a} = (a_x, a_y)^{\mathsf{T}}$ with variances of $\sigma_a x^2$ and $\sigma_a y^2$ on x-axis and y-axis, respectively. In this case, the $\mathbf{w}$ noise matrix, which is the motion EB for the tracked model, can be expressed as

$$\mathbf{w} = \begin{pmatrix} w_{px} \\ w_{py} \\ w_{vx} \\ w_{vy} \end{pmatrix} = \begin{pmatrix} \frac{a_x \Delta t^2}{2} \\ \frac{a_y \Delta t^2}{2} \\ a_x \Delta t \\ a_y \Delta t \end{pmatrix} \tag{5.18}$$

From (14), the noise is described by a zero mean and a covariance matrix $\mathbf{Q}$. When referring to (15) and (18), with variances of $\sigma_a x^2$ and $\sigma_a y^2$ applied, the $\mathbf{Q}$ matrix can be estimated by taking expectations of the $\mathbf{w}$ and $\mathbf{w}^{\mathsf{T}}$ matrix as

$$\begin{aligned} \mathbf{Q} &= \mathbb{E}[\mathbf{w}\mathbf{w}^{\mathsf{T}}] \\ &= \begin{bmatrix} \frac{\sigma_{ax}^2 \Delta t^4}{4} & 0 & \frac{\sigma_{ax}^2 \Delta t^3}{2} & 0 \\ 0 & \frac{\sigma_{ay}^2 \Delta t^4}{4} & 0 & \frac{\sigma_{ay}^2 \Delta t^3}{2} \\ \frac{\sigma_{ax}^2 \Delta t^3}{2} & 0 & \sigma_{ax}^2 \Delta t^2 & 0 \\ 0 & \frac{\sigma_{ax}^2 \Delta t^3}{2} & 0 & \sigma_{ax}^2 \Delta t^2 \end{bmatrix} \end{aligned} \tag{5.19}$$

Hence $\mathbf{Q}$ depends on Δt and the random acceleration variances. Δt changes from time to time because of different frames-per-second (fps) of camera and radar. With high Δt, the uncertainty and the large EB are generated, and vice versa.

Similarly, we apply this to measurement noise matrix $\mathbf{R_{radar}}$. Combining (1)–(5), the result $\mathbf{R_{radar}}$ is

$$\begin{aligned}\mathbf{R}_{\mathbf{radar}} &= \mathbb{E}[\mathbf{y}\mathbf{y}^{\mathsf{T}}] \\ &= \begin{bmatrix} \sigma_\rho^2 & 0 & 0 \\ 0 & \sigma_\theta^2 & 0 \\ 0 & 0 & \sigma_{v_d}^2 \end{bmatrix}\end{aligned} \tag{5.20}$$

5.3.4 EKF update

Similar to the prediction state vector definition, we need an update measurement vector $\mathbf{z}$ that consists of radar's measurement:

$$\mathbf{z}_{\mathrm{radar}} = (\rho, \theta, v_d)^{\mathsf{T}} \tag{5.21}$$

The update consists of a residual $\mathbf{y}$ term, which is the error of the measurement from the prediction:

$$\mathbf{z}_k = \mathbf{H}_k\mathbf{x}_k + \mathbf{y}_k \tag{5.22}$$

where

$\mathbf{H}_k$ measurement function matrix.

The $\mathbf{y}$ part obeys:

$$\mathbf{y} \sim \mathcal{N}(0, \mathbf{R}) \tag{5.23}$$

where the $\mathbf{R}$ is the inherent measurement noise cause by sensor, e.g., introduced in (5.20). The random variable also has a normal distribution. Thus, the measurement error covariance matrix becomes:

$$\mathbf{S}_k = \mathbf{H}_k\mathbf{P}_k\mathbf{H}_k^{\mathsf{T}} + \mathbf{R}_k \tag{5.24}$$

The optimal KF gain $\mathbf{G}$ can be obtained as

$$\mathbf{G}_k = \mathbf{P}_k\mathbf{H}_k^{\mathsf{T}}\mathbf{S}_k^{-1} \tag{5.25}$$

The state vector update $\mathbf{x}_{k(\mathrm{update})}$ is based on prediction:

$$\mathbf{x}_{k(\mathrm{update})} = \mathbf{x}_k + \mathbf{G}_k\mathbf{y}_k \tag{5.26}$$

$\mathbf{x}_{k(\mathrm{update})}$ is then used for the next state prediction step as $\mathbf{x}_{k-1}$. Similarly, it is also applied to the uncertainty covariance matrix $\mathbf{P}_{k(\mathrm{update})}$:

$$\mathbf{P}_{k(\mathrm{update})} = (\mathbf{I} - \mathbf{G}_k\mathbf{H}_k)\mathbf{P}_k \tag{5.27}$$

The uncertainty covariance is often used to scale the response of the KF. Commonly, in terms of representing the response, the KF gain $\mathbf{G}$ is used. It represents the relative weight of the measurements compared to the current state estimate, and is typically changing over time as the noise keeps changing throughout time. Rewriting the (5.25) with (5.24):

$$\mathbf{G}_k = \frac{\mathbf{P}_k\mathbf{H}_k^{\mathsf{T}}}{\mathbf{H}_k\mathbf{P}_k\mathbf{H}_k^{\mathsf{T}} + \mathbf{R}_k} \tag{5.28}$$

Thus, we can conclude that $\mathbf{G}$ as depends on $\mathbf{H}_k\mathbf{P}_k\mathbf{H}_k^{\mathsf{T}}$ and $\mathbf{R}_k$ terms, with $\mathbf{R}_k$ is fixed in a single measurement and not changing throughout the k steps. When $\mathbf{H}_k\mathbf{P}_k\mathbf{H}_k^{\mathsf{T}}$ is much higher than the $\mathbf{R}_k$ term, $\mathbf{G}$ is smaller. The low $\mathbf{G}$ will result in the model relying more on predictions. Hence, it is less responsive to measurements. On the other hand, when $\mathbf{H}_k\mathbf{P}_k\mathbf{H}_k^{\mathsf{T}}$ is much smaller than the $\mathbf{R}_k$ term, $\mathbf{G}$ is larger, and the EKF puts more weights on the current measurement. Thus, it gets more responsive to measurements.

5.3.5 Nonlinearity

The prominent part of this radar-EKF is to exploit the nonlinear property in EKF to deal with the radar's Doppler data, because Doppler is the range rate – the partial derivative of the range with time.

Because radar has detections in polar coordinate system, the conversion to Cartesian coordinate system is needed for every incoming measurement. Additionally, from (5.22), the $\mathbf{H}$ matrix should have a convertible formation to map radar detections from the state vector $\mathbf{x}$ to analyze $\mathbf{y}$. If we change (5.22) for radar observations:

$$\mathbf{z}_k = h(\mathbf{x}_k) + \mathbf{y}_k \tag{5.29}$$

The $h(\mathbf{x}_k)$ function represents a way to convert position and velocity of state vector to radar coordinates. Hence the $h(\mathbf{x}_k)$ function can be further updated as

$$h(\mathbf{x}_k) = \begin{pmatrix} \rho \\ \theta \\ v_d \end{pmatrix} = \begin{pmatrix} \rho \\ \theta \\ \frac{\partial \rho}{\partial t} \end{pmatrix} = \begin{pmatrix} \sqrt{p_x^2 + p_y^2} \\ \arctan(p_y/p_x) \\ \frac{p_x v_x + p_y v_y}{\sqrt{p_x^2 + p_y^2}} \end{pmatrix} \tag{5.30}$$

Heading back to the $\mathbf{H}$ matrix, the 3×4 matrix has higher dimensional components as the $h(\mathbf{x}_k)$ function indicates. Thus, it is a first-order nonlinear function in terms of radar-EKF. A way to deal with this is to use Jacobian matrix on $\mathbf{H}$:

$$\mathbf{H} = \left[\frac{\partial h(\mathbf{x}_k)}{\partial \mathbf{x}_k}\right]_{\mathbf{x}_{k-1}} = \begin{bmatrix} \frac{\partial \rho}{\partial p_x} & \frac{\partial \rho}{\partial p_y} & \frac{\partial \rho}{\partial v_x} & \frac{\partial \rho}{\partial v_y} \\ \frac{\partial \theta}{\partial p_x} & \frac{\partial \theta}{\partial p_y} & \frac{\partial \theta}{\partial v_x} & \frac{\partial \theta}{\partial v_y} \\ \frac{\partial v_d}{\partial p_x} & \frac{\partial v_d}{\partial p_y} & \frac{\partial v_d}{\partial v_x} & \frac{\partial v_d}{\partial v_y} \end{bmatrix} \tag{5.31}$$

Combining (30) and (32), the resulting $\mathbf{H}$ for radar is

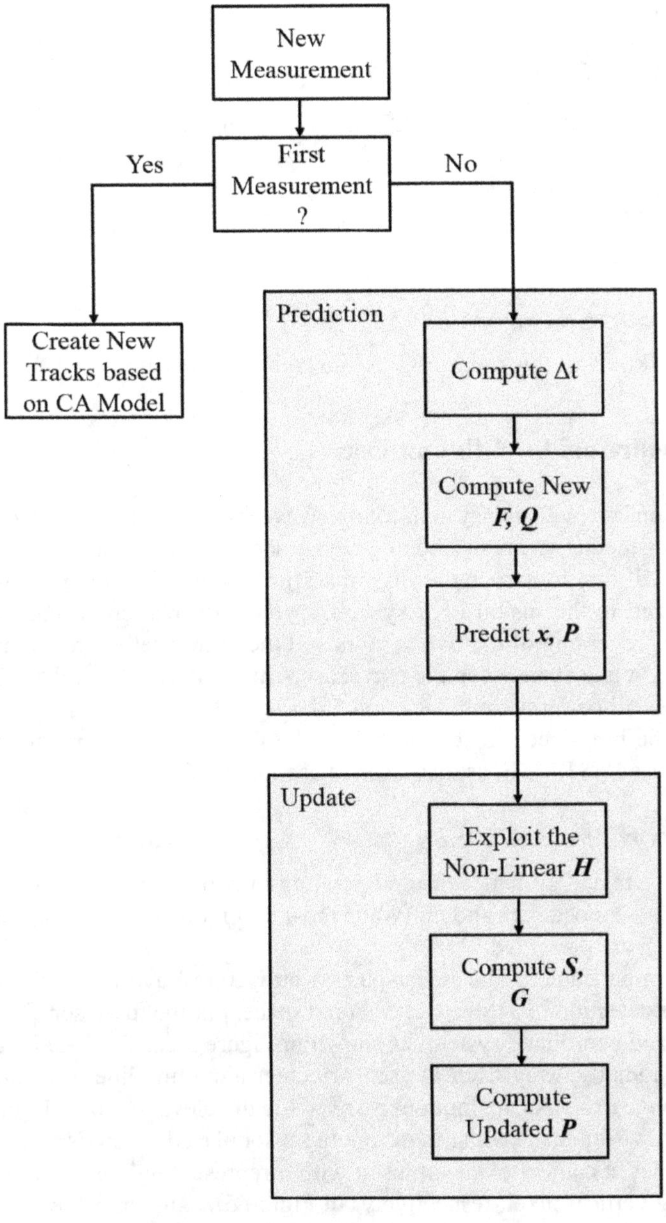

Figure 5.1 Radar-EKF workflow

$$\mathbf{H} = \begin{bmatrix} \frac{p_x}{\sqrt{p_x^2+p_y^2}} & \frac{p_y}{\sqrt{p_x^2+p_y^2}} & 0 & 0 \\ -\frac{p_y}{p_x^2+p_y^2} & \frac{p_x}{p_x^2+p_y^2} & 0 & 0 \\ \frac{p_y(v_xp_y-v_yp_x)}{(p_x^2+p_y^2)^{3/2}} & \frac{p_x(v_yp_x-v_xp_y)}{(p_x^2+p_y^2)^{3/2}} & \frac{p_x}{\sqrt{p_x^2+p_y^2}} & \frac{p_y}{\sqrt{p_x^2+p_y^2}} \end{bmatrix} \tag{5.32}$$

5.3.6 Radar-EKF workflow

The full workflow of this radar-EKF is shown in Figure 5.1.

5.4 Sensor fusion with camera

In this section, a new fusion system for mmWave radar and camera is proposed. And other heterogeneous sensors or homogeneous sensors can follow the same methodology. We will first introduce the EBs in section 5.4.1. It is a fundamental concept that is utilized in the fusion-EKF system for data association. Then, we will discuss the preprocessing of the two sensors and the HEM method for finding out the transformation matrix between the two sensors in sections 5.4.2 and 5.4.3. The new fusion-EKF will be discussed in section 5.4.4. The data association and synchronization will be introduced in section 5.4.5. The fusion system performance will be evaluated based on RMSE, as discussed in section 5.4.6..

5.4.1 Coordinates and EBs

In this sensor fusion system, sensors operating in two different coordinate systems are used, viz., monocamera and mmWave radar. In Figure 5.2, the coordinates in this fusion system are presented.

With a monocamera and linear-phased array mmWave radar, the fusion can improve the detection to a three-dimensional space. For the fused sensor system, we use a left-hand coordinate system, as shown in Figure 5.2a. The x-axis is the range axis that is radially away from the sensor center along the line of sight. y-axis is the cross-range axis, aka, azimuth axis. z-axis is the elevation axis. In Figure 5.2b, monocamera vision plane with monocamera sensor placed at zero is presented. (u, v) pixel variables at camera plane are used with u representing horizontal pixel and v representing vertical pixel, respectively. In Figure 5.2c, the mmWave radar sensor is conducting measurements with (ρ, θ, v_d), which are radar measurement range, azimuth angle and Doppler velocity (aka range rate), respectively.

If both sensors are presented, with a slight vertical difference in the positions of the camera and radar, as shown in Figure 5.2d, a fused system reconstructs the coordinate system with planes of vision and mmWave radar detections perpendicular to each other. As vision data are not totally limited to one plane, blockage of some targets could exist on different range and azimuth locations. Thus, the vision plane

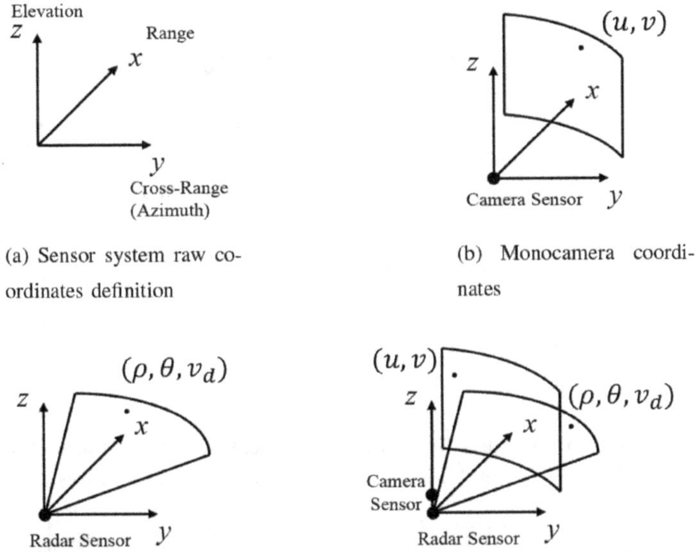

(a) Sensor system raw coordinates definition

(b) Monocamera coordinates

(c) Radar coordinates

(d) Fused sensor system coordinates

Figure 5.2 Coordinates definition

is penetrating throughout range and azimuth. In this case, the fusion can be constructed with correlations of (u, v) and (ρ, θ, v_d). Additionally, Cartesian and polar conversions are made throughout the fusion process. Radar's angular θ-angle axis and azimuth axis are used in following statements.

EBs of heterogeneous sensors have different coordinate systems as stated above. Here we need to define how the fused system will behave compared to before and after the sensor fusion.

Definition 1: Measurement point of the next inquiry lays within a certain region of approximation, from either radar or camera sensors. That region is then the theoretical EBs of the sensor.

EBs of the sensor fusion system is defined as following:

Definition 2: The actual localization of the detecting target lays within the approximate prediction region. The approximation is estimated by two or more sensors' system, and the region takes advantages from the sensors' EBs. The prediction region is then the fused EBs from the multi-sensor system.

With Definitions 1 and 2, the EBs of fused sensor system can be plotted as Figure 5.3. For vision error, the detecting target is difficult to solve along the range axis (depth). For radar error, the detecting target is limited as solving along θ-angle axis, which is caused by 4×2 MIMO transceivers configuration (15 degrees θ resolution at center). Fusion takes advantages from both the sensors, which results in better range and angle resolutions. However, in this application, we do not solve for the elevation (height) due to the constraint of having a linear antenna array from the

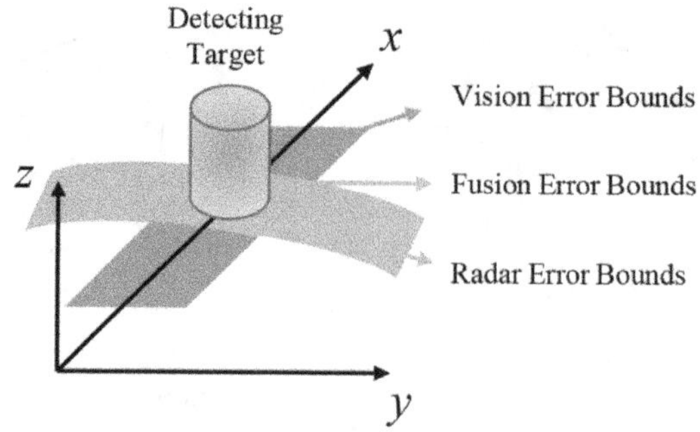

Figure 5.3 Error bounds of fused sensor system

mmWave radar sensor. With better mmWave radars with antenna channels along the elevation, a similar approach can be used to obtain the elevation EBs.

5.4.2 Camera preprocessing

Cameras, including stereo and monocameras, have intrinsic and extrinsic parameters, which are inherent due to the physical build of the pinhole and lens [40–43]. The extrinsic parameters with rotation and translation matrices are intended to map the world coordinates of the objects into camera coordinates. On the other hand, intrinsic parameters deal with the camera coordinates into the image pixel coordinates (u, v).

Here, we generally follow the four-step procedure from [43] and calibrate the monocular camera on the robotic operating system. In this case, intrinsic, extrinsic, and lens distortion parameters of the camera are prepared for warping and distortion image processing [44]. The camera calibration for this fusion project is shown in Figure 5.4. Figure 5.4a shows the sample of detected checkerboard points in image coordinates. With help from the extrinsic visualization in Figure 5.4b, multiple samples are collected and used to estimate intrinsic parameters.

(a) Checkerboard calibration for intrinsic parameters

(b) Extrinsic parameters visualization

Figure 5.4 Camera calibration

5.4.3 HEM

The transformation between heterogeneous sensors assumes that there exists a relationship between the targets detected in different sensors' domain. An HEM is used to associate radar's detection plane and camera's image plane:

$$c\begin{pmatrix} u \\ v \\ 1 \end{pmatrix} = \mathbf{T}\begin{pmatrix} x \\ y \\ 1 \end{pmatrix} \tag{5.33}$$

where c is a non-zero constant and x, y are the coordinates of the warped image in world coordinates. $\mathbf{T}$ is the transformation matrix that we are interested in, in order to warp the image plane to world plane. It is a 3×3 matrix and can be decomposed as

$$\mathbf{T} = \begin{bmatrix} \mathbf{T}_1 & \mathbf{T}_2 & \mathbf{T}_3 \end{bmatrix}^{\mathrm{T}} \tag{5.34}$$

We collect vision data with a calibrated camera, and try to associate the radar's localization of a specific point target like corner reflector. With Moore–Penrose pseudo inverse [45], the computing of the Euclidean norm solution can be obtained by linking several measured pairs. By applying least squares to the HEM, the transformation matrix $\mathbf{T}$ can be obtained as

$$\begin{cases} \mathbf{T}_1 = (\mathbf{J}^{\mathrm{T}}\mathbf{J})^{-1}\mathbf{J}^{\mathrm{T}}\mathbf{U} \\ \mathbf{T}_2 = (\mathbf{J}^{\mathrm{T}}\mathbf{J})^{-1}\mathbf{J}^{\mathrm{T}}\mathbf{V} \\ \mathbf{T}_3 = (\mathbf{J}^{\mathrm{T}}\mathbf{J})^{-1}\mathbf{J}^{\mathrm{T}}\mathbf{I} \end{cases} \tag{5.35}$$

$$\text{where}, \mathbf{J} = \begin{bmatrix} x_1 & y_1 & 1 \\ x_2 & y_2 & 1 \\ \vdots & \vdots & \vdots \\ x_n & y_n & 1 \end{bmatrix}, \mathbf{U} = \begin{bmatrix} cu_1 & cu_2 & \cdots & cu_n \end{bmatrix}, \mathbf{V} = \begin{bmatrix} cv_1 & cv_2 & \cdots & cv_n \end{bmatrix},$$

$\mathbf{I}$: identity matrix, n: number of measurements.

When collecting n samples of pairs from world coordinates to the camera image plane, the warped BEV can be best estimated by the transformation matrix $\mathbf{T}$.

After warped view from image toward BEV is applied, the BEV of warped image can be used to associate with the radar's detection point cloud map. The warped BEV image with BBox is consistently used in fusion-EKF for further processing. With camera's (x, y) coordinates for objects, the radar's detections with range, azimuth angle and Doppler velocities are then fused and provide tracking from fusion-EKF.

The BBox generation is performed using a background subtraction algorithm (BSA) [46]. In future processing within the EKF, the BBoxes and radar clusters are associated according to the algorithm introduced in section 5.4.5.

5.4.4 Fusion-EKF

In append to Radar-EKF introduced in section 5.3, the fusion-EKF takes camera's data into EKF as well. Note that fusion-EKF is not consisting two standalone EKFs and combining the tracks at a later stage. It is updating itself with both radar and camera, or any other sensors.

In prediction stage, state prediction does not depend on whether it is from the radar or camera. However, it will predict the updated state from whichever sensor it gets the current state from at the sensor incoming timestamp. Thus, the predictions run every time a sensor outputs.

In update stage, the camera's update measurement vector $\mathbf{z}_{\text{cam}}$ is expressed as

$$\mathbf{z}_{\text{cam}} = (u, v)^{\mathrm{T}} \tag{5.36}$$

And both measurements follow the same procedure, as introduced in (5.22).

For the EKF gain introduced in (5.28), fusion-EKF has the ability to adjust sensor's weights to obtain better EBs on different dimensions, e.g., camera's angular resolution and radar's range resolution. The **G** term also affects the region query and therefore the EBs of both sensors, which will be discussed in section 5.4.5.

The full workflow of this radar–camera fusion-EKF is shown in Figure 5.5.

The radar, camera, and fusion updates are shown in yellow, blue, and green, respectively. Radar raw data are fed with an ultra-short-range configuration on the digital signal processing chip module on TI radar. With output through a USB port to a host computer, the point cloud with Doppler information is extracted. Additional filtering like moving target indicator is applied to collaborate with the camera's BSA. A DBSCAN [47] or REDBSCAN [48] clustering algorithm is then applied to obtain clusters and potential target shape from radar scans. These are important parts to identify targets and similarly create "BBox" for radar targets before the radar–camera fusion-EKF. After the first stage, the noise deductions and the query data are sent to radar–camera fusion-EKF for further association and tracking.

Similarly, camera raw data are fed with intrinsic and extrinsic parameters with the exact locations and directions measured during the experiment. The calibration stage provides the BEV from warping the image onto the perspective view. The BSA is then used for image plane BBox generation. The clustering in radar preprocessing and the BBox generation in vision preprocessing have an inherent association. But without applied machine learning or segmentation, the data association needs to be done within the radar–camera fusion-EKF. Further details are discussed in section 5.4.5.

The fusion-EKF uses the CA maneuvering model in our setup. The CA model is a simplified model of Singer model [49] but without the acceleration factor, thus lowering orders in time deviation and further improve the processing speed. For tracking multiple targets with different velocities and positions, the CA model can handle different preset acceleration noises.

The key difference in our setup compared to other relevant work in this area is that (1) EBs are updated by both sensors, and (2) radar–camera fusion-EKF updates both vision BBox plane and radar localization BEV continuously. In this case, the time stamps of the prediction and update are different based on each sensor's fps.

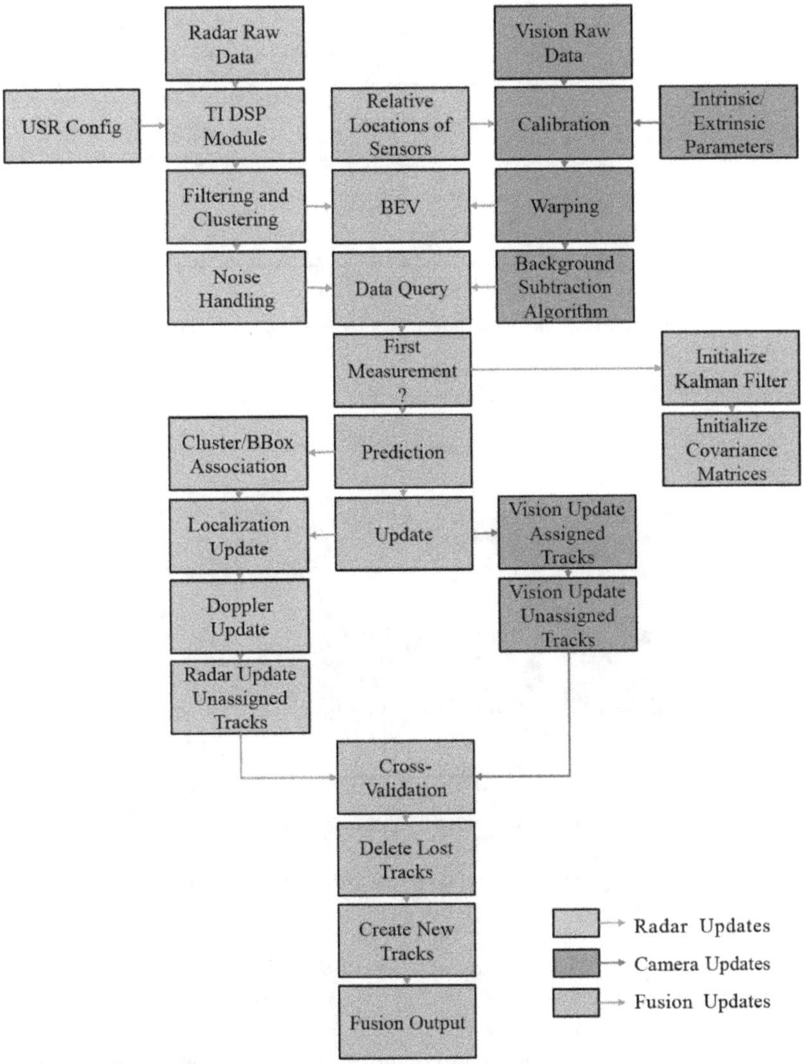

Figure 5.5 Radar–camera fusion-EKF workflow

This will be discussed in section 5.4.5. In the next set of relations, we denote the current state with k and the previous state with $k - 1$.

5.4.5 Data association and sensor synchronization

The track-oriented data association is implemented using timeline alignment of sensors to (1) get updates from heterogeneous sensor's perspective, and (2) synchronization over certain regions of fusion-EKF output to obtain the tracked target localization. This process is overlying on the radar–camera fusion-EKF in Figure 5.5. With HEM to transform

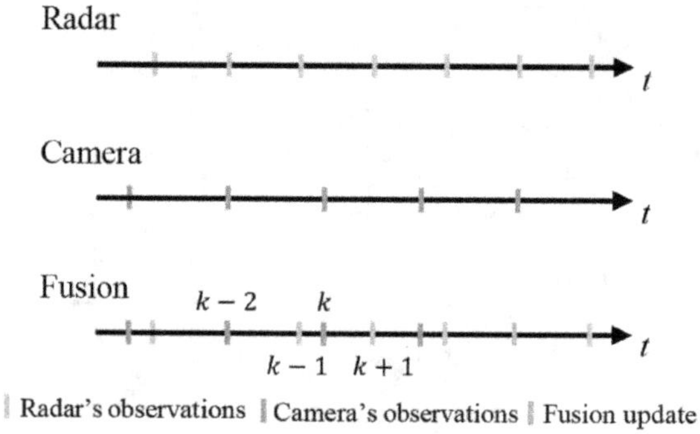

(a) Radar-camera association timeline. A sample state steps $k-2, k-1, k, k+1$ are shown in the fusion timeline.

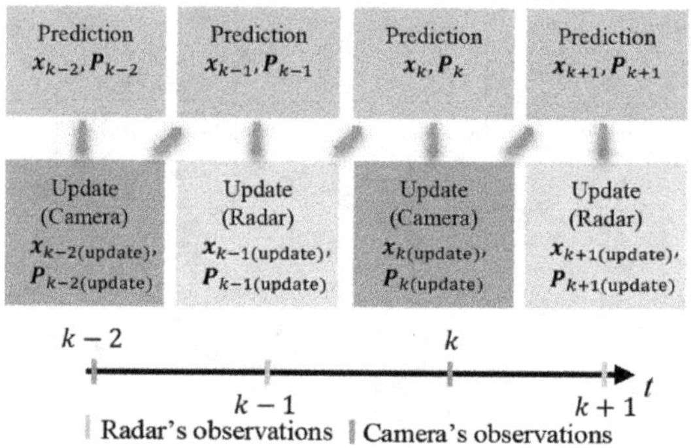

(b) Association actions based on timeline alignment of state steps $k-2, k-1, k, k+1$.

Figure 5.6 Radar–camera association timeline alignment

between coordinates, EBs are recalled to optimize the association of radar clusters and camera BBoxes. The fusion-EKF track is created or deleted according to the existence from both sensors.

1. *Timeline alignment:* The timeline alignment for radar–camera fusion system can be seen in Figure 5.6. With different frame rates from both sensors, observations of different sensor type on different time stamps are collected on the processing unit. And fusion association is shown with different Δt with different state steps k in Figure 5.6a. The green-colored step on the fusion timeline actually means both sensors output their observations at the same time ($\Delta t \leq 0.005s$). In this case, predictions and updates follow the same rule as the radar–camera fusion-EKF.

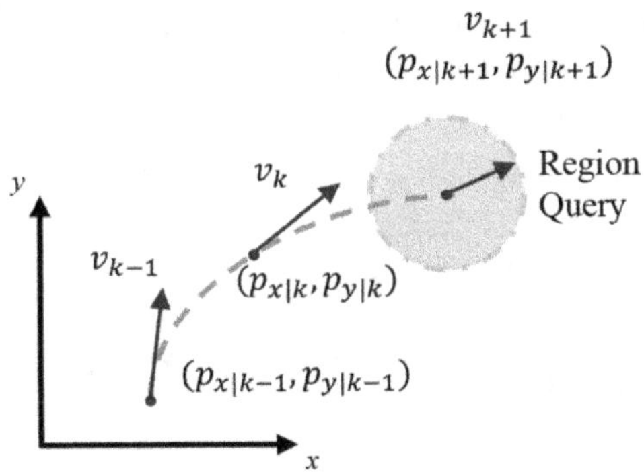

Figure 5.7 Radar–camera association region search. Blue dash line: EKF projected trajectory; yellow region: search region for next state for radar clusters and camera bounding box BEV projection.

In Figure 5.6b, as we have two tuned updates for each step depending on which sensor data are obtained from, the prediction and update for each observation from radar or camera are shown. As stated in the previous section, the updated $\mathbf{x}_{k(\text{update})}$ and $\mathbf{P}_{k(\text{update})}$ matrices are used for the next prediction step. In other words, the measurement update of $\mathbf{x}_{k(\text{update})}$ and $\mathbf{P}_{k(\text{update})}$ is only based on the prediction within the same step. And the prediction relies on the previous update step.

2. *Region search within EBs:* From 525. (25) and section 5.4, we have an idea of the fusion-EKF gain **G**. The association of radar clusters and camera BBoxes based on fusion-EKF gain **G** using HEM is shown in Figure 5.7. The fusion-EKF trajectory in blue dash line is shown with steps $k-1, k, k+1$. The different predictions after update measurements are shown with positions and velocities. The yellow region query for the upcoming measurement is set in Figure 5.7. Here are the EBs resulting from Definitions 1 and 2:

1. EB along x-axis of radar is

$$\text{EB}|x_{\text{radar}} = \sigma_\rho \cos\theta_i \tag{5.37}$$

where θ_i is the angle of interest from target, which is the target angular location. EB along azimuth (y-axis) of radar is

$$\text{EB}|y_{\text{radar}} = \sigma_\rho \sin\theta_i \tag{5.38}$$

The assumption is that the radar has even spacing MIMO antenna with even angular resolution. Thus, we can conclude that at the center, radar has better range EB and less accurate when it is on the side. On the other hand, radar has relative bad azimuth EB and needs other sensors to compensate for the angular EB.

2. EB along range of camera is

$$\mathrm{EB}|x_{\mathrm{cam}} = c(\mathbf{T}^{\mathrm{T}}\mathbf{T})^{-1}\mathbf{T}^{\mathrm{T}}\mathbf{U} \tag{5.39}$$

   And EB along azimuth of camera is

$$\mathrm{EB}|y_{\mathrm{cam}} = c(\mathbf{T}^{\mathrm{T}}\mathbf{T})^{-1}\mathbf{T}^{\mathrm{T}}\mathbf{V} \tag{5.40}$$

   For camera pixels using HEM, warping along the vertical axis to x-axis is typically limited compared to the horizontal axis to y-axis. Thus, the range resolution is restricted especially for long ranges. The azimuth resolution is consistent when looking at horizontal pixels to angle pair. So camera's EB along the range is worse compared to radar's. And EB along the azimuth is better with adequacy horizontal pixels.

3. Combining radar and camera EBs. There exist two standalone query regions to search for cluster or BBox from heterogeneous sensors.
4. Fusion EBs use EKF output of the new prediction to get the estimated localization of the cluster/BBox. This results in missing clusters/BBoxes or multiple clusters/BBoxes.

Even with the fusion EBs, there still exists multiple clusters/BBoxes or missing clusters/BBoxes. Reasons include:

1. Concealed by front object using the camera.
2. Sensor failure by radar or camera.
3. Radar's multipath effect.
4. Lighting condition (overexposed or no lighting) by the camera.
5. Others.

Algorithm 1 shows the region search and track updates for tracking multiple targets. Different tracks share the incoming clusters/BBoxes measurement updates from sensors. Fusion-EKF deals with the region search based on EBs estimations. The fusion-EKF provides invisible counts for heterogeneous sensors. If either sensor fails temporary, the tracking will continue using the other sensor's update. The tracking will still be valid even if misses several clusters/BBoxes. However, a track will be terminated after a certain duration of missing detections from both sensors after cross-validation. With updates for fusion localization, the improved EBs can provide a better result for the system and thus greatly decreases uncertainty errors from either sensor. In addition, new tracks are initialized based on cross-validation from different sensors with a certain threshold.

Therefore, the association is done with timeline alignment together with EBs region search within tracks. With help from different advantages from heterogeneous sensors, fusion EBs are greatly reduced: along range using the radar as it provides better localization; and along azimuth using the camera as it provides better localization. The experiment result will further approve of this association and EBs' deduction.

Algorithm 1 Pseudo-code of the tracking and association for multi-targets

```
Require: EKF tracks: tracks at current time stamp
Initialize: i = 0
After radar-camer fusion-EKF, get number of tracks: N
while i < N do
  if Track valid then
    Find EBs for next step from prediction
    Find radar clusters and BBoxes within EBs
    if radarclusters not equal to 1 then
      radarInvisibleCount ++
    else
      Update radar's localization
    end if
    if BBoxes not equal to 1 then
      camInvisibleCount ++
    else
      Update camera's localization
    end if
  end if
  if radarInvisibleCount > threshold_radar then
    Cross-validation failed.
    Track invalid, valid until camera updates
  end if
  if camInvisibleCount > threshold_cam then
    Cross-validation failed.
    Track invalid, valid until radar updates
  end if
  i ++
end while
Return: EBs with localization for each track
```

5.4.6 EB evaluation

From Figure 5.7 and Definition 2, the EB of the next query is the region of the prediction and association. With RMSE, the EB evaluation can be estimated as:

$$\begin{aligned}\mathrm{RMSE}(\hat{o}) &= \sqrt{\mathbb{E}((\hat{o}-o)^2)} \\ &= \sqrt{\frac{1}{N}\sum_{n=1}^{N}(\hat{o}-o)^2}\end{aligned} \tag{5.41}$$

where o is the operator of the estimation variable.

With a window function to estimate the EB, (5.41) is modified with:

$$\mathrm{RMSE}(\hat{o}) = \sqrt{\frac{1}{w}\sum_{n=k-(w-1)/2}^{k+(w-1)/2}(\hat{o}-o)^2} \tag{5.42}$$

where w is the window size of the estimation on the kth step. As introduced before, the EBs of radar are the radar range, angle, and Doppler variances, $\begin{pmatrix}\sigma_\rho^2 & \sigma_\theta^2 & \sigma_{v_d}^2\end{pmatrix}^\mathsf{T}$. Here the sliding window applied RMSE is used to measure the EB on where the prediction region is. In this case, the EB evaluates radar's fusion performance with camera's vision.

In the experiment, the EB of radar is measured and evaluated with RMSEs in range, Doppler and angle domains, respectively.

5.5 Sensor tracking and fusion experimental results

The experiment is implemented by two humans' walking inside an indoor environment. The application implemented sensor fusion in this experiment used ELP USB8MP02G-L75 monocamera with 640 × 480 YUY2 video picture encoding format [50], transmitting at 30 fps, and a TI AWR1642 mmWave radar working at 77 GHz at 30 fps, which is a 1D 4 RX by 2 TX phased array antenna. AWR1642 is only capable of measuring target in range and azimuth plane.

The vision output and radar output from EKF can be seen from Figure 5.8. In the vision's output, BBoxes with positions and velocities on x-axis and y-axis are shown, along with the radar's point cloud clusters. Different colors stand for different clusters with the searching radius of a human. The clusters provide median filtered centroids with a green marker and black marker edge. Red and blue markers with black marker edge are the EKF output of the tracking. In the radar's output, radar point clusters' centroid as well as fusion-EKF prediction localization for different tracks are presented. These figures are combinations of tracks thus showing multiple targets' multiple sensors' tracking results.

As we stated in the section 5.4.4, we only have a linear array radar, i.e., the elevation information is out of scope, and thus only points on the ground (elevation equals 0) are shown. This approach is also applicable when applying to the elevation information obtained from a compatible sensor to have tracking more reliable.

Additionally, in case of blockage between targets, the output BBox shows "radar only" in Figure 5.8p and Figure 5.8r. As invisible from the camera's BSA. The "radar only" label still keeps tracking of different tracks until a camera's vision validates the track. Otherwise, after certain invisible counts of the track, the track is deleted because of lack of cross-validation.

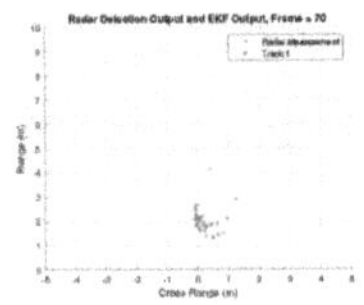
(a) Frame number 70 radar EKF output

(b) Frame number 70 camera EKF output

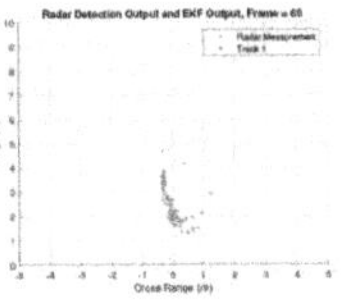
(c) Frame number 80 radar EKF output

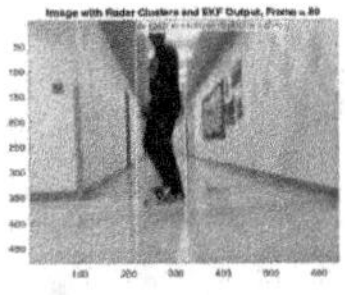
(d) Frame number 80 camera EKF output

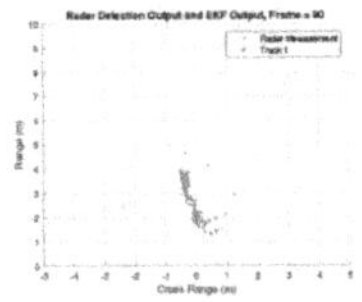
(e) Frame number 90 radar EKF output

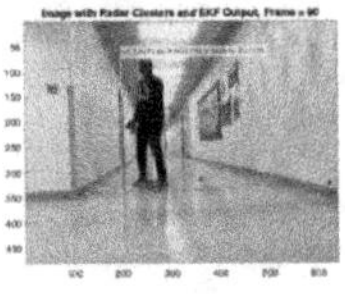
(f) Frame number 90 camera EKF output

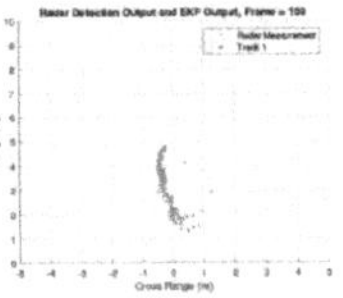
(g) Frame number 100 radar EKF output

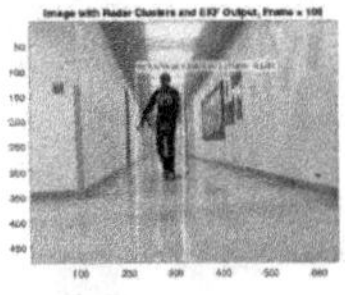
(h) Frame number 100 camera EKF output

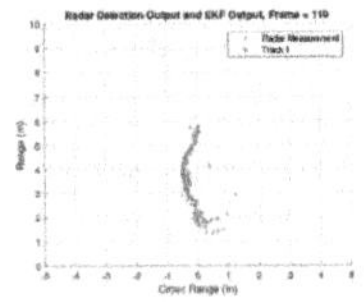
(i) Frame number 110 radar EKF output

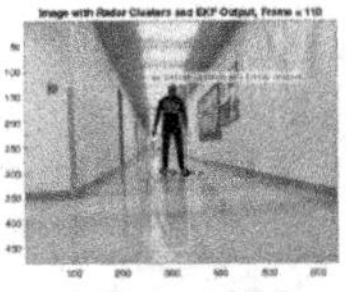
(j) Frame number 110 camera EKF output

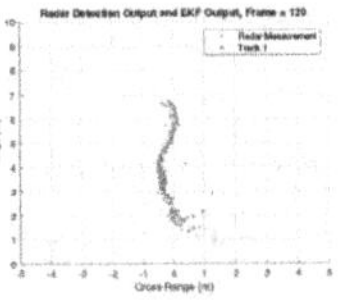
(k) Frame number 120 radar EKF output

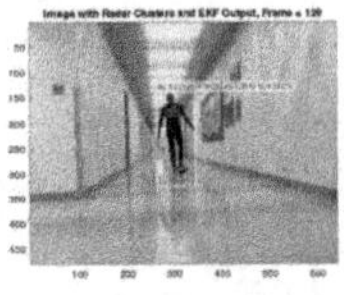
(l) Frame number 120 camera EKF output

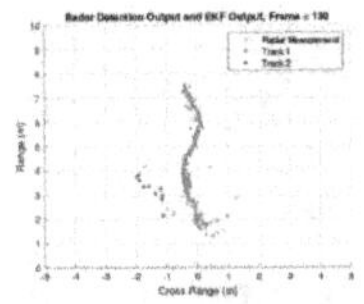
(m) Frame number 130 radar EKF output

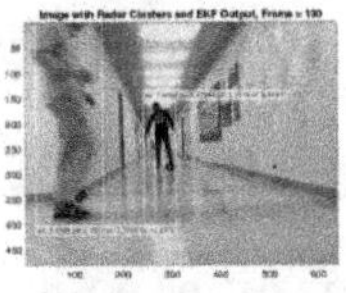
(n) Frame number 130 camera EKF output

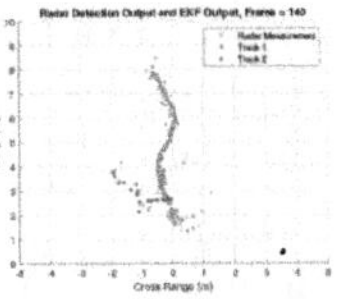
(o) Frame number 140 radar EKF output

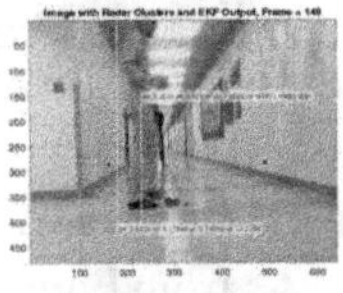
(p) Frame number 140 camera EKF output

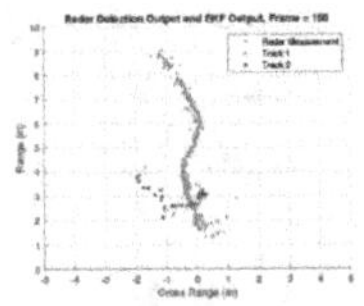
(q) Frame number 150 radar EKF output

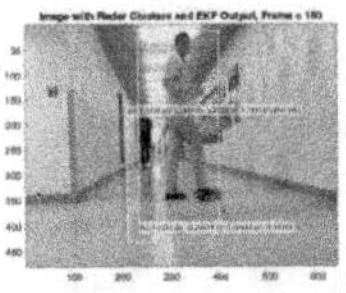
(r) Frame number 150 camera EKF output

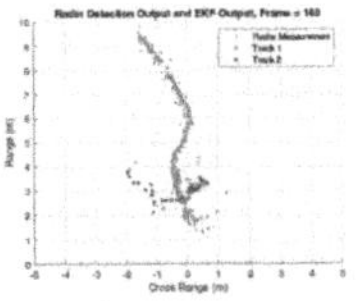
(s) Frame number 160 radar EKF output

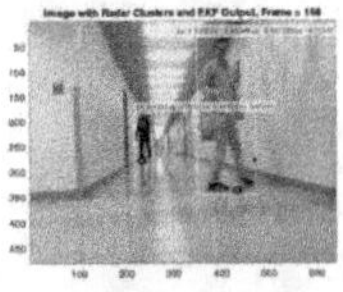
(t) Frame number 160 camera EKF output

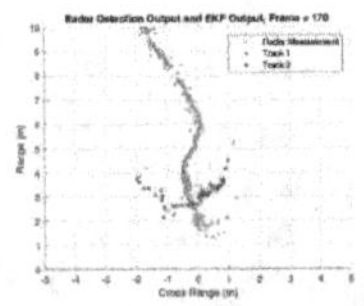
(u) Frame number 170 radar EKF output

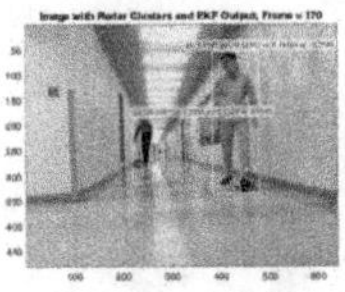
(v) Frame number 170 camera EKF output

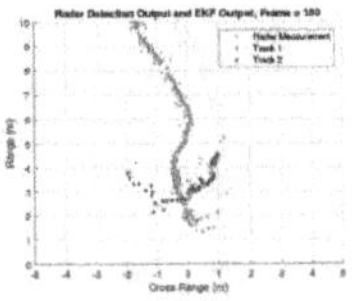
(w) Frame number 180 radar EKF output

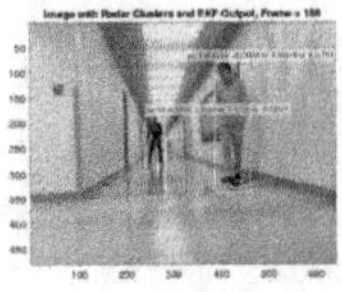
(x) Frame number 180 camera EKF output

Figure 5.8 EKF output

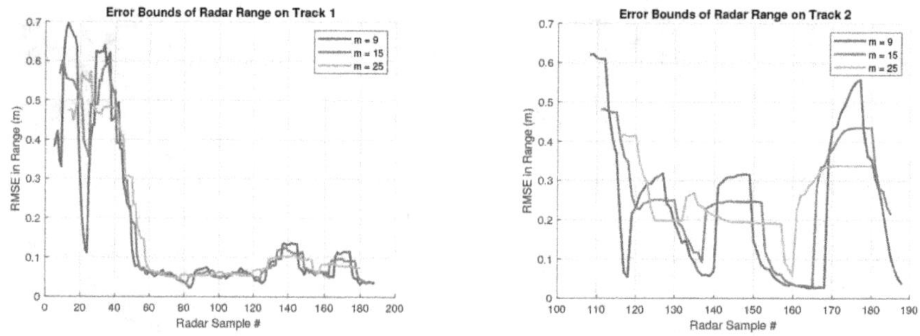

(a) Radar's range RMSE of EKF estimates with memory of 9, 15 and 25 measurements for track 1

(b) Radar's range RMSE of EKF estimates with memory of 9, 15 and 25 measurements for track 2

Figure 5.9 EKF RMSE of radar's range

The radar's tracking, shown in Figure 5.8m,o,q,s,u,w, provides multiple targets' tracking. If camera vision data were unavailable, the output may be unreliable until the fusion EBs' update. The region search and association avoids the missed assignment of tracks and improves tracking results.

To evaluate the performance of the radar–camera fusion-EKF system, we use the RMSE metrics, and also to further validate if the fusion-EKF takes advantages from both sensors. From 5.42, the range, angle, and Doppler RMSE plots for different Tracks 1 and 2 are shown in Figures 5.9–5.11, respectively. The memory used for RMSE is 9, 15, and 25, which are fast, median, and slow windows, respectively. With fps of 30 Hz, the window is estimating a state in about 0.3, 0.5 and 0.83 seconds, respectively.

As from Figure 5.9a, after initializing the track, the position RMSE drops rapidly to around 0.05 m. From 5.1, the range resolution of the fusion system is 0.0436 m, which is the theoretical range resolution. After the fusion-EKF, the practical range

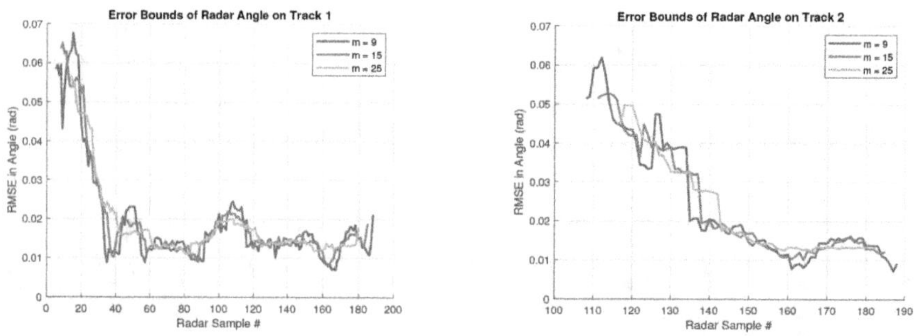

(a) Radar's angle RMSE of EKF estimates with memory of 9, 15 and 25 measurements for track 1

(b) Radar's angle RMSE of EKF estimates with memory of 9, 15 and 25 measurements for track 2

Figure 5.10 EKF RMSE of radar's angle

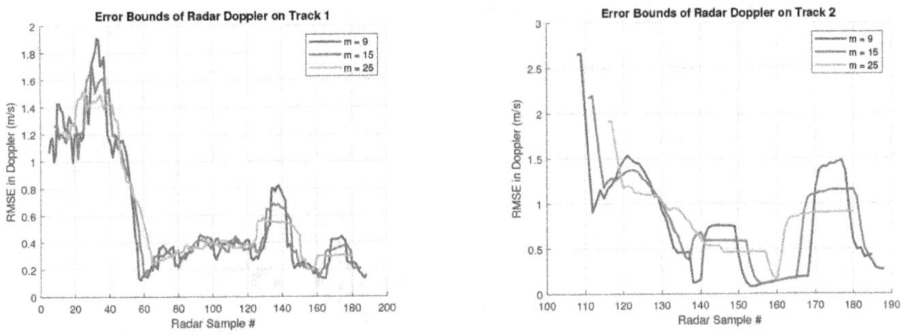

(a) Radar's Doppler RMSE of EKF estimates with memory of 9, 15 and 25 measurements for track 1

(b) Radar's Doppler RMSE of EKF estimates with memory of 9, 15 and 25 measurements for track 2

Figure 5.11 EKF RMSE of radar's Doppler

resolution achieved is around 0.05 m after stabilizing from Figure 5.9a. The weakness of radar's detecting cross-range objects is overcome from Figure 5.9b. As for radar's cross-range objects in a typical radar scan, the cross-range velocities are close to zero because Doppler velocities are essentially range rates. This problem is solved by adding the camera's vision in the proposed fusion-EKF.

As from Figure 5.10a, after initializing the track, the angular RMSE drops rapidly to around 0.015 rad, which is 0.859 degrees. From 5.2, the angular resolution of radar at center field-of-view (FOV) is 0.25 rad, which is the theoretical angular resolution at center FOV. Outside, that is $1/\cos\theta_i$. After the fusion-EKF, the angular resolution achieved is around 0.015 rad after stabilizing from Figure 5.10a. This track is along about the center FOV so that it verifies that the camera improves fusion's track significantly. In Figure 5.10b, a similar result is achieved. This confirms the improvements on radar's EB on the azimuth axis by using the camera data.

As from Figure 5.11a, after initializing the track, the Doppler RMSE drops rapidly to around 0.3 m/s. From 5.3, the theoretical Doppler resolution of radar is 0.616 m/s. After the fusion-EKF, the practical Doppler resolution achieved is around 0.3 m/s after stabilizing from Figure 5.11a. The fusion-EKF improves Doppler estimations on both range rate and cross-range rate. In Figure 5.11b, a resolution of only 0.5 m/s is achieved. Fusion-EKF greatly improves radar's cross-range rate detections, and thus allows for reliable cross-range objects detections.

From Figures 5.9–5.11, two motion maneuvering models are handled with different tracks and different CA models from fusion-EKF. The result shows different CA models can handle tracking for different targets with different maneuvering behaviors.

Additional experiment analysis on this EB-based sensor fusion is conducted that uses the sensor fusion result of radar and camera data and compares

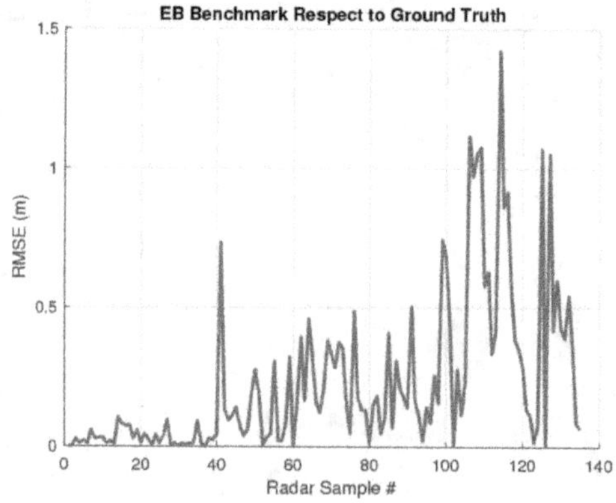

Figure 5.12 *Fusion-EKF benchmark on EB respect to ground truth*

Table 5.1 *Average localization error of radar–camera fusion-EKF*

	Average localization error
Range	0.2902 m
Azimuth angle	0.0134 rad
Velocity	0.7864 m/s

the localization result with ground truth. The benchmark of radar–camera fusion-EKF tracking result respect to ground truth can be seen in Figure 5.12. The RMSE of fusion localization deviation from ground truth is 0.3664 m. This is a tolerable deviation that is good enough to separate human targets, and is a reasonable region search EB from Definitions 1 and 2. The average localization error comes from the track association from heterogeneous sensors and the performance of the proposed fusion-EKF can be seen in Table 5.1. With EBs reduced in range, angle, and Doppler dimensions, the radar–camera fusion-EKF provides a solution to heterogeneous sensor fusion. Additionally, the cumulative distributions of the localization errors in range, azimuth angle, and velocity are presented in Figure 5.13.

Some comparison of errors or EBs between the proposed radar–camera fusion-EKF and other radar tracking algorithms is shown in Table 5.2. Because mmWave radar has better range resolution than traditional radars, range EB is much superior than the other lower frequency radars. Meanwhile, azimuth angle EB is also improved greatly due to the fusion approach with the camera. Therefore, the proposed fusion-EKF provides extra information from the camera to improve the detection and tracking of the mmWave radar sensor.

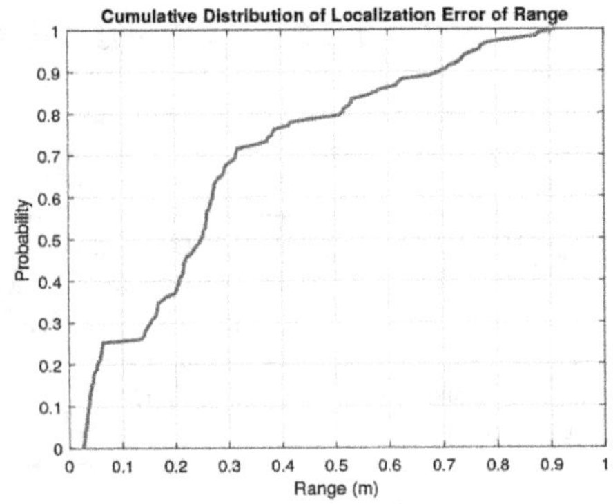

(a) Cumulative distribution of localization error of range

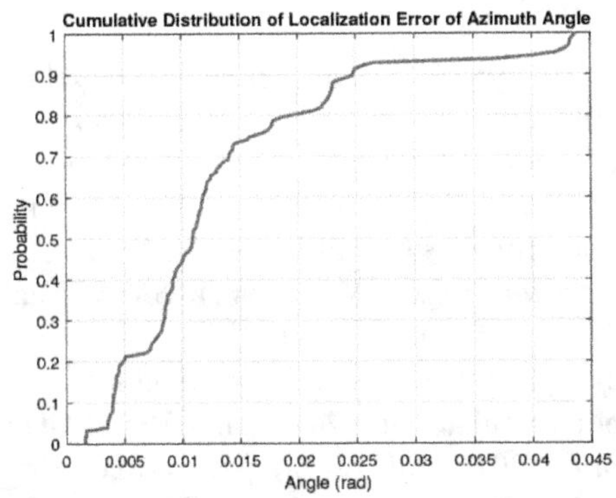

(b) Cumulative distribution of localization error of azimuth angle

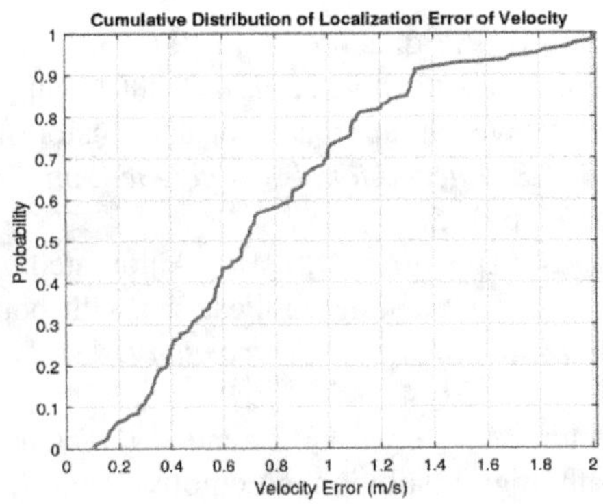

(c) Cumulative distribution of localization error of velocity

Figure 5.13 Cumulative distribution of localization errors

Table 5.2 Comparison of error/EBs using radar–camera fusion EKF with other radar tracking algorithms

Method	Error or EBs
CA model of radar–camera fusion-EKF	0.2902 m, 0.0134 rad (short-range configuration with camera fusion)
CCA model from [30]	2.49 m (Euclidean distance without fusion)
600 m airport surveillance system from [51]	5 m, 0.15°
New track association algorithm from [52]	0.5°

References

[1] Clark S., Durrant-Whyte H. 'Autonomous land vehicle navigation using millimeter wave radar'. *Proceedings of IEEE International Conference on Robotics and Automation*; Leuven, Belgium, 1998. pp. 3697–702.

[2] Motomura T., Uchiyama K., Kajiwara A. 'Measurement results of vehicular RCS characteristics for 79 ghz millimeter band'. *IEEE Topical Conference on Wireless Sensors and Sensor Networks (WiSNet)*; Anaheim, CA, 2018. pp. 103–06.

[3] Tan Q.J.O., Romero R.A. 'Ground vehicle target signature identification with cognitive automotive radar using 24–25 and 76–77 GHz bands'. *IET Radar, Sonar & Navigation*. 2018, vol. 12(12), pp. 1448–65.

[4] Wei J., Snider J.M., Kim J., Dolan J.M., Rajkumar R., Litkouhi B. 'Towards a viable autonomous driving research platform'. *Proceedings of IEEE Intelligent Vehicles Symposium (IV)*; Gold Coast City, QLD, 2013. pp. 763–70.

[5] Shima N., Babasaki M., Akidzuki Y., *et al.* 'Fusion sensor for driving assistance system'. *Fujitsu Ten Technical Journal*. 2001, vol. 17, pp. 35–44.

[6] Wu J.K., Wong Y. 'Bayesian approach for data fusion in sensor networks'. *Proceedings of IEEE International Conference on Information Fusion*; Waikoloa, HI, 2006. pp. 1–5.

[7] Scheiner N., Haag S., Appenrodt N, *et al.* 'Automated ground truth estimation for automotive radar tracking applications with portable gnss and imu devices'. *20th International Radar Symposium (IRS)*; Ulm, Germany, 2019. pp. 1–10.

[8] Thormann K., Baum M., Honer J. 'Extended target tracking using gaussian processes with high-resolution automotive radar'. *21st International Conference on Information FUSION (FUSION)*; Cambridge, 2018. pp. 1764–70.

[9] Xia Y., Wang P., Berntorp K. 'Extended object tracking with automotive radar using learned structural measurement model'. *IEEE Radar Conference (RadarConf20)*; Florence, Italy, 2020. pp. 1–6.

[10] Zhang R., Cao S. 'Real-time human motion behavior detection via CNN using mmwave radar'. *IEEE Sensors Letters*. 2019, vol. 3(2), pp. 1–4.

[11] Sengupta A., Jin F., Zhang R., Cao S. 'mm-Pose: real-time human skeletal posture estimation using mmWave radars and CNNS'. *IEEE Sensors Journal*. 2020, vol. 20(17), pp. 10032–44.

[12] Major B., Fontijne D., Ansari A, *et al.* 'Vehicle detection with automotive radar using deep learning on range-Azimuth-doppler tensors'. *Proceedings of the IEEE/CVF International Conference on Computer Vision Workshops*; Seoul, South Korea, 2019.

[13] Zhang R., Cao S. 'Extending reliability of mmWave radar tracking and detection via fusion with camera'. *IEEE Access*. 2019, vol. 7, pp. 137065–79.

[14] Kim J., Han D.S., Senouci B. 'Radar and vision sensor fusion for object detection in autonomous vehicle surroundings'. *Tenth International Conference on Ubiquitous and Future Networks (ICUFN)*; Prague, Czech Republic, 2018. pp. 76–78.

[15] Kim J., Kim Y., Kum D. 'Low-level sensor fusion network for 3D vehicle detection using radar range-Azimuth heatmap and monocular image'. *Proceedings of the Asian Conference on Computer Vision*; Virtual Kyoto, 2020.

[16] Dubrofsky E. Homography estimation. [Diplomová práce]. Vancouver: Univerzita Britské Kolumbie; 2009.

[17] Song Y., Wang X., Zhu J., Lei L. 'Sensor dynamic reliability evaluation based on evidence theory and intuitionistic fuzzy sets'. *Applied Intelligence*. 2018, vol. 48(11), pp. 3950–62.

[18] Eltrass A., Khalil M. 'Automotive radar system for multiple-vehicle detection and tracking in urban environments'. *IET Intelligent Transport Systems*. 2018, vol. 12(8), pp. 783–92.

[19] Aditya P., Apriliani E., Arif D.K., Baihaqi K. 'Estimation of three-dimensional radar tracking using modified extended Kalman filter'. *Journal of Physics: Conference Series*. 2018, vol. 974(1), p. 012071.

[20] Ljung L. 'Asymptotic behavior of the extended Kalman filter as a parameter estimator for linear systems'. *IEEE Transactions on Automatic Control*. 1979, vol. 24(1), pp. 36–50.

[21] Angelov A., Robertson A., Murray-Smith R., Fioranelli F. 'Practical classification of different moving targets using automotive radar and deep neural networks'. *IET Radar, Sonar & Navigation*. 2018, vol. 12(10), pp. 1082–9.

[22] Hajri H., Rahal M.-C. 'Real time lidar and radar high-level fusion for obstacle detection and tracking with evaluation on a ground truth'. *International Journal of Mechanical and Mechatronics Engineering*. 2018, vol. 12(8), pp. 821–27.

[23] Kwon S.K., Hyun E., Lee J.-H., Lee J., Son S.H. 'A low-complexity scheme for partially occluded pedestrian detection using lidar-radar sensor fusion'. *Proceedings of IEEE 22nd International Conference on Embedded and Real-Time Computing Systems and Applications (RTCSA)*; Daegu, South Korea, 2016. pp. 104–104.

[24] Debes C., Merentitis A., Heremans R., *et al.* 'Hyperspectral and lidar data fusion: outcome of the 2013 GRSS data fusion contest'. *IEEE Journal of Selected Topics in Applied Earth Observations and Remote Sensing*. 2014, vol. 7(6), pp. 2405–18.

[25] Ulrich M., Maile F., Löcklin A., Yang B., Kleiner B., Ziegenspeck N. 'A model for improved association of radar and camera objects in an indoor environment'. *Proceedings of IEEE Sensor Data Fusion: Trends, Solutions, Applications (SDF)*; Bonn, 2017. pp. 1–6.

[26] Sikdar A., Cao S., Zheng Y.F., Ewing R.L. 'Radar depth association with vision detected vehicles on a highway'. *Proceedings of IEEE Radar Conference*; Cincinnati, OH, 2014. pp. 1159–64.

[27] Bunyak F., Palaniappan K., Nath S.K., Seetharaman G. 'Flux tensor constrained geodesic active contours with sensor fusion for persistent object tracking'. *Journal of Multimedia*. 2007, vol. 2(4).

[28] Zhong Z., Liu S., Mathew M., Dubey A. 'Camera radar fusion for increased reliability in adas applications'. *Electronic Imaging*. 2018, vol. 30(17), pp. 1–4.

[29] Muresan M.P., Nedevschi S. 'Multimodal sparse lidar object tracking in clutter'. *Proceedings of IEEE International Conference on Intelligent Computer Communication and Processing (ICCP)*; Cluj-Napoca, Romania, 2018. pp. 215–21.

[30] Schubert R., Richter E., Wanielik G. 'Comparison and evaluation of advanced motion models for vehicle tracking'. *Proceedings of IEEE International Conference on Information Fusion*; Cologne, 2008. pp. 1–6.

[31] Nishigaki M., Rebhan S., Einecke N. 'Vision-based lateral position improvement of radar detections'. *Proceedings of IEEE International Conference on Intelligent Transportation Systems (ITSC)*; Anchorage, AK, 2012. pp. 90–97.

[32] Folster F., Rohling H. 'Lateral velocity estimation based on automotive radar sensors'. *Proceedings of IEEE International Conference on Radar*; Kraków, Poland, 2006. pp. 1–4.

[33] Wilson S.L., Carlson B.D. 'Radar detection in multipath'. *IEE Proceedings - Radar, Sonar and Navigation*. 1999, vol. 146(1), pp. 45–54.

[34] Ryu I.-hwan., Won I., Kwon J. 'Detecting ghost targets using multilayer perceptron in multiple-target tracking'. *Symmetry*. 2018, vol. 10(1),16.

[35] Schmidt R. 'Multiple emitter location and signal parameter estimation'. *IEEE Transactions on Antennas and Propagation*. 1986, vol. 34(3), pp. 276–80.

[36] Roy R., Kailath T. 'ESPRIT-estimation of signal parameters via rotational invariance techniques'. *IEEE Transactions on Acoustics, Speech, and Signal Processing*. 1989, vol. 37(7), pp. 984–95.

[37] Kailath T., Sayed A.H., Hassibi B. *Linear Estimation*. Upper Saddle River, NJ: Prentice Hall; 2000.

[38] Smith J.O. *Mathematics of the discrete fourier transform (DFT): with audio applications*. http://www.w3k.org/: W3K Publishing; 2007.

[39] AWR1642 Evaluation Module (AWR1642BOOST) *Single-Chip mmWave Sensing Solution* [online]. 2020.Texas Instruments. Available from http://www.ti.com/lit/ug/swru508b/swru508b.pdf [Accessed 7/28/2022].

[40] Mirzaei F.M., Kottas D.G., Roumeliotis S.I. '3D LIDAR–camera intrinsic and extrinsic calibration: identifiability and analytical least-squares-based initialization'. *The International Journal of Robotics Research*. 2012, vol. 31(4), pp. 452–67.

[41] Zhang Q., Pless R. 'Extrinsic calibration of a camera and laser range finder (improves camera calibration)'. *IROS*. 2004, vol. 3, pp. 2301–6.

[42] Zhang Z. 'A flexible new technique for camera calibration'. *IEEE Transactions on Pattern Analysis and Machine Intelligence*. 2000, vol. 22(11), pp. 1330–4.

[43] Heikkila J., Silven O. 'A four-step camera calibration procedure with implicit image correction'. *Proceedings of IEEE Computer Society Conference on Computer Vision and Pattern Recognition*; San Juan, PR, 1997. pp. 1106–12.

[44] Zhou Wang., Bovik A.C. 'A universal image quality index'. *IEEE Signal Processing Letters*. 2002, vol. 9(3), pp. 81–4.

[45] Moore E.H. 'On the reciprocal of the general algebraic matrix'. *Bulletin of the American Mathematical Society*. 1920, vol. 26, pp. 394–5.

[46] Barnich O., Van Droogenbroeck M. 'Vibe: a universal background subtraction algorithm for video sequences'. *IEEE Transactions on Image Processing*. 2011, vol. 20(6), pp. 1709–24.

[47] Birant D., Kut A. 'ST-DBSCAN: an algorithm for clustering spatial–temporal data'. *Data & Knowledge Engineering*. 2007, vol. 60(1), pp. 208–21.

[48] Robust and adaptive radar elliptical density-based spatial clustering and labelling for mmWave radar point cloud *Proceedings of Asilomar Conference on Signals, Systems, and Computers*; Pacific Grove, CA, 2019. pp. 919–24.

[49] Li X.R., Jilkov V.P. *Proceedings of SPIE International Conference on Signal and Data Processing of Small Targets*; Orlando, FL, 2000. pp. 212–36.

[50] *Yuv pixel formats* [online]. Available from http://fourcc.org/yuv.php.

[51] Garcia J., Molina J.M., Berlanga A., de Miguel G. 'Data fusion alternatives for the integration of millimetre radar in airport surveillance systems'. *Proceedings of IEEE International Radar Conference*; Arlington, Virginia, 2005. pp. 796–801.

[52] Peng B., Guan X. 'A new track association algorithm of radar and esm'. *Proceedings of CIE International Conference on RADAR (RADAR)*; IEEE, 2016. pp. 1–5.

Chapter 6

Automotive radar target classification

Xiuzhang Cai[1] and Kamal Sarabandi[1]

6.1 Introduction

The technology behind self-driving cars or autonomous cars is considered as one of the most influential emerging technologies in the near future. One of the main challenges in the autonomous driving or advanced driver assistance systems is to reliably detect and identify all targets in complex traffic environment. Correctly identifying targets and their attributes is essential to make autonomous cars.

Traffic target classification using cameras or lidars has been studied extensively [1–3], and due to their high resolutions and the intuitive nature of data interpretation, significant progress has been made. However, the performance of cameras or lidars could be largely degraded by poor visibility conditions such as at night or in foggy or raining conditions. By contrast, radar data is, for the most part, unaffected by the atmospheric conditions and the time of day [4, 5]. Hence, it is important to develop solutions for target classification using data obtained from millimeter-wave (MMW) radar.

Numerous studies have been reported for the classification of aircrafts, military vehicles, and terrestrial targets for various remote sensing applications with radar data. Since aircrafts must be detected at distant ranges, they are usually modeled as targets of single scatterer, and the frequency or time (range) information of the radar response is used to classify the targets [6, 7]. Terrestrial [8, 9] and military vehicles [10, 11] are detected by airborne or satellite synthetic aperture radar (SAR) systems, and therefore the SAR images of targets are applied in the target classifications. For the scenarios related to traffic, MMW automotive radars can encounter different targets at varying ranges and sizes that may only be partially illuminated and thus cannot always be considered classical point targets. SAR methods are obviously not applicable due to the difficulties in estimating the platform.

[1]Department of Electrical Engineering and Computer Science, University of Michigan, Ann Arbor, MI, USA

The vulnerable road users like pedestrians and bicyclists are considered to be the most important objects to be correctly identified. Since a pedestrian will frequently have different velocities of their arms and legs when they are moving, it is observed that moving pedestrians often exhibit very different Doppler radar response patterns compared to vehicles. This feature is described and utilized to distinguish pedestrians from vehicles in many studies [12–17].

Recent advances in machine learning (ML) techniques, including convolutional neural network (CNN) [18, 19] and convolutional recurrent neural network (RNN) [20], enable new methods for radar target classifications. In [21], multiple frames for the range Doppler images of a continuous movement of a target are put into target classifications models based on RNN. Similar to other Doppler-based target classification methods, this approach can only be applied to moving targets. In [22], the phase and magnitude of each antenna element is used as input for a deep neural network to distinguish pedestrians from vehicles. While greater than 90% accuracy is achieved, the data samples used in the paper is very limited (only based on one vehicle and one pedestrian), which limits the scope and applicability of the neural network model. For some advanced radars that are capable of imaging, the targets then can be represented by radar images in target classification [23, 24]. In [23], the proposed algorithm applies CNN to classify static targets only into two broad categories: parked vehicles and nonvehicles. In [24], targets are visualized as radar point clouds, where radar cross section (RCS) information is completely discarded, and zero-Doppler data are excluded as well, which limits the target classification to only dynamic targets. For zero-Doppler target classification problem, the RCS and their statistical information, and RCS radar images are utilized for target classification [25, 26].

The radar target classification relies on both the classification algorithms and the differentiable radar data for different targets. In this chapter, some ML approaches used in target classification are introduced in Section 6.2. Unlike camera or Lidar data, radar data interpretation is less intuitive and more challenging for developing the pertinent signal processing algorithms. Besides, as various types of radars can be used for automotive radars, different forms of radar data can be generated to represent the different features of the targets. We will present the classification with the following forms of radar data: micro-Doppler signature is presented in Section 6.3, statistical features of RCS are presented in Section 6.4, and radar images are presented in Section 6.5. The basic principle of generating different forms of radar data will be presented, and the pros and cons of each will be discussed in details.

6.2 Machine learning approaches for classification

The history of creating intelligent computer maybe is as long as that of computers. In the middle of twentieth century when the first computer was invented, Turing raised a question: Can machines think? To define the problem more clearly, he proposed the famous imitation game also known as the Turing test. Since then, tremendous progress has been achieved in computer hardware, software, and algorithms;

however, the study of bringing intelligence to machine will never stop. Nowadays still no machines can fully replace human yet, but the discipline of ML is progressing rapidly, and its applications are expanding to an unprecedented wide range of research disciplines including radar technology.

One sign of intelligence is the ability to learn, which means the capability to acquire new knowledge or to build proper connection between new knowledge and old knowledge or experience, and this process is referred to ML [27]. For the past decays, numerals ML algorithms like neural networks have been developed and successfully applied to a wide variety of problems traditionally considered unsolvable by machine, for example, picture classification, voice recognition, and Go game play [28].

In our application, the radar target classification problem can be considered as a data-driven classification problem where the mature ML algorithms can be directly applied. The objective of this section is to give the readers the ML applied in the radar classification in a simple language. For those readers who are interested in a deeper understanding of ML or deep learning with other applications, we refer them to several excellent textbooks [27, 29–31].

6.2.1 Basic concept of machine learning

Radar target classification problem can be described as follows: given a set of radar data as inputs and a set of target types as an output, find a classifier function or ML model to map the input radar data to the output target types. One example of the radar target classification is depicted in Figure 6.1(a). Please note the objective is not just mapping the radar data from the training set to their corresponding target types, but to find the model that can recognize the similarity of the radar data with the same target type and know the difference of the radar data between different categories. Therefore, the trained model should be able to work with the radar data other than the training set with reasonably good accuracy. For example, a model trained with some radar data from various young pedestrians should be able to identify a walking old man as a pedestrian once presented with the associated radar data. This is illustrated in Figure 6.1(b). An input radar dataset with M element can be denoted as $X = \{x_1, x_2, \ldots, x_M\}$, and since this classification problem is a case of supervised learning, each input radar data corresponds to an output target type or a label, that is included in a label vector $\vec{y} = \{y_1, y_2, \ldots, y_M\}$. Both X and $\vec{y}$ are needed to train the ML model.

In the ML algorithm, the general structure of the ML model can be specified before training, and the parameters of the model are determined by the ML algorithms in the training stage. Training an ML model is an optimization process to maximize the performance of the model for both training and test datasets. The performance of the classification model is usually evaluated based on its accuracy, which is defined as the number of correct outputs over the total number of tests. However, in some scenarios only measuring the accuracy is not sufficient. For example, in one problem of classifying the radar data between vehicles and pedestrians, if 95% training and test samples are collected for vehicles, then suppose an

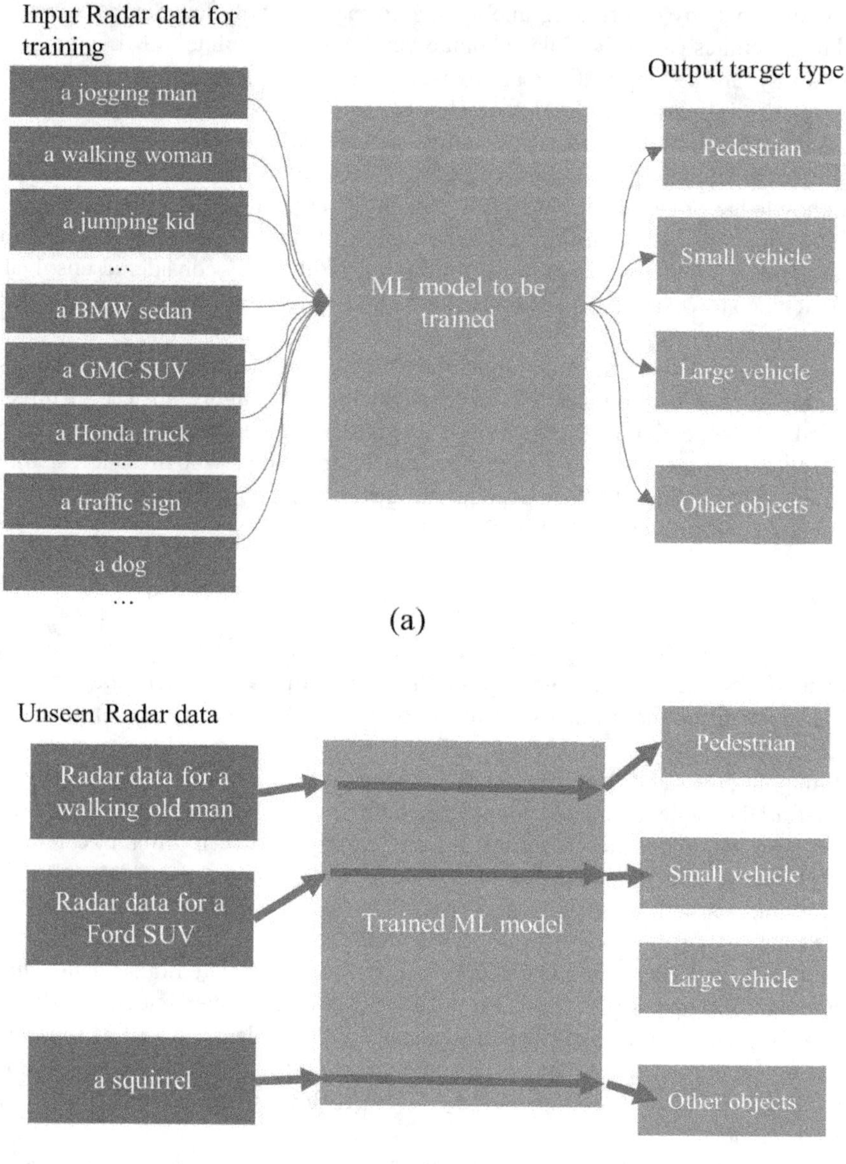

Figure 6.1 *(a) Illustration of the radar target classification problem with ML model; (b) expected performance for the trained ML model with the input of some radar data that are not used in the training stage*

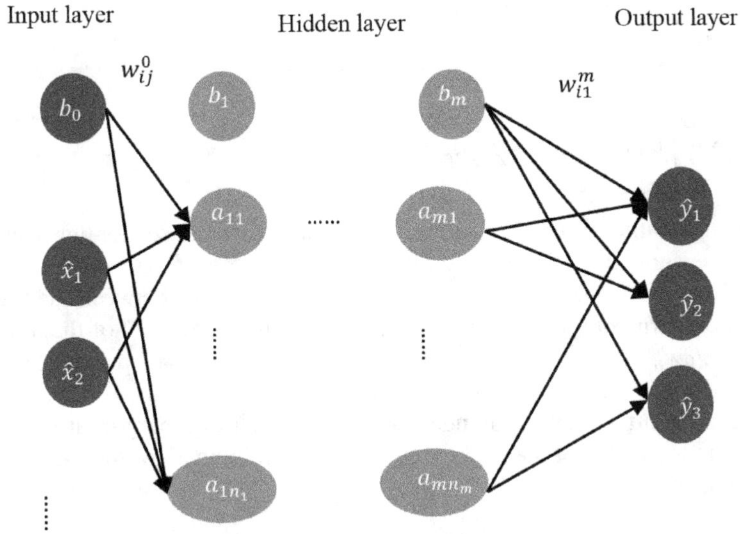

Figure 6.2 The structure of one MLP used in classification problem with three categories

ML model considers all data as vehicles, it will produce 95% accuracy. We know it is a poor model since the true positive rate for pedestrians' data is 0. To avoid such issue, we can either try to balance the number of samples for different categories or give more weights for the true positive rate of the categories with less samples.

The classification problem is similar to a curve fitting problem, and in general, the more complex is the ML model, the more accurately it can fit the training data. However, the increased complexity of ML model sometimes leads to worse performance for the test data due to the overfitting with the training data. On the other hand, if the performance of an ML model can be further optimized for both training and test dataset, this model is underfitting. In general, the goal of training an ML model is to obtain the best accuracy while minimize the performance difference between training dataset and test dataset and keep the model as simple as possible.

6.2.2 Multilayer perceptron

The multiplayer perceptron (MLP), also known as artificial neural network (ANN), is one of the simplest ML models and is very useful in solving many ML problems including data classification. Originally, this mathematical model was developed to mimic the biological neural network, which is used by living beings including humans to learn knowledge. The structure of one example of MLP can be seen in Figure 6.2. It takes the normalized input $X = \{\hat{x}_1, \hat{x}_2, \ldots, \hat{x}_M\}$ and outputs the probability of three predefined categories $y = \{\hat{y}_1, \hat{y}_2, \hat{y}_3\}$. The hidden layers are composed of some constants b_m and variables a_{mn}. The variables in the hidden layer or the output layer are determined by the variables and constant in the previous layer and some weighting factor w_{ij}^{m}, and are given by,

$$a_{1i} = \tanh(b_0 w_{0i}^0 + \sum_{j=1}^{M} \hat{x}_j w_{ji}^0) \tag{6.1}$$

$$a_{ki} = \tanh\left(b_{k-1} w_{0i}^{k-1} + \sum_{j=1}^{n_{k-1}} a_{(k-1)j} w_{ji}^{k-1}\right) \tag{6.2}$$

$$\hat{y}_i = \frac{1}{2}\left[\tanh\left(b_m w_{0i}^m + \sum_{j=1}^{n_m} a_{mj} w_{ji}^m\right) + 1\right] \tag{6.3}$$

where tanh() is the hyperbolic tangent function to introduce nonlinearity to the model and ensures the value is between −1 and 1. The performance is measured by an error function E, which usually is the mean square error between the predicted output and the true output. To optimize the weighting factors w_{ij}^m of the model, the gradient $\partial E/\partial w_{ij}^m$ is computed, which is also known as the backpropagation of the neural network. Many training algorithms are used to obtain the weighting factors including gradient descent, conjugate gradient, and Levenberg–Marquardt algorithm, and more details of the backpropagation and training algorithms can be found in the textbook [32].

6.2.3 Convolutional neural network

The MLP model usually is applied for the problem with relatively small size of uncorrelated input data as the complexity of MLP grows dramatically when the number of inputs increase. In the radar target classification application, many types of radar data are in the form of images including Range-Doppler images, micro-Doppler signature, and radar images. For the classification with image data where patterns can be found between adjacent pixels, the CNNs are more appropriate and leads to better results.

The typical geometry of a CNN is shown in Figure 6.3. Compared to the MLP, three more operations are performed to the data: *convolution, ReLU activation,* and *pooling.* In the convolution layer, part of input variables or the neuron from previous layer is convolved with a trained matrix called feature map. Suppose the input neurons are a 2D matrix with size of $J \times K$ represented by a matrix: $A = \{a_{11}, a_{12}, \ldots, a_{1K}; a_{21}, \ldots a_{2K}; \ldots, a_{JK}\}$, and the feature map is a $L \times M$ matrix

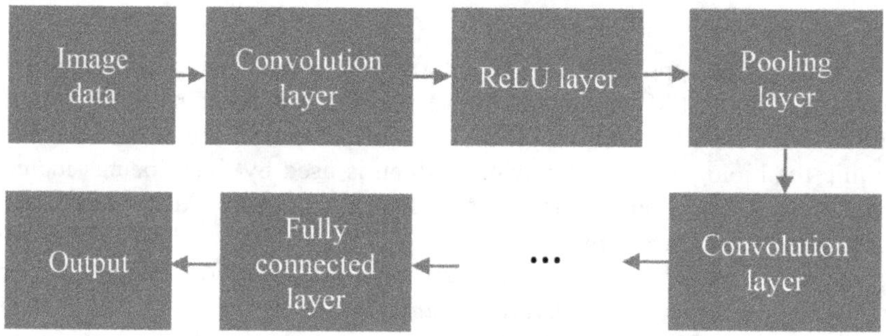

Figure 6.3 The structure of one example of CNN

0	1	2	0	1	2
2	0	0	0	2	1
1	0	1	1	0	0
3	2	1	0	0	1
0	0	0	0	0	2
0	1	0	0	0	0

conv.

Feature map

1	1	2
2	0	0
1	0	1

=

11	4	5	6
8	2	7	7
9	7	4	3
7	4	1	2

Figure 6.4 Convolution of a 6 × 6 matrix and one 3 × 3 feature map

represented by $W^{(i)} = \left\{ w_{11}^{(i)}, w_{12}^{(i)}, \ldots, w_{1M}^{(i)}; w_{21}^{(i)}, \ldots, w_{LM}^{(i)} \right\}$, then the jth, k^{th} element of the hidden neurons in the convolution layer can be expressed as [32]:

$$c_{jk}^{(i)} = \sum_{l=1}^{L} \sum_{m=1}^{M} w_{lm}^{(i)} a_{j+l,k+m} \tag{6.4}$$

One example of the convolution operation is given in Figure 6.4. It shows the convolution results of a 6 × 6 matrix and a 3 × 3 feature map. The values of the hidden neurons indicate how well the feature map matches the corresponding portion of the input matrix. The feature maps serve vital roles in CNN, noting that there could be more than one distinguishable feature on the images for different categories, so we may use several feature maps in the convolution layer. Besides, the features may be revealed in different levels and thus a deep CNN is often used in the image recognition problem. The ReLU is short for rectified linear unit and is commonly used in CNN. Its function is to create nonlinearity for the system, and it can be simply expressed as:

$$x_{ReLU\_out} = \begin{cases} x_{in}, & x_{in} > 0 \\ 0, & x_{in} \le 0 \end{cases} \tag{6.5}$$

The pooling layer is used to simplify the data after ReLU layer. For example, by taking 2 × 2 max-pooling, the matrix after convolution in Figure 6.4 is reduced to a 2 × 2 matrix, as shown in Figure 6.5.

The convolution layer, ReLU layer, and pooling layer can be repeated several times to create a deep CNN, and finally, the data will be rearranged into a 1D array and feed to the fully connected layer. The fully connected layer has the same structure as the MLP, and the output is the same as that of MLP as well.

6.2.4 Recurrent neural network

The last type of ML model introduced in this section is the RNN. It is a type of neural network that is usually dedicated for sequential input data. For example, in the

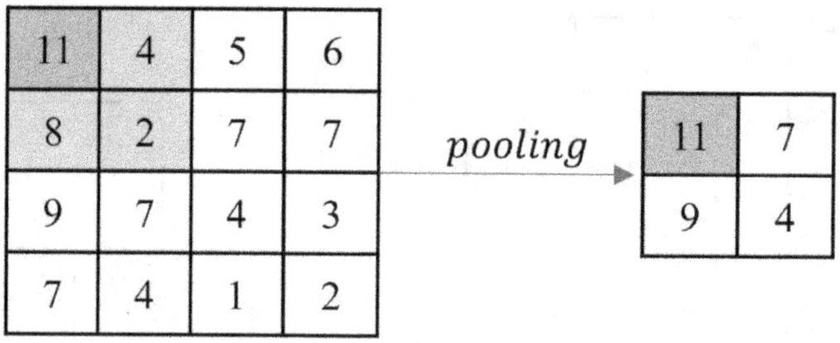

Figure 6.5 The 2 × 2 maximum pooling for a 4 × 4 matrix

speech recognition problem, the length of words is not fixed and the words highly depend on the context. It will be hard to apply MLP or CNN as they usually require fixed size input data and the output does not rely on the previous input data. RNN is designed to resolve this issue. It uses both the new input data and the neural network fed with the old input data to generate output, and the neural network from previous state can be considered as memory of the RNN. In the radar target classification application, the radar data can be collected for a short period of time and apply RNN to classify the data.

One example of a RNN is given in Figure 6.6 . It shows the recurrent graph and the corresponding unfolded graph of a simple RNN in the forward propagation mode. In the graph, x denotes the input vector, and U, V, W are the weight matrix optimized after training the RNN, and y denotes the predicted output. At different time steps, the weight matrix U, V, W will remain the same, and the output is updated by the equations [29]:

$$h^{(t)} = \tanh\left(b + Wh^{(t-1)} + Ux^{(t)}\right) \tag{6.6}$$

$$\hat{y}^{(t)} = \text{softmax}\left(c + Vh^{(t)}\right) \tag{6.7}$$

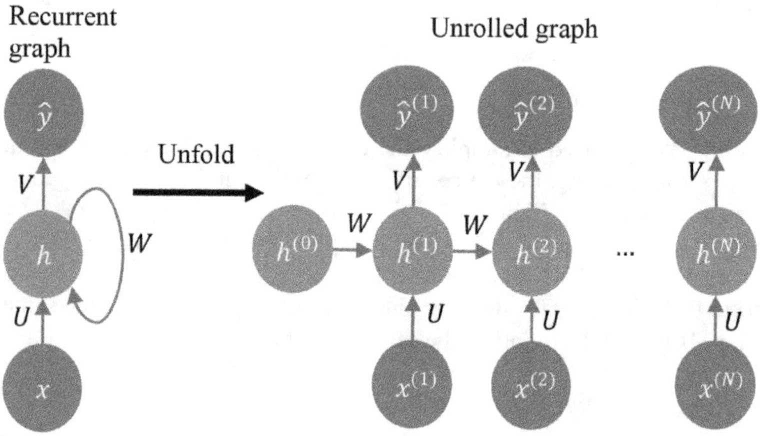

Figure 6.6 The recurrent and the unrolled graph of an RNN

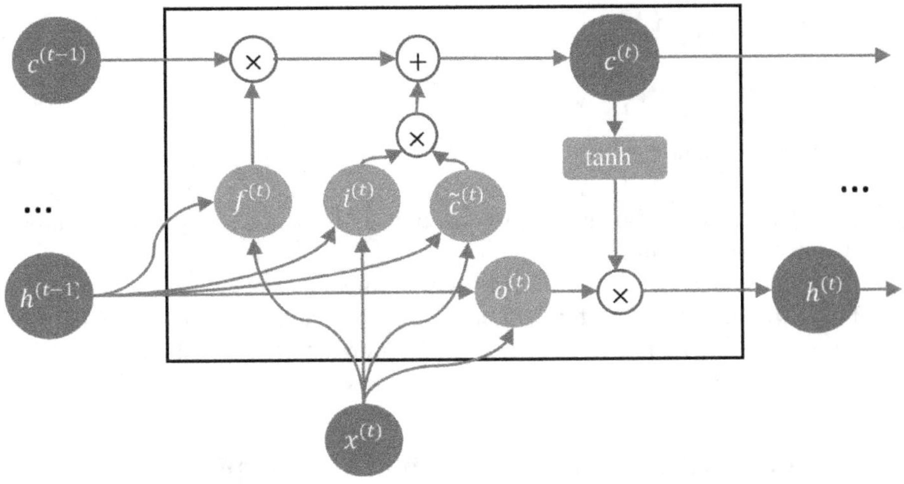

Figure 6.7 The unrolled graph of the LSTM at one time step t

where b and c denote the bias vectors.

The output $\hat{y}^{(N)}$ in Figure 6.6 depends on the inputs from all previous time steps, but sometimes the information from the too early steps may not be necessary. Therefore, researchers proposed some new networks that are able to "forget" the information from the previous time steps. One widely used RNN network with this feature is the long short-term memory (LSTM) [32]. The geometry of the LSTM is depicted in Figure 6.7. The state vectors $h^{(t)}$ are also the output of LSTM network. Besides $h^{(t)}$, the cell $c^{(t)}$ will also propagate to the next time step in the LSTM. The internal variables $f^{(t)}$, $i^{(t)}$, and $o^{(t)}$ are the output of sigmoid functions with values between 0 and 1, and $\tilde{c}^{(t)}$ is the output of hyperbolic tangent function with values between −1 and 1.

LSTM introduces three gates to control the weights of information from current and the previous time steps. $f^{(t)}$ denotes the forget gate's activation vector, and if it equals 0, then $c^{(t-1)}$ is "forgotten" and not used to evaluate the output. $i^{(t)}$ is the activation vector for input gate, and the product of $i^{(t)}$ and $\tilde{c}^{(t)}$ will be added to the cell to generate $c^{(t)}$. The symbol $o^{(t)}$ denotes the output gate's activation vector, and the output $h^{(t)}$ is determined by both $o^{(t)}$ and $c^{(t)}$.

The relations between those variables are given by [32]:

$$f^{(t)} = \sigma\left(b_f + W_f x^{(t-1)} + U_f h^{(t-1)}\right) \tag{6.8}$$

$$i^{(t)} = \sigma\left(b_i + W_i x^{(t-1)} + U_i h^{(t-1)}\right) \tag{6.9}$$

$$o^{(t)} = \sigma\left(b_o + W_o x^{(t-1)} + U_o h^{(t-1)}\right) \tag{6.10}$$

$$\tilde{c}^{(t)} = \tanh\left(b_c + W_c x^{(t-1)} + U_c h^{(t-1)}\right) \tag{6.11}$$

$$c^{(t)} = f^{(t)} \circ c^{(t-1)} + i^{(t)} \circ \tilde{c}^{(t)} \tag{6.12}$$

$$h^{(t)} = o^{(t)} \circ \tanh\left(c^{(t)}\right) \tag{6.13}$$

where $\sigma\,()$ denotes the logistic sigmoid function, and $\circ$ denotes the Hadamard product (element-wise product).

There are other variants of LSTM used in different applications, and it can be combined with CNN as well for the image-like input data. More details and examples of the RNN or LSTM can be found in the aforementioned textbooks or online sources.

6.3 Radar target classification with micro-Doppler signature

6.3.1 Description of micro-Doppler signature

One of the main advantages of radar sensors compared to other sensors is the high accuracy in speed detection and target discrimination based on speed. The relative radial velocity of one target can be perceived by the well-known Doppler effect, and the Doppler frequency shift is given by,

$$f_D = \tfrac{2f}{c} v_r, \tag{6.14}$$

where f denotes the radar's operating frequency, c denotes the speed of light, and v_r is the relative radial velocity of the target with respect to the radar.

This Doppler frequency shift can be detected in different ways for different radars. In automotive radar, Frequency-modulated continuous-wave (FMCW) waveform is commonly used because of its low-power and low-cost feature. For FMCW radars, the Doppler shift can be evaluated by measuring the phase difference across chirp to chirp of target's response. The simplified phase term of the radar signal model can be expressed as [33]:

$$d(n,p) = \exp\left\{i2\pi\left[\left(\tfrac{2KR}{c} + f_D\right)\tfrac{n}{fs} + f_D p T_0 + \tfrac{2f_cR}{c}\right]\right\} \tag{6.15}$$

where K denotes the frequency modulation constant and is equal to $(f_2 - f_1)/t_c$. R denotes the range of the target, and f_s is the sampling frequency of one chirp and can be expressed by the number of frequency samples N divided by t_c. f_c denotes the carrier frequency, and n and p refer the index of frequency samples and chirps. The waveform of the FMCW radar is illustrated in Figure 6.8. It shows the transmitted radar signals for one frame and one channel (one transmitter and one receiver).

The Doppler shift of targets can be obtained by taking fast Fourier transform (FFT) to the all-chirp data after range FFT, and the velocity resolution is determined by the number of chirps in one frame P and the chirp period T_c,

$$dv = c/\left(2f_c P T_c\right) \tag{6.16}$$

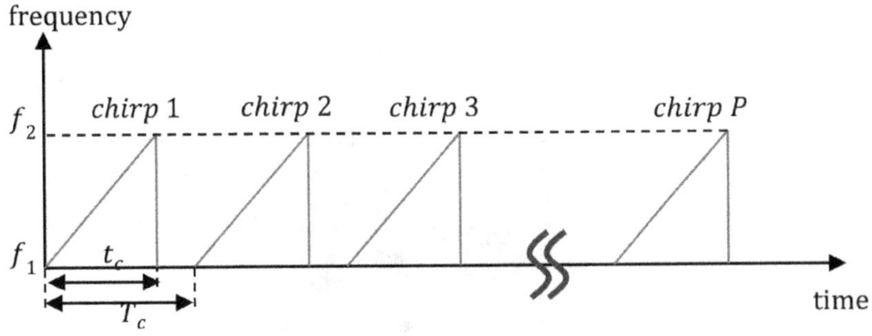

Figure 6.8 A typical FMCW radar waveform, with starting frequency f_1, ending frequency f_2, chirp duration time t_c, and chirp period T_c

Then, the maximum unambiguous velocity is simply given by:

$$v_m = c/\left(2f_cT_c\right) \tag{6.17}$$

One important safety feature of automotive radars is detection and identification of pedestrians. Compared to other moving objects like vehicles, the motional patterns of pedestrians are more complex. For example, the arms or legs may move back and forth and exhibit different radial velocity relative to the radar, as illustrated in Figure 6.9. The average speed of a pedestrian is usually below 5 m/s, and the cycle time of one frame of the automotive radars usually is required to be less than 50 ms, which corresponds to the velocity resolution of 0.04 m/s at 77 GHz according to (6.16). Therefore, the radar should be able to discriminate the motion of arms or legs from the main human body through the Doppler spectrum. This phenomenon captured by the radar is known as micro-Doppler effect, since those Doppler shift components are usually weaker than the main body.

Since the movement of pedestrian will change over time, it is important to examine the time dependence of the micro-Doppler behavior. The Doppler spectrum as a function of time is also called the micro-Doppler signature. This feature was first introduced by Chen [34, 35] and has been further studied and applied in the target classification by many other researchers [12–17, 36–38]. One example of

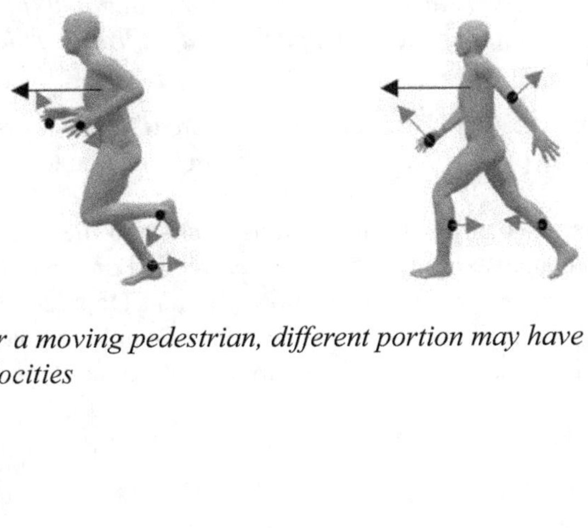

Figure 6.9 For a moving pedestrian, different portion may have different velocities

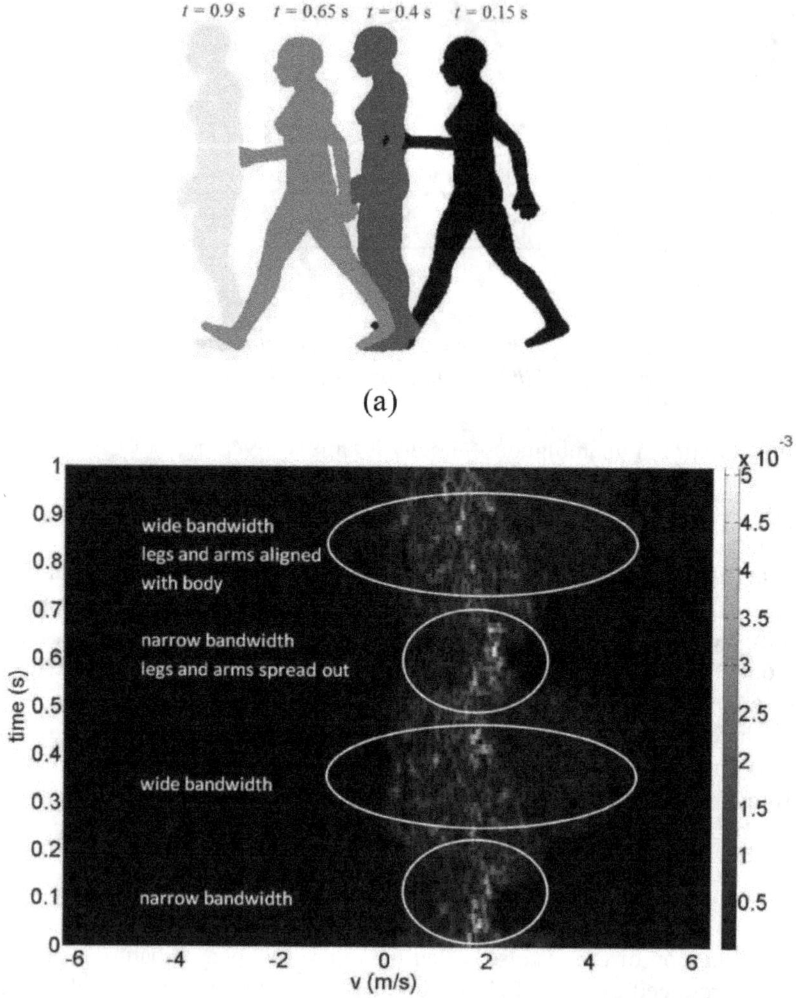

Figure 6.10 (a) Different stage of one period of walking and (b) the corresponding micro-Doppler signature from simulation [15]

the micro-Doppler signature [15] is shown in Figure 6.10. It shows that at different stages of one walking period, the Doppler spectrum (relative radial velocity) spreads differently.

In fact, any target that has small motions relative to the main body like rotations and vibrations may exhibit micro-Doppler effect. For example, some portions of the wheels of a car or a bicycle may show different Doppler shifts from the car or bicycle when it is moving [39, 40]. The micro-Doppler signature of a bicycle is more complex than that of a car as the motions of human legs are involved and

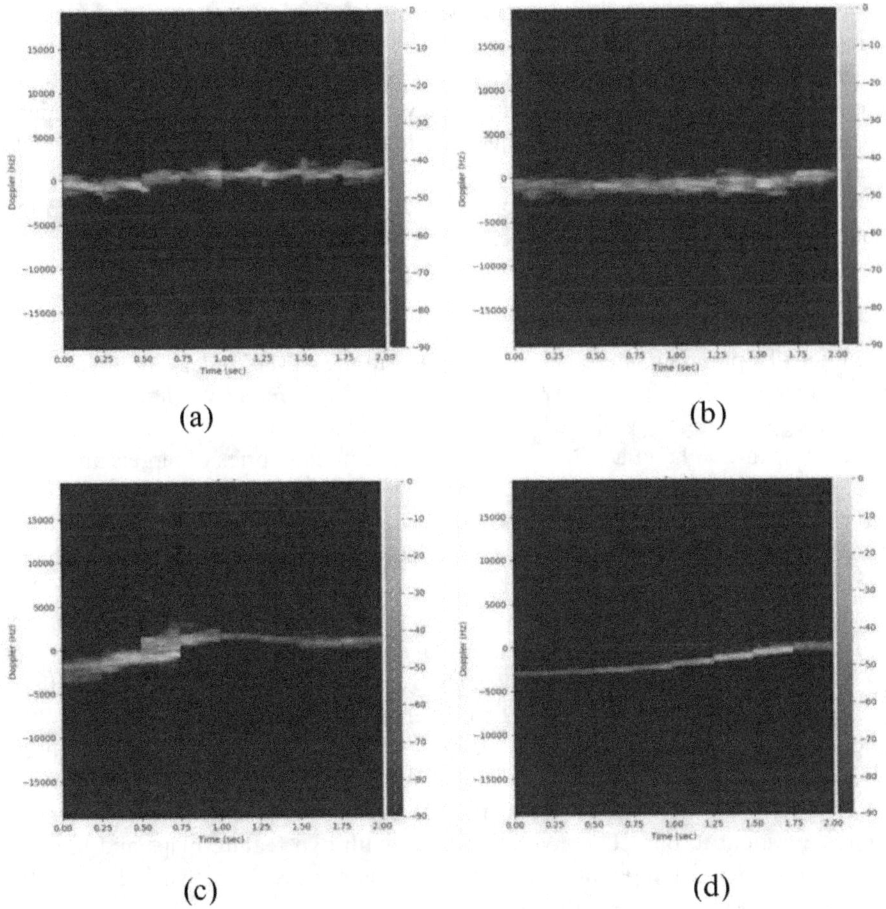

Figure 6.11 The measured micro-Doppler signature examples for (a) a walking person, (b) two people walking together, (c) a bicycle, and (d) a car [17]

create additional Doppler shift other than the wheels. One example of the MMW band micro-Doppler signature comparison between different traffic targets including one pedestrian, two pedestrians walking together, a bicycle, and a car is depicted in Figure 6.11 [16]. In this study, a 76.5 GHz automotive radar is used to measure the micro-Doppler signatures of the aforementioned targets. The radar has the velocity resolution of 0.3 m/s and the maximum unambiguous velocity of 76.6 m/s. As we can see the micro-Doppler spectrum for one pedestrian changes periodically, which matches the pedestrian's movement pattern. The variance on the spectrum becomes less in the two pedestrians' scenario as their movements are asynchronized. The Doppler spectrum of the bicycle is affected by the movements of the legs, which result in the micro-Doppler spectrum spreads widely at some points. By contrast, the

Doppler spectrum of a car is quite narrow since the dominant scattering mechanism of radar signals is from the car body.

The motional Doppler spectrum patterns may be quite different for the different types of traffic targets (e.g. pedestrians, bicycles, vehicles). Therefore, the micro-Doppler signature can serve as a unique feature to distinguish different traffic targets. The micro-Doppler effect has many other applications as well, such as pedestrian fall detection and hand gesture recognition. For readers who are interested in this topic, more studies and discussions can be found in the textbook [36].

6.3.2 Classification example

The micro-Doppler signature has been applied in the MMW radar target classification in many studies [12–17]. While there are many different ways to classify the micro-Doppler data, we will use [17] as an example to demonstrate the classification approach with CNN and LSTM in this chapter. Four categories of targets are measured: single pedestrian (60 samples), a car (60 samples), a bicycle (22 samples), and two pedestrians together (44 samples).

It utilizes both CNN and CNN-LSTM network in the classification problem. The structure of the CNN proposed in [17] is similar to the standard deep CNN visual geometry group. In the forward propagation model, the 2 s micro-Doppler signatures are fed to the 2D convolution layer with 30 feature maps with size of 5×5. After maximum pooling, the data go to another 2D convolution layer with 15 feature maps whose size is 3×3. Then, after another maximum pooling and 20% dropout, the data are put into the three-layer fully connected layers with size of [128, 50, 2] (three layers with 128 neurons, 50 neurons, and 2 neurons in each layer).

Regarding the CNN-LSTM network used in [17], the 0.25 s micro-Doppler signatures are input to one 2D convolution layer with two feature maps and then the output are flattened and go into a following 50-neuron LSTM network. Finally, the data are processed with the fully connected layer.

The ML models are applied in several classification problems: classification between (1) car and pedestrian; (2) car, pedestrian, and bike; and (3) car, single pedestrian, and two pedestrians. In the two-category classification, thanks to the good differentiability of the micro-Doppler signature between pedestrians and cars, nearly 100% accuracy is achieved for both CNN and CNN-LSTM. The accuracy is defined by the number of correct predictions over the total number of test cases (the true positive rate for each category is not provided). Table 6.1 demonstrates the classification accuracy with different models. It shows the LSTM model with 0.5 s long data sample performs better than that with 2 s long data sample, which indicates the LSTM model has overfitting issue with 2 s long data sample.

For the three-category classification problem, the performance for the test data is given in Table 6.1. It shows that the accuracy for most cases is downgraded to around 80%. This is because some of the micro-Doppler signatures for bike are similar to those for a car, and some of those for two pedestrians are not too much different from those for a single pedestrian. Overall, the CNN-LSTM shows better performance than just CNN in this example.

Table 6.1 Test accuracy of the target classification based on micro-Doppler signature [17]

Classification	**CNN (2 s long data)**	**CNN (0.5 s long data)**	**CNN-LSTM (2 s long data)**	**CNN-LSTM (0.5 s long data)**
Car-pedestrian classification	98.33%	–	92%	99%
Car-pedestrian-bike classification	79%	83%	93%	83%
Car-1 pedestrian-2 pedestrians classification	81%	78%	80%	84%

6.4 Radar target classification with statistical RCS information

6.4.1 Physical meaning of RCS

RCS of a target is a parameter solely attributed to the target's geometry, aspect angle and polarization of the radar, but it is independent of radar power, distance to the target, and antenna aperture size. A target's RCS indicates the reflectivity and its detectability for a radar operating at a specific frequency. To better utilize RCS values of targets in a traffic scene for target classification, we should start from the radar equation relating the received power to the transmitted power and the radar antenna parameters. The wave propagation and scattering of radar signals can be divided into two parts: (1) the transmitter part where the electromagnetic (EM) wave with power P_t is sent from the transmitter and reaches a target at a certain direction and range, and (2) the wave scattered from the target that propagates back to the receiver of the radar. It is known that for a point source in the 3D space, the power density of the scattered wave decays as $1/r^2$, so the incident power density on the target can be expressed as:

$$S_i = \frac{P_t G_t}{4\pi r_t^2} \tag{6.18}$$

where G_t denotes the antenna gain of the transmitter in the direction of the target with respect to the transmitter, and r_t denotes the distance from the transmitter to the target. Radar cross section denoted by σ is defined as a fictitious area of the target that can collect the incident power density at the target location and radiate the collected power isotropically in such a way that the power density at the receiver is equal to that of the scattered field at the receiver location. The mathematical form is given by:

$$\sigma_{pq} = \lim_{r\to\infty} 4\pi r^2 \left| \frac{E_s^p}{E_i^q} \right|^2, \; p,\, q = v \; or \; h \tag{6.19}$$

where v denotes vertical polarization, and h denotes horizontal polarization; E_s^p denotes the scattered field at the receiver with p polarization and E_i^q refers to the incident field with q polarization.

Hence, the power density at the receiver can be computed from:

$$S_r = \frac{S_i\sigma}{4\pi r_r^2} \tag{6.20}$$

where r_r denotes the distance from the receiver to the target. The receiver antenna can collect the incident power density using its effective area denoted by, that is

$$P_r = S_r A_{eff} \tag{6.21}$$

where A_{eff} is given by:

$$A_{eff} = \frac{G_r\lambda^2}{4\pi} \tag{6.22}$$

Here, G_r is the antenna gain of the receiver and λ denotes the wavelength. Combining equations from (6.18) to (6.22), we can relate the received power with transmitted power through:

$$P_r = \frac{P_t G_t G_r \lambda^2 \sigma}{(4\pi)^3 r_t^2 r_r^2} \tag{6.23}$$

Equation (6.23) is known as the radar equation [41].

Many factors affect the RCS of a target, such as the dimension, geometry, material, angle of view, frequency, and polarization. For some targets with simple geometry like sphere, cylinder, and corner reflector, the RCS in the backscattering direction can be calculated analytically. For example, the backscattering RCS of a large metallic sphere with radius of a is just the area of the great circle πa^2. The RCS of complex targets, however, can only be obtained from measurement or numerical simulation.

The RCS can be measured from the target directly regardless its motion; therefore, one advantage of using RCS data compared to the micro-Doppler data is that it can be applied to both moving and stationary target for this radar target classification problem. Generally, a comprehensive dataset for an exhaustive list of targets is required to obtain a convincing ML-based classification model. Radar measurement can provide reliable data; however, gathering them can be very expensive, time consuming, and sometimes impractical. For example, it is difficult to take radar measurements of a heavy truck or bus for all azimuth and elevation angles as it is almost impossible to put a large vehicle on a turn table for accurate angular variation. On the other hand, numerical simulation provides the ability to be performed for any kind of target of any distance or any direction with much lower cost. Furthermore, some targets like pedestrian models with all ages, weights, and heights can easily be generated and examined in simulation, whereas such measurements usually have very limited samples and can be time consuming and uncomfortable for the test subjects.

6.4.2 RCS simulation in MMW band

There are many different EM numerical simulation methods, and in general, they can be divided into two categories: full-wave methods and asymptotic methods. Both categories have their own applicable regions. The full-wave methods including method of moment, finite element method, and finite-difference time-domain method can generate accurate results; however, they are computational very expensive (with time complexity of $O(n^3)$ or $O(n^2 \log n)$) and could only simulate objects with dimensions less than 10 wavelengths in most modern computers. The current automotive radars operate at around 77 GHz, and traffic targets like vehicles and pedestrians have dimensions on the order of hundreds to thousands of wavelengths at this frequency. Therefore, it is almost impossible to apply full-wave methods to this problem.

On the other hand, asymptotic methods require much less computational resources and work well. They have been successfully used to simulate the backscattering RCS of traffic targets at MMW frequency [15, 26, 42]. The asymptotic methods used in RCS simulation for electrically large targets include physical optics method (PO), raytracing-based method such as shooting and bouncing rays, and uniform geometric theory of diffraction. Each of them has their own advantages and best applicable scenarios. PO is the most efficient approach, and it is accurate for convex targets, but it is inefficient in calculating multi-scattering or higher order solutions. Shooting and bouncing rays treat EM wave propagation as rays, and on boundary the rays get reflected, refracted, or diffracted according with the geometric optics and geometric theory of diffraction rules. It can solve the multi-scattering problem and thus appropriate for concave geometry like cavities. Geometric theory of diffraction or uniform geometric theory of diffraction account for the diffracted fields that are neglected in the geometric optics solution and makes it more accurate for targets with sharp edges or vertices. In this chapter, the PO method is presented and used to simulate RCS of desired targets since most traffic targets do not often present sharp edges and are for the most part convex.

The PO method evaluates the scattered fields from targets for any given incident fields and using (6.19), RCS can be calculated. Usually, the RCS for co-polarization is much larger than that for the cross polarization, and for electrically very large objects, the co-polarized RCS σ_{vv} and σ_{hh} are very similar to each other, therefore the RCS in the following discussion refers only to vertical polarization.

The scattered field of a current source can be obtained from the Huygens principle:

$$\vec{E}_s = \frac{ik_0 e^{i\vec{k}_s\cdot\vec{r}}}{4\pi r}\left(\bar{I} - \hat{k}_s\hat{k}_s\right)\iint\left[Z_0\vec{J}_s\left(r'\right) - \hat{k}_s \times \vec{M}_s\left(r'\right)\right]e^{-i\vec{k}_s\cdot\vec{r}'}\,ds' \tag{6.24}$$

where $\bar{I}$ is the dyadic idemfactor, $\hat{k}_s$ is the unit vector of scattering direction, $\vec{r}$ is the observation point, and $\vec{r}'$ is on the surface of target. $\vec{J}_s$ and $\vec{M}_s$ are the equivalent electric and magnetic surface currents.

In the PO approximation, only line-of-sight surface portion of the target is illuminated by the radar, and the surface is considered locally flat. This approximation is also known as tangent plane approximation. The equivalent electric and magnetic surface current $\vec{J}_s$ and $\vec{M}_s$ can be simply calculated from the tangential E field and H field on the surface of the scatterer [15]:

$$\vec{J}_s = \left[\vec{H}_i \cdot \hat{t}\left(1 + R_{TM}\right)\left(\hat{n} \times \hat{t}\right) + \tfrac{1}{Z_0}\vec{E}_i \cdot \hat{t}\left(\hat{n} \cdot \hat{k}_i\right)\left(R_{TE} - 1\right)\hat{t}\right] e^{i\vec{k}_i \cdot \vec{r}}, \tag{6.25}$$

$$\vec{M}_s = \left[Z_0\vec{H}_i \cdot \hat{t}\left(\hat{n} \cdot \hat{k}_i\right)\left(R_{TM} - 1\right)\hat{t} - \vec{E}_i \cdot \hat{t}\left(1 + R_{TE}\right)\left(\hat{n} \times \hat{t}\right)\right] e^{ik_i \cdot \vec{r}} \tag{6.26}$$

where Z_0 is the characteristic impedance of free space, $\vec{E}_i$ and $\vec{H}_i$ are the incident electric and magnetic fields, $\hat{k}_i$ is the unit vector along the incident direction, $\hat{t}$ is the local tangential vector on the surface, defined as $\hat{t} = \left(\vec{k}_i \times \hat{n}\right) / \left|\vec{k}_i \times \hat{n}\right|$, and $\hat{n}$ is the unit normal vector of the surface. R_{TM} and R_{TE} are the Fresnel reflection coefficients for TM and TE polarizations, respectively, and are given by,

$$R_{TE} = \left(\eta_2 cos\theta_1 - \eta_1 cos\theta_2\right) / \left(\eta_2 cos\theta_1 + \eta_1 cos\theta_2\right), \tag{6.27}$$

$$R_{TM} = \left(\eta_1 cos\theta_1 - \eta_2 cos\theta_2\right) / \left(\eta_1 cos\theta_1 + \eta_2 cos\theta_2\right). \tag{6.28}$$

Here, η_1 and η_2 are the characteristic impedance of air and target medium; θ_1 and θ_2 are the incident and refracted angles with respect to the surface normal of the target. For metallic targets, $R_{TE} = -1$ and $R_{TM} = 1$, and they are independent of the incident angle and $\vec{M}_s = 0$.

The scattered E field in (6.23) is only evaluated for a single frequency, however, the MMW automotive radar operates in a wide frequency band with up to 4 GHz bandwidth. In order to simulate the radar response over a wide frequency range in a computationally efficient manner, we can rearrange the integral in (6.23) as a frequency-independent term and a phase integral term:

$$\iint \left[Z_0\vec{J}_s\left(r'\right) - \hat{k}_s \times \vec{M}_s\left(r'\right)\right] e^{-i\vec{k}_s \cdot \vec{r}'} ds' = \left[Z_0\vec{J}_{sm} - \hat{k}_s \times \vec{M}_{sm}\right] \iint e^{i\vec{k}_i \cdot \vec{r}' - i\vec{k}_s \cdot \vec{r}'} ds' \tag{6.29}$$

where $\vec{J}_s\left(r'\right) = \vec{J}_{sm} e^{i\vec{k}_i \cdot r'}$ and $\vec{M}_s\left(r'\right) = \vec{M}_{sm} e^{i\vec{k}_i \cdot r'}$. The modified surface currents $\vec{J}_{sm}$ and $\vec{M}_{sm}$ have almost no dependence on the position r', over a relatively small area, or the frequency over the desired bandwidth, and therefore they can be pulled out from the integral. These are calculated once at discrete and distant frequency points. Since the raytracing or shadowing process is frequency independent, the only term needed to be evaluated for different frequencies is the phase integral term $\iint e^{i\vec{k}_i \cdot \vec{r}' - i\vec{k}_s \cdot \vec{r}'} ds'$. The phase integral is over the area of one triangular facet, which can be evaluated analytically based on Gordon's method or large-element PO approach [43],

$$\iint e^{i(\vec{k}_{in} - \vec{k}_{sn}) \cdot \vec{r}'} ds' = e^{i\vec{k} \cdot \vec{o}'} \sum_1^3 \vec{e}_i \cdot \hat{t} \frac{e^{ik_\alpha (\vec{v}_i + \vec{v}_{i+1}) \cdot \hat{\alpha}/2} \sin\left(\frac{1}{2} k_\alpha \vec{e}_i \cdot \hat{\alpha}\right)}{\frac{1}{2} k_\alpha \vec{e}_i \cdot \hat{\alpha}}, \tag{6.30}$$

where $\vec{k} = \vec{k}_i - \vec{k}_s$, $\vec{O}'$ denotes one point on the facet, $\hat{t}$ is a tangential unit vector on the surface perpendicular to $\vec{k}$, $\vec{k}_\alpha$ is the tangential component of $\vec{k}$, and is given by $\vec{k}_\alpha = \vec{k} - \left(\vec{k} \cdot \hat{n}\right)\hat{n}$; k_α equals the length of $\vec{k}_\alpha$, $\hat{\alpha}$ denotes the unit vector along $\vec{k}_\alpha$ direction, $\vec{v}_i$ denotes the coordinate of the vertex i and $\vec{e}_i$ denotes the edge vector,

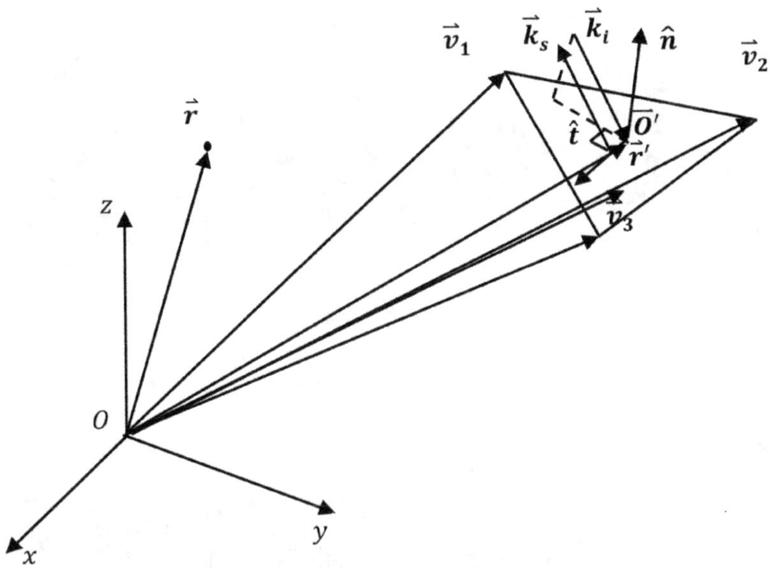

Figure 6.12 Coordinate system of the large-element PO method for a triangular facet

and is given by $\vec{e}_i = \vec{v}_{i+1} - \vec{v}_i$. The coordinates of all the vectors are illustrated in Figure 6.12.

6.4.3 Statistical representation of RCS

The geometry of traffic targets is usually complex, and their dimensions compared to the wavelength is very large. This results in large RCS fluctuation as a function of the incident angle or frequency. This phenomenon has been observed by both experiment [44, 45] and through simulations [25]. Before RCS data can be applied in the target classification, the RCS needs further processing to present differentiability for different types of target. One idea is to extract the statistical information of the RCS. We take the example in [25] to demonstrate the classification using the statistical information of RCS. Figure 6.13(a) shows the computer-aided design (CAD) models of a pedestrian, a sedan, and a palm tree. The simulated RCS for all three targets changing with the frequency is depicted in Figure 6.13(b), and Figure 6.13(c) illustrates the frequency averaged RCS as a function of incident angle. In both cases, the RCS fluctuates rapidly and has large variations.

A large complex target will demonstrate multiple scattering centers at MMW band, and the scattered fields from different scatterer points on the target can add constructively or destructively. This is the main reason that causes the large fluctuation in RCS over frequency or incident angle. To apply the RCS in classification problem, we have to stabilize the data to achieve better differentiability. The study of utilizing statistics to analyze the randomness of RCS dated back to 1950s by

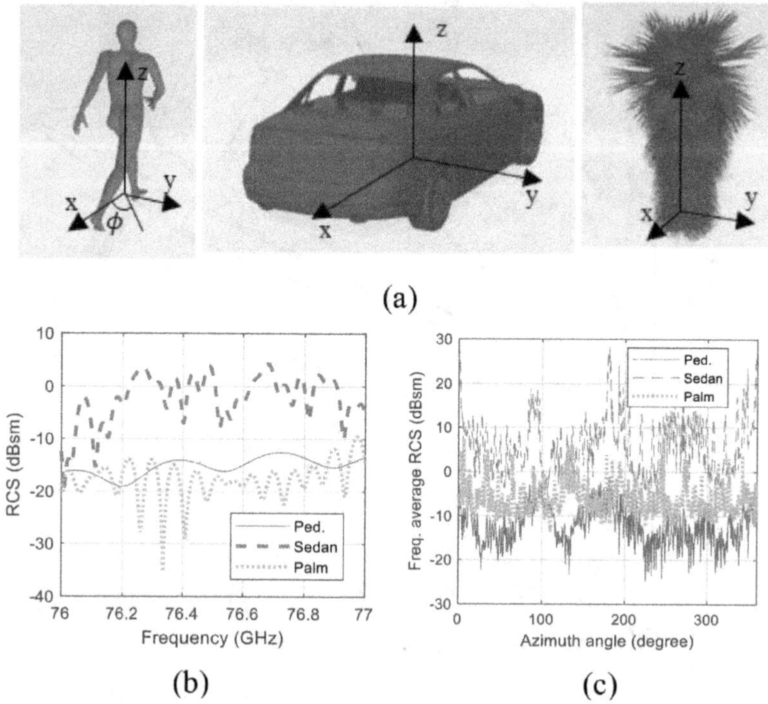

Figure 6.13 *(a) The CAD models used in simulation (a 1.8 m tall pedestrian, a sedan, and a dwarf palm tree); (b) the simulated RCS as a function of frequency for the three targets from $\phi = 0°, 35°$, and $105°$, respectively, in the range of 50 m; and (c) the frequency averaged RCS as a function of azimuth angle in the range of 50 m [25]*

Swerling. In this work, several statistical models were developed and applied to many different radar applications successfully [45, 46].

In this study, to obtain reliable statistical models for traffic targets in MMW band, we simulate the RCS data for more than 80 different kinds of traffic targets for all azimuth angles (ϕ) with 0.5° step and 2 GHz bandwidth with 100 MHz steps (21 frequency points). Each target is simulated to create more than 15,000 RCS samples that are then reduced into 72 groups of statistical data (5° per group) for different aspect angles. By fitting the data to different known distributions, it is found that the Weibull distribution is the best fit for the RCS values of traffic targets in MMW or sub-MMW bands [47]. The probability density function of the Weibull distribution is given by:

$$f(x|A,B) = \frac{B}{A}\left(\frac{x}{A}\right)^{B-1} \exp\left\{-\left(\frac{x}{A}\right)^{B}\right\}, \tag{6.31}$$

where A is the scale parameter, and B is the shape parameter. If $B = 1$, the Weibull distribution reduces to the exponential distribution, which coincides with the

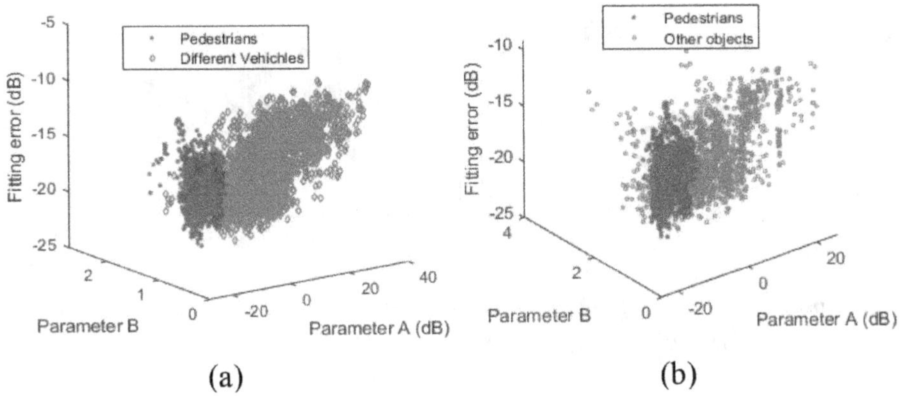

Figure 6.14 The parameters A *and* B *and the fitting error of Weibull distribution for different targets with categories in (a) pedestrians and vehicles and (b) pedestrians and other objects [25]*

Swerling I model. Different B parameters can be used to represent how the statistical distribution deviates from the exponential distribution. In target classification problem, it is always desired to have more distinguishable features. We also notice that not every target RCS values can be well matched to the Weibull distribution. This is because different types of targets like pedestrians and cars have different dimensions and may have different scattering behaviors. For example, some portions of a car have large flat surface that can dominate the RCS while the scatterers on a pedestrian may distribute more evenly. Different scattering behaviors may lead to different fitting error for the Weibull distribution. Therefore, the fitting error is considered as an additional feature in this classification problem. The fitting error is measured by the Kolmogorov–Smirnov test, which is defined as the root-mean-square (RMS) error between the empirical cumulative density function of the raw data and the fitted cumulative density function of Weibull distribution with best fitting parameters [48]. Figure 6.14 shows the statistical information scatter plot for some targets with different categories. In the figures, we can imagine that some artificial boundaries can be created to separate most part of the data in different categories, and the ML approach is applied to find such boundaries.

6.4.4 Classification with ANN

In this traffic target classification example, targets of interest are divided into three categories: pedestrians, vehicles, and other objects. About 90 diverse targets are simulated and processed to generate more than 6000 samples of the statistical parameters described in the previous subsection. For the target categories, the pedestrians include people with different genders, ages, poses (standing, walking, and jogging), and body types (skinny, average, and overweight); vehicles include bikes, motorcycles, sedans, supercars, SUVs, buses, and trucks; other objects include some animals like dogs, deer, and horse, stationary objects like traffic sign, lamp posts, mailboxes,

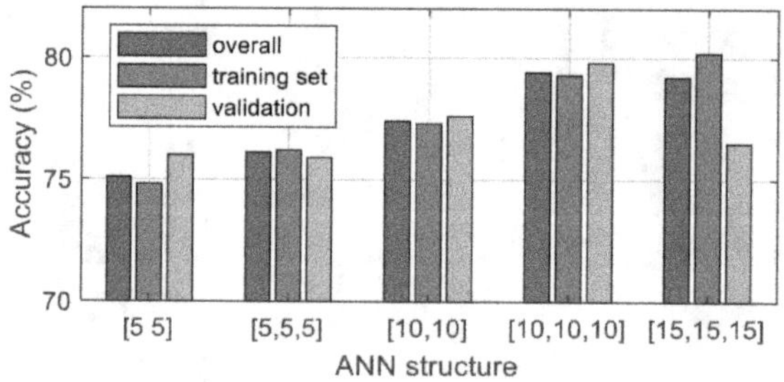

Figure 6.15 The performance of classification with statistical RCS data using different ANN structures [25]

trees, trash bins, etc. The number of samples for the three categories are close to each other to avoid bias issue in the classification.

In this example, because the number of inputs is small (only three statistical parameters), the ANN (MLP) is selected as the ML model. The ANN is constructed and trained in MATLAB® [49]. About 70% of data are randomly selected as training set, and 15% of data are used for validation and the rest is used for testing. To select the optimal structure, different structures of the ANN are trained and compared. The performance comparison is shown in Figure 6.15. In this figure, [5, 5, 5] means the ANN has three layers and each layer has five neurons. Here, it is shown that [10, 10, 10] is the optimal ANN structure since it has one of the best overall accuracies and the accuracy for validation dataset is almost the same as that for training set.

The performance of the target classification based on statistical information of RCS is given in Table 6.2. The true-positive rate of the classification model is marked in bold. It shows that if we want to classify the data for all three categories, more than 90% accuracy is achieved for pedestrians and vehicles, but the correct rate for the prediction with other objects data is only 29%. This is due to the fact that the scope of other objects category is broad, some small targets like a dog or a lamp post may exhibit similar statistical RCS as that for pedestrians, and the statistical RCS for some large target like a tree or bus station may be close to that for vehicles. If we only consider the targets near the sidewalk so that we can exclude the vehicles

Table 6.2 Overall performance of the target classification based on statistical information of RCS [25]

		Predict target			**Predict target (Veh. excluded)**	
True target		*Ped.*	*Veh.*	*Other Obj.*	*Ped.*	*Other Obj.*
	Ped.	**95.3%**	0.3%	4.4%	**95.8%**	4.2%
	Veh.	1.3%	**92.9%**	5.8%	–	–
	Other Obj.	19.7%	51.2%	**29%**	26.4%	**73.6%**

Table 6.3 Performance of the target classification model based on statistical information of RCS with measured data [25]

		Pred. target			Pred. target (Veh. excluded)	
True target		*Ped.*	*Veh.*	*Other Obj.*	*Ped.*	*Other Obj.*
	Mann.	**80.2%**	0.9%	18.9%	**95.6%**	4.4%
	Veh.	5.2%	**86.0%**	8.8%	–	–
	Sign Post	41.0%	20.1%	**38.9%**	43.0%	**57.0%**

data from the training model, then we can achieve more than 95% accuracy for pedestrian and more than 70% accuracy for other objects.

The performance of the models trained with simulated data are validated with the measured data as well. The measured targets include a mannequin covered in a reflective coating in three different poses, two traffic sign posts (one square shape and one U shape), and three vehicles (two sedans and one sport utility vehicle (SUV)). The measurements are performed with a 77 GHz instrumental radar, and the targets are on a turntable and measured for all azimuth angles. The statistical RCS are collected and input into the trained models. The classification performance with the measured data can be found in Table 6.3. It shows that more than 80% data for the mannequin and vehicles can be correctly identified with the three-category model, and more than 95% data of mannequin are correctly predicted as pedestrian, while 57% accuracy is achieved for the measured data from sign posts. Though statistical RCS can provide more differentiability than the raw RCS, different types of targets with similar dimensions may still demonstrate similar statistical features for the RCS. For example, the statistical RCS for the sign posts are similar to those for the pedestrians. This limits the performance of the target classification model based on statistical information of RCS data.

6.5 Radar target classification using radar images

6.5.1 Radar images simulation

To achieve better performance of the classification model, more features are needed. The radar images (range-azimuth/elevation angles) can provide the dimension or shape information of targets, and like the RCS of targets, the radar images are independent of the target's movement. Therefore, it is desired to develop classification model based on radar images. Radar images can be generated in many ways including analog beamforming, digital beamforming, and multiple-input multiple-output (MIMO) antenna arrays. In this example, we use typical MIMO antenna arrays used in automotive radars to create radar images. The MIMO antenna array selected is equivalent to 1D or 2D uniform linear arrays (ULA), and by applying different phase and weighting factors, the synthesized main beam can be scanned.

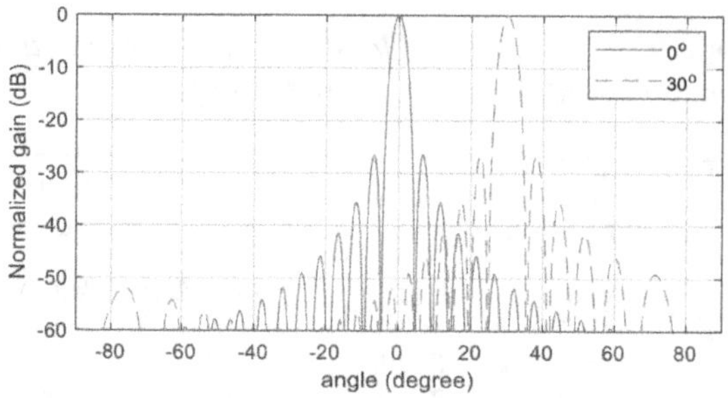

Figure 6.16 The radiation pattern of a 50-element ULA with λ/2 spacing and triangular weighting function [25]

To ensure the best quality of the radar image, the sidelobe level of the synthesized antenna pattern should be kept as low as possible. The sidelobe level can be controlled by different weighting factors for the antenna elements with the tradeoff of achieving a wider main beam. Tapered weighting factors include triangular, Hamming, Chebyshev, and Blackman weighting factor. In the example presented here, we choose the triangular weighting function for its simplicity, which is given by,

$$W_i = 1 - \left|\frac{2i}{N+1} - 1\right|, \quad 1 \le i \le N \tag{6.32}$$

where N denotes the number of channels in the ULA. By altering the phase of each element, the main beam's direction can be steered. The radiation patterns for a 50-element ULA with main beam at 0° (boresight) or 30° are shown in Figure 6.16.

The simulation process for the RCS in the previous subsection is only for one transmitter and one receiver. To generate the radar images with MIMO radars, one idea is to simulate the radar signals for all channels (all combination of transmitters and receivers), and then applying different weights and phase (e.g., taking Fourier transform) to the signals from all channels to obtain the radar response from each direction. In one example, a radar with 50-element ULA with 0.7 wavelength spacing is considered in the simulation. The radar response is collected for 512 frequency samples from 76 to 77 GHz, so the corresponding range resolution is 0.15 m. After processing, the range-azimuth angle (r-ϕ) radar images for some traffic targets are shown in Figure 6.17. The main beam is scanning with step of 1° and the radar images are threshold to have 30 dB dynamic range. In Figure 6.17, the images are normalized to their maximum RCS values. It can roughly illustrate the dimension or shape of the targets, and good differentiability is shown in the r-ϕ radar image for targets with different dimensions. But it is hard to distinguish the targets with similar dimensions like pedestrians and traffic signpost.

Each r-ϕ radar image is obtained from simulating 25,600 or 51,200 RCS data (50 or 100 channels with 512 frequencies). Repeat the RCS simulation for more than 25,000 times is rather inefficient. It is noted that ray tracing or shadowing process should be the same for all channels, and therefore the ray tracing algorithm only needs to compute once. Besides, the RCS for different frequencies can be obtained

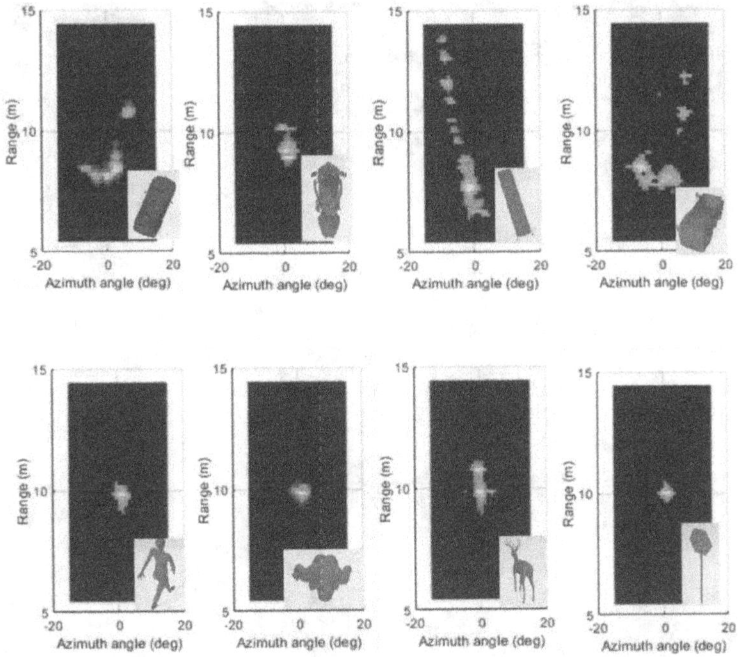

Figure 6.17 *Simulated range-azimuth angle radar images for various traffic targets in the range of 10 m with a 2° beamwidth radar in azimuth direction [25]*

from the wideband PO method described in the previous subsection. For one channel, the scattered fields from one current source is calculated by (6.23), and we can simply add some phase compensation based on the angle of arrival to it to compute the scattered fields for other channels with the far field approximation:

$$E_s^{(j)} = E_s^{(1)} e^{jd_j(\sin\theta_r)}, \quad 2 \leq j \leq N \tag{6.33}$$

where $E_s^{(1)}$ denotes the scattered field at the first array element from one current source, d_j denotes the distance between the jth array element to the first one, and θ_r denotes the angle of arrival from boresight direction. The total scattered fields will be the summation of the scattered fields from all current sources. Moreover, by utilizing parallel computing with graphic processing unit (GPU), the simulation can be done within reasonable time.

Some more advanced radars can scan their main beams in both azimuth and elevation directions, and therefore can produce azimuth-elevation angle (ϕ-θ) radar images and 3D (xyz) radar images. Such radar images can be generated with simulation as well. The simulation process is similar to that for the r-ϕ radar image except for the fact that much more channels are needed. For example, in an MIMO configuration, the radar has 50 transmitters in the elevation direction and 50 receivers in the

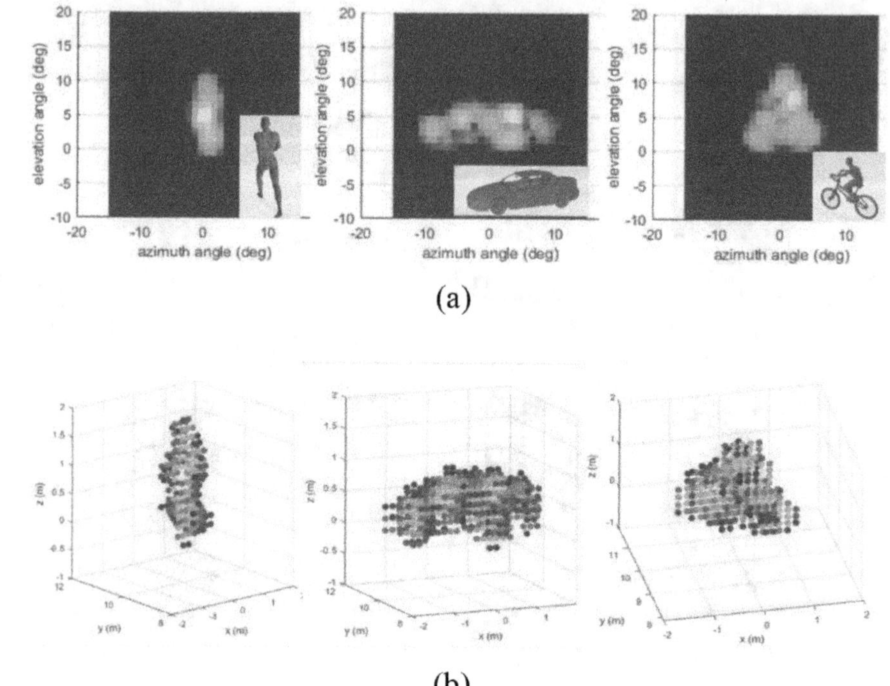

Figure 6.18 The simulated (a) azimuth-elevation angle radar images and (b) 3D radar images for a pedestrian, a sedan, and a man riding a bike in the range of 10 m and with the 2° beam width radar [25]

azimuth direction, and both have 0.7 λ spacing. The resulting beamwidth is about 2° in both elevation and azimuth direction, and there are 2,500 channels in total. Thus, a 3D radar image requires about 1.3 million simulated RCS data (2,500 channels with 512 frequency samples) to create. Some examples of the ϕ-θ radar images and the 3D radar images are shown in Figure 6.18. The scale in the images is normalized to their maximum RCS values, and to avoid the error caused by the sidelobe (26 dB lower than the main beam), the pixel with RCS 26 dB lower than the maximum RCS is not shown. We can clearly observe the profiles of targets and the strong and weak scattering positions from the ϕ-θ radar images and the 3D radar images, which gives us confidence on achieving good performance in target classification.

6.5.2 *Classification of radar images with CNN*

The quality of the radar images depends on both the range of the target and the radar's beamwidth. To study how this will affect the classification performance, the r-ϕ image simulations are performed for targets in different ranges and by varying the number of elements or antenna spacings (resulting in different antenna beamwidths). In the example in [25], simulations are performed for the ranges from 10

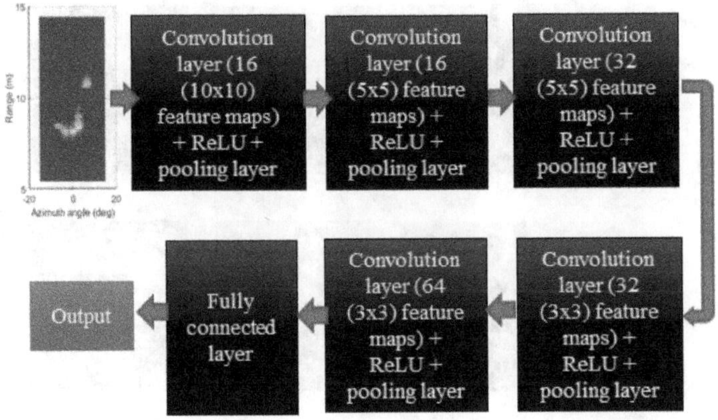

Figure 6.19 The CNN structure used for target classification with rϕ radar images [25]

to 30 m with the boresight beamwidth of 1°, 2°, and 3°. The 1° beamwidth can be obtained by 100 element ULA with 0.7 λ inter-element spacing. In each range and radar configuration, more than 6,000 radar images for about 90 different targets from 72 azimuth directions (5° step) are generated for classification.

CNN is the natural solution in the classification problem based on radar images. After optimization, the CNN used in the *r-ϕ* image has five convolutional layers, and the detailed structure of the CNN is shown in Figure 6.19. The CNN is trained using the MATLAB® Deep Learning toolbox [49] as well. About 70% of the data are randomly selected as training dataset and the rest is used as validation dataset. The classification performance of the models for different ranges and beamwidths are given in Table 6.4. All models have the same CNN structures, and the radar

Table 6.4 Performance of the target classification model based on rϕ RCS images [25]

	True positive rate with 3° beamwidth		
Range	*Pedestrians*	*Vehicles*	*Other Obj.*
10 m	91.9%	99.5%	91.6%
20 m	93.3%	98.5%	87.0%
30 m	90.8%	98.8%	81.5%
	True positive rate with 2° beamwidth		
10 m	97.4%	99.6%	91.7%
20 m	93.8%	98.1%	81.4%
30 m	96.4%	97.8%	80.1%
	True positive rate with 1° beamwidth		
10 m	94.9%	99.1%	93.2%
20 m	94.0%	99.7%	88.4%
30 m	87.0%	96.7%	93.5%

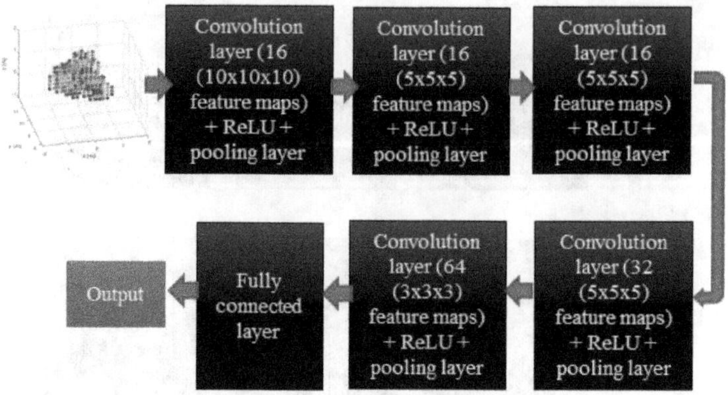

Figure 6.20 The CNN structure used for target classification based on 3D radar images [25]

scanning angle step is 1° for all radars for better comparison. It is shown that the performance is improved for the radar images that obtained from near range or with narrow beamwidth. This result is expected since the quality of radar images is better for targets at closer range or narrower radar beamwidth.

Similarly, the 3D radar images are utilized in the classification problem. The performance of classification is compared for different range values (10, 20, and 30 m) and radar beamwidths (2° and 3°) as well. More than 6000 3D radar images are generated using the simulation procedure described before using the same targets in the r-ϕ radar images classification. The classification models for 3D radar images are also based on CNN. Because the input of the CNN has three dimensions, the feature maps are set to be 3D as well. The proposed CNN structure for 3D radar images is illustrated in Figure 6.20. The CNNs have five convolution, ReLU, and pooling layers, and they are trained using the MATLAB® Deep Learning toolbox as well.

Table 6.5 Performance of the target classification model based on 3D RCS images [25]

	True positive rate with 3° beamwidth		
Range	*Pedestrians*	*Vehicles*	*Other Obj.*
10 m	99.5%	99.5%	99.4%
20 m	99.3%	99.2%	97.8%
30 m	98.0%	98.6%	96.5%
	True positive rate with 2° beamwidth		
10 m	100%	99.9%	99.7%
20 m	100%	99.7%	94.5%
30 m	98.9%	99.0%	97.8%

Table 6.5 shows the classification performance based on 3D radar images. Excellent performance (>98% accuracy for pedestrians and vehicles and >94.5% accuracy for other objects) is achieved for both 3° beamwidth and 2° beamwidth for range values from 10 to 30 m. Since more features can be detected by 3D radar images, it is not surprising that the radar target classification based on 3D radar images outperforms the other models in terms of accuracy. The drawback of the 3D radar images-based target classification is that it requires a more advanced radar and more effort to process the data than other approaches.

Notably, the radar images are generated from simulation with ideal conditions, for example, no noise or clutters are included, and no combined targets cases (e.g., a car very close to a pedestrian) are considered. Besides, additional radar signal processing like targets isolation is needed to produce radar images for single target from measurement data, and such process may introduce more error for the classification. Overall, it is still promising to utilize radar images in the target classification, but keeping in mind the excellent classification performance based on radar image simulation could be overoptimistic.

6.6 Conclusion

As an emerging application of the modern radar technology, the automotive radar target classification is demonstrated in this chapter. This is a cross-disciplinary research that involved ML and radar scene simulation. Some commonly used ML models including MLP, CNN, and RNN (LSTM) are introduced. Besides these classification algorithms, a good classification model requires the differentiable input data as well. Depending on the types of radar and signal processing approaches, the radar data can be represented in many forms. Radar can provide excellent velocity detections, and some examples utilizing the micro-Doppler signature for dynamic target classification are provided. This approach shows good performance for distinguishing pedestrians from vehicles. In a more general scenario regardless the movement of targets, the classification models based on the statistical information of RCS are demonstrated. The models are trained with a large high-fidelity simulation dataset and are validated by measurement. Though the RCS-based classification models can be applied to both dynamic and static scenarios, for the targets in both near and far range, their performance is limited by the small number of features. For radars with imaging capabilities, the 2D or 3D radar images can be generated and utilized in the radar target classification. Excellent performance can be achieved, but the model based on radar images may only be applied very successfully only for targets that are not too far from the radar due to the limited angular resolution of the radar.

It should be pointed out that not all models or radar data for target classification are covered in this chapter. Also, some technical challenges related to the target classification are not discussed, for example, the algorithms to isolate and clean up the radar data for each target are of such method not discussed here. Although not all aspects of radar target classification are covered in this chapter, it is hoped that the reader is able to gain some sense of the available methods for radar target classification with ML.

References

[1] Eykholt K., Evtimov I., Fernandes E, *et al.* 'Robust physical-world attacks on deep learning visual classification'. *Proceedings of the IEEE Conference on Computer Vision and Pattern Recognition*; Salt Lake City, UT, IEEE, 2018. pp. 1625–34.

[2] Kidono K., Miyasaka T., Watanabe A., Naito T., Miura J. 'Pedestrian recognition using high-definition LIDAR'. *IEEE Intelligent Vehicles Symposium (IV)*; Baden-Baden, Germany, IEEE, 2011. pp. 405–10.

[3] Choi J., Ulbrich S., Lichte B., Maurer M. 'Multi-target tracking using a 3D-lidar sensor for autonomous vehicles'. *16th International IEEE Conference on Intelligent Transportation Systems (ITSC 2013)*; The Hague, Netherlands, IEEE, 2013. pp. 881–86.

[4] Van Brummelen J., O'Brien M., Gruyer D., Najjaran H. 'Autonomous vehicle perception: the technology of today and tomorrow'. *Transportation Research Part C: Emerging Technologies*. 2018, vol. 89(4), pp. 384–406.

[5] Bagloee S.A., Tavana M., Asadi M., Oliver T. 'Autonomous vehicles: challenges, opportunities, and future implications for transportation policies'. *Journal of Modern Transportation*. 2016, vol. 24(4), pp. 284–303.

[6] Zyweck A., Bogner R.E. 'Radar target classification of commercial aircraft'. *IEEE Transactions on Aerospace and Electronic Systems*. 1996, vol. 32(2), pp. 598–606.

[7] Kim K.-T., Kim H.T. 'One-Dimensional scattering centre extraction for efficient radar target classification'. *IEE Proceedings - Radar, Sonar and Navigation*. 1999, vol. 146(3), pp. 147–58.

[8] Ferrazzoli P., Guerriero L., Schiavon G. 'Experimental and model investigation on radar classification capability'. *IEEE Transactions on Geoscience and Remote Sensing*. 1999, vol. 37(2), pp. 960–8.

[9] Pierce L.E., Sarabandi K., Ulaby F.T. 'Application of an artificial neural network in canopy scattering inversion'. *International Journal of Remote Sensing*. 1994, vol. 15(16), pp. 3263–70.

[10] Ding J., Chen B., Liu H., Huang M. 'Convolutional neural network with data augmentation for SAR target recognition'. *IEEE Geosci Remote Sens Lett*. 2016, vol. 13(3), pp. 364–68. IEEE.

[11] Chen S., Wang H. 'SAR target recognition based on deep learning'. *International Conference on Data Science and Advanced Analytics (DSAA)*; Shanghai, China, IEEE, 2014. pp. 541–47.

[12] Heuel S., Rohling H. 'Pedestrian classification in automotive radar systems'. *13th International Radar Symposium*; Warsaw, Poland, IEEE, 2012. pp. 39–44.

[13] Heuel S., Rohling H. 'Two-stage pedestrian classification in automotive radar systems'. *12th International Radar Symposium (IRS)*; Leipzig, Germany, 2011. pp. 477–84.

[14] Bartsch A., Fitzek F., Rasshofer R.H. 'Pedestrian recognition using automotive radar sensors'. *Advances in Radio Science*. 2012, vol. 10, pp. 45–55.

[15] Vahidpour M., Sarabandi K. 'Millimeter-wave doppler spectrum of walking bodies'. *Transactions on Geoscience and Remote Sensing*. 2012, vol. 50(7), pp. 2866–79.

[16] Lee S., Yoon Y., Lee J., Kim S. 'Human–vehicle classification using feature-based SVM in 77-GHz automotive FMCW radar'. *IET Radar Sonar & Navigation*. 2017, vol. 10(10), pp. 1589–96.

[17] Angelov A., Robertson A., Murray-Smith R., Fioranelli F. 'Practical classification of different moving targets using automotive radar and deep neural networks'. *IET Radar, Sonar & Navigation*. 2018, vol. 12(10), pp. 1082–9.

[18] Krizhevsky A., Sutskever I., Hinton G. 'Imagenet classification with deep convolutional neural networks'. *Proceedings of the Advances in Neural Information Processing Systems (NIPS)*. 2012, pp. 1097–115.

[19] Long J., Shelhamer E., Darrell T. 'Fully convolutional networks for semantic segmentation fully convolutional networks for semantic segmentation'. *Proceedings of the IEEE Computer Society Conference on Computer Vision and Pattern Recognition*; 2015.

[20] Girshick R., Donahue J., Darrell T., Malik J. 'Rich feature hierarchies for accurate object detection and semantic segmentation'. *IEEE Conference on Computer Vision and Pattern Recognition*; Columbus, OH, IEEE, 2014.

[21] Kim S., Lee S., Doo S., Shim B. 'Moving target classification in automotive radar systems using Convolutional recurrent neural networks'. *26th European Signal Processing Conference (EUSIPCO)*; Rome; 2018. pp. 1482–6.

[22] Lim S., Lee S., Yoon J., Kim S. 'Phase-based target classification using neural network in automotive radar systems'. *IEEE Radar Conference (RadarConf)*; Boston, MA, IEEE, 2019. pp. 1–6.

[23] Lombacher J., Laudt K., Hahn M., Dickmann J., Wohler C. 'Semantic radar grids'. *IEEE Intelligent Vehicles Symposium (IV)*; Los Angeles, CA, IEEE, 2017. pp. 1170–75.

[24] Schumann O., Hahn M., Dickmann J., Wohler C. 'Semantic segmentation on radar point clouds'. *21st International Conference on Information FUSION (FUSION)*; Cambridge, 2018. pp. 2179–86.

[25] Cai X., Giallorenzo M., Sarabandi K. 'Machine learning-based target classification for MMW radar in autonomous driving'. *IEEE Transactions on Intelligent Vehicles*. 2021, vol. 6(4), pp. 678–89.

[26] Cai X., Sarabandi K. 'A machine learning based 77 GHz radar target classification for autonomous vehicles'. *IEEE International Symposium on Antennas and Propagation and USNC-URSI Radio Science Meeting*; Atlanta, GA, USA; 2019. pp. 371–2.

[27] Goodfellow I., Bengio Y., Courville A. *Deep learning*. Cambridge, MA: MIT Press; 2016.

[28] DeepMind, *DeepMind AlphaGo [online]*. Available from https://deepmind.com/research/case-studies/alphago-the-story-so-far.

[29] Gurbuz S. *Deep neural network design for radar applications*. Institution of Engineering and Technology, 2020.

[30] Russell S., Norvig P. 'Artificial intelligence: a modern approach'. *Series in Artificial Intelligence*. Englewood Cliffs, NJ: Prentice-Hall; 1995.

[31] Nielsen M. *Neural networks and deep learning*. Vol. 25. Determination Press; 2015.

[32] Gers F.A., Schmidhuber J., Cummins F. 'Learning to forget: continual prediction with LSTM'. *Neural Computation*. 2000, vol. 12(10), pp. 2451–71.

[33] Patole S.M., Torlak M., Wang D., Ali M. 'Automotive radars: a review of signal processing techniques'. *IEEE Signal Processing Magazine*. 2017, vol. 34(2), pp. 22–35.

[34] Chen V.C., FayinL., Shen-ShyangH., Wechsler H., Li F. 'Micro-doppler effect in radar: phenomenon, model, and simulation study'. *IEEE Transactions on Aerospace and Electronic Systems*. 2006, vol. 42(1), pp. 2–21.

[35] Chen V. *The micro-doppler effect in radar*. second edition. Norwood, MA: Artech; 2019.

[36] Smith G.E., Woodbridge K., Baker C.J. 'Radar micro-doppler signature classification using dynamic time Warping'. *IEEE Transactions on Aerospace and Electronic Systems*. 2010, vol. 46(3), pp. 1078–96.

[37] Kim Y., Moon T. 'Human detection and activity classification based on micro-Doppler signatures using deep convolutional neural networks'. *IEEE Geoscience and Remote Sensing Letters*. 2016, vol. 13(1), pp. 8–12.

[38] Cao P., Xia W., Ye M., Zhang J., Zhou J. 'Radar-ID: human identification based on radar micro-Doppler signatures using deep convolutional neural networks'. *IET Radar, Sonar & Navigation*. 2018, vol. 12(7), pp. 729–34.

[39] Kellner D., Barjenbruch M., Klappstein J., Dickmann J., Dietmayer K. 'Wheel extraction based on micro Doppler distribution using high-resolution radar'. *IEEE MTT-S International Conference on Microwaves for Intelligent Mobility (ICMIM)*; 2015. pp. 1–4.

[40] Belgiovane D., Chen C.-C. 'Micro-doppler characteristics of pedestrians and bicycles for automotive radar sensors at 77 ghz'. *11th European Conference on Antennas and Propagation (EUCAP)*; IEEE, 2017. pp. 2912–16.

[41] Ulaby F.T., Long D.G., Blackwell W.J, *et al. Microwave Radar and Radiometric Remote Sensing*. Vol. 5 (4). Ann Arbor, MI: University of Michigan Press; 2014.

[42] Obelleiro F., Araújo M.G., Rodríguez J.L. 'Iterative physical-optics formulation for analyzing large waveguides with lossy walls'. *Microwave and Optical Technology Letters*. 2001, vol. 28(1), pp. 21–6.

[43] Gordon W. 'Far-Field approximations to the Kirchoff-Helmholtz representations of scattered fields'. *IEEE Transactions on Antennas and Propagation*. 1975, vol. 23(4), pp. 590–2.

[44] Schipper T., Fortuny-Guasch J., Tarchi D., Reichardt L., Zwick T. 'RCS measurement results for automotive related objects at 23-27 ghz'. *Proceedings*

of the 5th European Conference on Antennas and Propagation (EUCAP); IEEE, 2011. pp. 683–86.

[45] Swerling P. 'Probability of detection for fluctuating targets'. ASTIA Document Number AD 80638, 1954.

[46] Shnidman D.A. 'Expanded swerling target models'. *IEEE Transactions on Aerospace and Electronic Systems*. 2003, vol. 39(3), pp. 1059–69.

[47] Alaqeel A.A., Ibrahim A.A., Nashashibi A.Y., Shaman H.N., Sarabandi K. 'Experimental characterization of multi-polarization radar backscatter response of vehicles at j-band'. *IEEE Transactions on Intelligent Transportation Systems*. 2019, vol. 20(9), pp. 3337–50.

[48] Massey F.J. 'The kolmogorov-smirnov test for goodness of fit'. *Journal of the American Statistical Association*. 1951, vol. 46(253), pp. 68–78.

[49] *The Mathworks, MATLAB Statistics and Machine Learning Toolbox. Natick, Mi*. Available from https://www.mathworks.com/products/statistics.html

Chapter 7

Road condition recognition with radar

Xiuzhang Cai[1] *and Kamal Sarabandi*[1]

7.1 Introduction

Weather-related road conditions severely impact the driving safety of vehicles. According to Federal Highway Administration, in the United States, each year more than 1,300 people are killed and more than 116,800 people are injured due to icy, slushy, or snowy pavement [1] and nearly 5,700 people are killed and more than 544,700 people are injured in crashes on the wet pavement [2]. Many studies show that snowy or wet pavement can largely increase the crash probability [3, 4], and the cost of weather-related crashes is about $22–$51 billion each year [5].

Any controllable motion of vehicles is dependent on the friction between the tires and the road. Wet, snowy, or icy conditions can largely reduce the tire–road friction coefficient. Typically, the friction coefficient will drop by 20–30% with wet pavement [6], by more than 75% on the snowy road and by more than 90% on the icy road [7]. Many advanced vehicle safety systems including anti-lock braking system (ABS), autonomous emergency braking system, and autonomous driving rely on the correct estimation of tire–road friction [8–10]; therefore, it is also important to detect the road conditions with vehicle's sensors to enable the best functionality of those advanced systems.

The detection of road conditions can be roughly divided into two categories: effect-based methods and cause-based methods [11]. The effect-based methods utilize sensors such as strain gauge, accelerometer, ABS wheel speed sensor, pressure sensor or microphone to measure tire deformation, rim deformation, wheel speed vibrations, tire noise or slip to estimate the road conditions, and the tire–road friction coefficient [12–18]. Those methods involve many sophisticated sensors and models, and good accuracy can be achieved. However, the effect-based methods usually require excitations (acceleration or brake) to trigger the measurable effects (deformation or slip), and they can only detect the road conditions that the vehicle is driving on.

[1]Department of Electrical Engineering and Computer Science, University of Michigan, Ann Arbor, MI, USA

In order to detect the road conditions in front of the vehicle, other sensors including optical sensors, LiDAR, acoustic sensors, cameras, and radar can be used. Those sensors utilize light, microwave, or acoustic wave to detect the wet, snowy, or icy road conditions directly, which are the cause of the friction drop of tires, so these approaches are also called cause-based methods. In one study, the infrared sensor is used to discriminate the dry, wet, icy, and snowy road surfaces [19]. The incident angle is about 45°, so it can only detect the road less than 1 m ahead of the vehicle. Another study uses LiDAR to detect different road conditions including dry, wet, slushy, and full snow coverage up to 40 m in range [20]. It measures the intensity of laser points falling on the road and applies machine learning techniques to classify the data from different road conditions. Since there is large overlap between the intensity for different road conditions, the overall accuracy of the LiDAR-based classification is about 67%. Intuitively, camera images can provide good differentiation for different road conditions just like human eyes and have been successfully applied in road condition recognition by many studies [21–24]. Excellent performance has been reported; however, the accuracy of models will be largely reduced with poor visibility, for example, in the night [25]. Besides LiDAR and camera, radar is also a commonly used sensor in the vehicles with advanced safety features, and compared to the other two types of sensors, radar is less affected by the weather or visibility condition. It is of great importance and interest to use radar to detect the road condition as well.

Several studies on the radar response of various road conditions at 24 [26], 77 [27, 28], and 94 GHz [29] are reported, and some promising results are shown in identifying road conditions with the radar data. Applications of radar for road surface recognition was first presented in [30]. To best understand and utilize the radar data, it will be good to investigate the road surface scattering phenomenon of radar signals. The process of mmWave radar signals scattering from roads under different conditions can be considered as a random rough surface and/or random media scattering problem. Depending on the road conditions, different scattering mechanisms such as surface or volumetric scattering may need to be considered. For example, for a dry road surface, the permittivity of the road surface is relatively small and with low loss, and therefore the volumetric scattering inside the road will be the major wave backscattering contribution; on the other hand, if the road surface is wet, whose permittivity is large and with high loss, the surface scattering may dominate the total wave scattering.

This chapter will be organized as follows: in Section 7.2, the ground truth information of the road surfaces used in our experimental investigation including the roughness, material properties of road, and the snow and ice in mmWave band will be discussed. A surface scattering model at 77 GHz based on full-wave analysis in conjunction with Monte-Carlo simulation will be presented in Section 7.3. In Section 7.4, a semi-empirical volumetric scattering model based on radiative transfer theory is covered. In Section 7.5, the radar measurement data for the road surface with various road conditions including wet, icy, and snowy will be demonstrated. Finally, Section 7.6 concludes this chapter.

7.2 Ground truth of road surfaces and radar metrics

7.2.1 Statistics of rough road surfaces

There are many types of roads with different materials, and some commonly seen types are asphalt, concrete, and dirt roads. Different types of roads may have different roughness, ingredients, and material properties. The pictures of some examples of asphalt, concrete, and dirt roads are shown in Figure 7.1. Intuitively, the dirt road has the roughest surface, and the asphalt road is rougher than the concrete road. Mathematically, the roughness of the surface can be evaluated by some statistical features such as the root-mean-square (RMS) height, autocorrelation function, and the correlation length [31–33]. The RMS height is defined as

$$s = \sqrt{\frac{1}{A}\iint_A \left(z(x,y) - \bar{z}\right)^2 dxdy} \tag{7.1}$$

where $z(x,y)$ denotes the measured height on the surface at a horizontal position (x,y), A is the measured area, dx, dy are the sampling space in x and y direction, and $\bar{z}$ is the average height of the surface.

Figure 7.1 The pictures for the (a) asphalt, (b) concrete, and (c) dirt roads

Besides the RMS height, autocorrelation function is often used in describing the roughness. It shows how rapidly the height of surface changes with positions and is defined as [33]:

$$C\left(x-x', y-y'\right) = \frac{1}{s^2}\left\langle z\left(x,y\right), z\left(x',y'\right)\right\rangle \tag{7.2}$$

where (x', y') are close to (x,y), and $\langle\cdot\rangle$ is the ensemble average between the two sets of data. The autocorrelation function is a function of Δx and Δy. Generally, the faster the correlation drops with Δx or Δy, the rougher the surface is. The autocorrelation function for rough surfaces is usually approximated by exponential and Gaussian autocorrelation functions [32]. The mathematical form for exponential autocorrelation function is given by [33]:

$$C\left(x-x', y-y'\right) = \exp\left(-\sqrt{\left[\left(x-x'\right)^2+\left(y-y'\right)^2\right]}/l\right) \tag{7.3}$$

where l denotes the correlation length, which means the distance between two points on the rough surface such that the correlation between them will drop to e^{-1}. Similarly, the Gaussian autocorrelation function is:

$$C\left(x-x', y-y'\right) = \exp\left(-\left[\left(x-x'\right)^2+\left(y-y'\right)^2\right]/l^2\right) \tag{7.4}$$

The rough surface with exponential autocorrelation function will be rougher than that with Gaussian correlation function even with the same RMS height and correlation length. To find out which autocorrelation function is more closely related to the road surface, we take some measurements on both asphalt and concrete road surfaces with a laser profilometer [34]. The laser profilometer can precisely measure the relative height of the road surface $z\left(x,y\right)$ with a resolution of 0.01 mm. It is attached to a linear stage to scan the road surface in one direction. Figure 7.2 depicts some examples of the measured height for concrete, weathered asphalt, and new asphalt road surfaces. The means of the height of all three surfaces are set to 0. It shows that the profiles of all three types of road surfaces are randomized and have different variances. By measuring more than 50 samples, the statistical features of the rough surfaces including the RMS height and autocorrelation function can be obtained. Figure 7.3 shows the comparison between the autocorrelation function generated from the measured data and the exponential/Gaussian autocorrelation function for the concrete and asphalt road surface. It shows that the exponential autocorrelation function provides a better fit to the empirical data.

The mean value and variance of the RMS height and correlation lengths for all measured samples with different types of road surfaces are summarized in Table 7.1. It shows that the concrete surface has a much smaller RMS height and a much larger correlation length than the asphalt surface, which means it is much smoother than the asphalt surface. The variance of the RMS height and correlation lengths indicates the homogeneity of the roughness across different samples. The concrete surface has a relatively low variance, which means that its roughness is more homogeneous. The variance for new asphalt is smaller than that for weathered asphalt, which is also a sign that the new asphalt is more homogeneous than the weathered asphalt as we

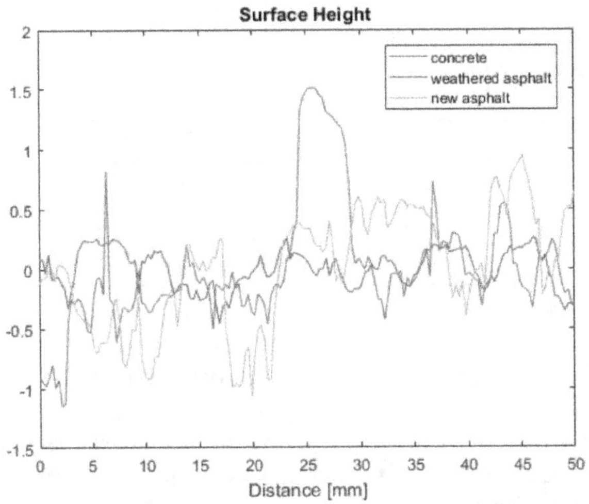

Figure 7.2 *The height of surface (in mm) as a function of scanning distance for different types of road surfaces*

expect. The RMS height s and correlation length l are reported by other studies as well. In Reference [35], the measured s for smooth concrete, smooth asphalt, and rough asphalt is 0.2, 0.34, and 0.9 mm, respectively; and the measured l for smooth concrete, smooth asphalt, and rough asphalt is 38.6, 4.2, and 6.3 mm.

7.2.2 Effective permittivity of asphalt and concrete roads

Both asphalt and concrete are composite materials with various ingredients. Asphalt contains aggregate particles (pebbles, rocks, and sand contribute to 85–90% vol), bitumen (about 5–7% vol), air (5–8% vol), and some other additives or fines [36].

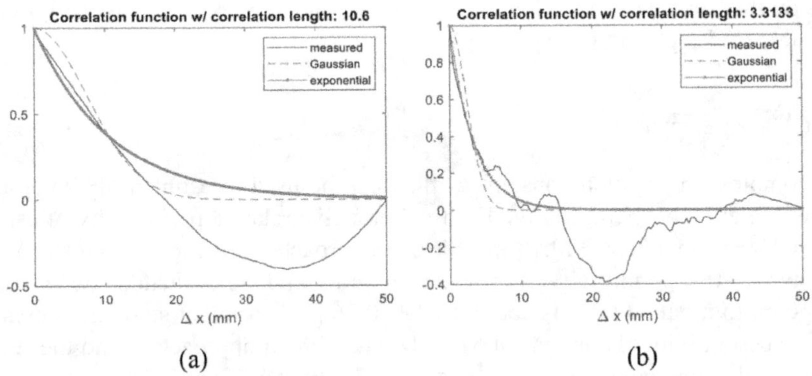

Figure 7.3 *The autocorrelation as a function of distance for (a) a measured sample of the concrete surface and (b) a measured sample of the asphalt surface [34]*

Table 7.1 RMS height and correlation length for measured concrete and asphalt surfaces [34]

Type	$\bar{s}$ **(mm)**	$\sigma(s)$ **(mm)**	$\bar{l}$ **(mm)**	$\sigma(l)$ **(mm)**
Weathered asphalt	1.24	1.17	2.71	2.48
New asphalt	0.81	0.49	1.80	1.66
Concrete	0.20	0.08	13.2	4.51

Concrete is mainly composed of cement, sand, rocks, and air [35]. In radar applications, one of the most important material properties is permittivity. Different ingredients may have different permittivity; however, since the shapes of ingredients are randomized and the particle sizes of ingredients usually are comparable to or larger than the wavelength for mmWave radar, the overall permittivity of asphalt or concrete cannot be directly computed by the permittivity of different ingredients. On the other hand, people can directly measure the EM response of asphalt or concrete mixtures to estimate the effective permittivity. Chang [37] utilizes a dielectric probe (Agilent 85070E) connected with a network analyzer to measure the permittivity of 69 asphalt samples with frequencies from 200 MHz to 13 GHz. The reported effective relative permittivity of asphalt in the X band is around $4 + i0.5$. In the mmWave band, the reflectivity of several polished asphalt samples at normal incidence is measured [35], and the real part of the effective relative permittivity is given by:

$$\varepsilon_e' = \left(\frac{1+\Gamma}{1-\Gamma}\right)^2 \tag{7.5}$$

where Γ is the reflection coefficient of a bulk asphalt sample. The imaginary part can be evaluated by measuring the reflectivity of a thin asphalt layer with a metal plate on the back. The reflection coefficient of a metal-backed asphalt sample with thickness of d is given by [35]:

$$\Gamma_m = \frac{\Gamma - e^{2ik_a d}}{1 - \Gamma e^{2ik_a d}} \tag{7.6}$$

The complex propagation constant k_a then can be used to estimate the imaginary part of the effective permittivity. The final effective relative permittivity of asphalt at 94 GHz is about $3.18 + i0.1$ [35]. Similar approaches are applied to estimate the effective relative permittivity of concrete mixtures, and the evaluated effective relative permittivity at 94 GHz is about $4.17 + i0.36$ [29]. Another study measures the relative permittivity of concrete at 51 GHz with similar approaches, and the results show that the relative permittivity of concrete is about $6.7 + i0.7$ [38].

In conclusion, the relative permittivity of the asphalt in the mmWave band is about $(3.1 \text{ to } 4) + i(0.1 \text{ to } 0.5)$, and that of the concrete is around $(4.1 \text{ to } 6.7) + i(0.3 \text{ to } 0.7)$.

7.2.3 Effective permittivity of water, ice, and snow

Under the wet, icy, or snowy road conditions, water, ice, or snow is involved and therefore their permittivity needs to be studied. The permittivity of pure water as a function of frequency below 100 GHz can be described by the Debye relaxation model [39]:

$$\varepsilon_w = \varepsilon_{w\infty} + \frac{\varepsilon_{w0} - \varepsilon_{w\infty}}{1 + j2\pi f \tau_w} \tag{7.7}$$

where f is the frequency in Hz, ε_{w0} denotes the static dielectric constant where $f = 0$, and $\varepsilon_{w\infty}$ denotes the high-frequency dielectric constant when $f \to \infty$, τ_w is the relaxation time constant in s. ε_{w0}, $\varepsilon_{w\infty}$, and τ_w are temperature-dependent, and Liebe [40] developed an empirical model to fit those parameters as a function of temperature. In the automotive radar frequency band (around 77 GHz), the relative permittivity of water in room temperature (20°C) is $8.84 + i15.17$.

Dry snow is composed of ice flakes and air. Because the volume fraction of ice flakes is usually low and the ice flake can be modeled as an ice sphere, the relative permittivity of dry snow can be calculated by the Polder–Van Santen mixing formula for spherical inclusions [35, 41]:

$$(1 - v)\frac{\varepsilon_{air} - \varepsilon_{snow}}{\varepsilon_{air} + 2\varepsilon_{snow}} + v\frac{\varepsilon_{ice} - \varepsilon_{snow}}{\varepsilon_{ice} + 2\varepsilon_{snow}} = 0 \tag{7.8}$$

where v denotes the volume fraction of ice. The relative permittivity of ice in the mmWave band has been measured and modeled in Reference [42]. The measured real part of ice's relative permittivity is $\varepsilon'_{ice} = 3.15$, and the measured imaginary part is temperature-dependent, and $\varepsilon''_{ice} = 0.002 \sim 0.01$ for the temperature ranging from −28°C to −3°C. Therefore, for an ice volume fraction of 0.0742, the effective relative permittivity of the snow is about $\varepsilon_{snow} = 1.11 + i0.00074$ [35].

7.2.4 Scattering coefficient

The road surface scattering problem can be considered as a random rough surface or random media scattering problem. The monostatic automotive radar usually illuminates the road at near grazing angles and measures the scattered fields in the backscattering direction. As the surface of the road, the shapes, sizes, and positions of the ingredients of the road are usually randomized, the scattered fields in the backscattering direction are stochastic as well. Therefore, the scattered power in the backscattering direction is usually added incoherently, which means the total scattered power from a large area is roughly the sum of the scattered power from all portions of it.

A very useful metric for the scattering ability of the road surface is the scattering coefficient. The scattering coefficient is defined as RCS per unit area [39]:

$$\sigma^0_{pq} = \frac{\langle \sigma_{pq} \rangle}{A} \tag{7.9}$$

where p and q are polarizations for scattered and incident wave, usually is specified as horizontal or vertical polarization, $\langle \sigma_{pq} \rangle$ denotes the ensemble-average operation over many measurements of the radar cross-section (RCS) of the distributed target, and A denotes the illuminated area of the distributed target. Similar to the RCS, the scattering coefficient is an attribute of the target, and it may change with the frequency of radar signals but is independent of the types of radar.

7.3 Surface scattering model based on full-wave simulation

Intuitively, the scattering coefficients of the random rough surface depend on the roughness of the surface, the dielectric constant of the material, and the incident angle. One advantage of numerical models or simulations is that in the simulations those parameters can be easily tuned accordingly and then the relations between the scattering coefficients and those parameters can be studied quantitatively.

In the random rough surface wave scattering problem, many electromagnetic models have been developed and applied such as the small perturbation method [43, 44], Kirchhoff approximation [45, 46], and full-wave solutions [34, 47–49]. The small perturbation method is usually applied to the low-frequency region where the RMS height and correlation length of the rough surface are much smaller than the wavelength. On the other hand, the Kirchhoff approximation is a high-frequency approximation that is only applicable to the rough surface whose RMS height is much larger than the wavelength. In our automotive radar applications, the previous section has shown that both the RMS height and correlation length are comparable to the wavelength; therefore, neither of the two aforementioned methods can be applied here.

Full-wave numerical methods can be applied to this random rough surface scattering problem as they do not take approximation for Maxwell's equations and the error purely comes from the numerical error and the geometric error of the rough surface. One of the major issues in full-wave methods for rough surface scattering problems is the edge effect. The rough surfaces in reality usually do not have edges, but in simulation, the rough surfaces have limited dimensions due to the limitation of computing ability, and the edges of the rough surfaces under simulation may introduce additional discontinuity for the current and cause large error [47].

One approach to eliminate the edge effect is to taper the incident wave such that the incident fields are nearly zero around the edges of the rough surfaces [47, 48]. However, if the incident angle is near grazing (>80°), it requires hundreds of wavelengths for the dimension of the rough surface to obtain accurate results [47], which is computationally very expensive and impractical. Another popular solution is to use the periodic boundary condition (PBC) with plane wave incidence [34, 49]. Theoretically, the PBC can work with any incident angle with good accuracy. Its drawback is that it can only generate the scattered fields for a few discretized scattering directions depending on the Bragg modes. Thereby, the dimensions of

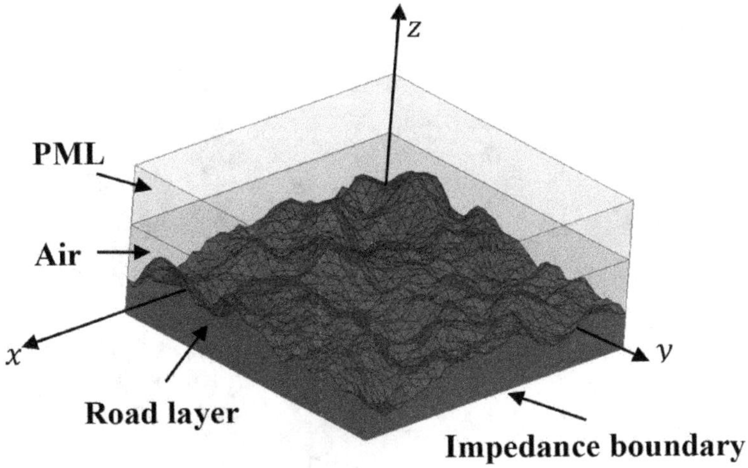

Figure 7.4 *The Simulated setup for a random rough surface in HFSS [33]*

the surface should be carefully chosen to ensure backscattering direction is one of the discretized scattering directions.

7.3.1 Full-wave simulation for the random rough surfaces

We choose the full-wave simulation software Ansys' High-Frequency Structure Simulator (HFSS) with PBC to compute the near-field scattered fields from random rough surfaces. HFSS is based on finite element method, which is proved to have good accuracy and has been widely used in electromagnetics simulation such as antennas. An example of the simulation setup is shown in Figure 7.4. The top layer is the perfect matched layer (PML) that prevents scattered fields reflecting back from the top boundary of the simulation objects. Please note some other applications such as antenna simulations usually use the radiation boundary condition instead of PML to absorb the scattered fields as radiation boundary requires less computational resources. However, the radiation boundary condition cannot be applied jointly with the PBC in HFSS. The middle layer is the air layer, and the bottom layer is the road layer with random rough surface. The random rough surface is generated with exponential autocorrelation function with RMS height $s = 0.8$ mm and correlation length $l = 2$ mm. The bottom of the simulation model is assigned as impedance boundary, which is carefully calculated to mimic infinitely thick road layer. PBC is applied on the four sides of the box under simulation. Notably, in order to apply PBC, the boundary of the road layer should be periodic as well. It means the height of the opposite boundaries of the rough surface should be identical,

$$z(x_0, y_0) = z(x_0 + nL_x, y_0 + nL_y) \tag{7.10}$$

where (x_0, y_0) are the positions on the boundary of the rough surface; L_x and L_y are the dimensions of the rough surface in x and y direction, $n = 0$ or 1.

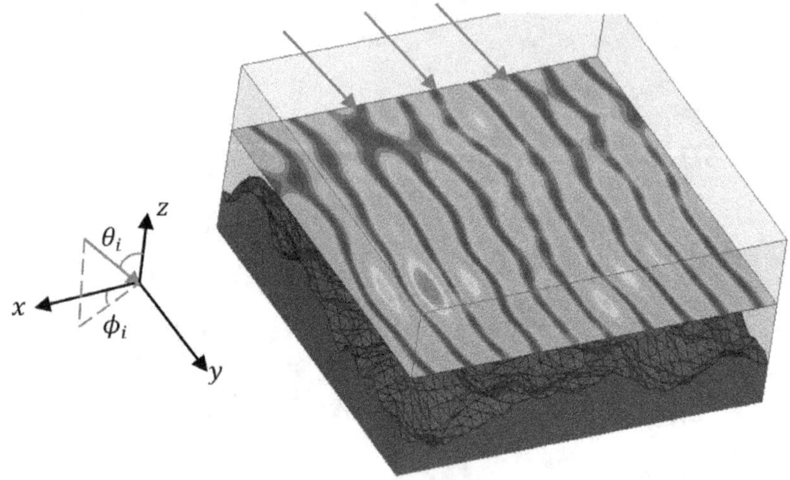

Figure 7.5 The simulated tangential scattered E fields on the top of air layer of the random rough surface model in HFSS [34]

Due to the available computational abilities, the dimensions of the entire model in Figure 7.4 are set to be around $5 \times 5 \times 2.5\lambda$ ($2\times 2 \times 1$ cm), and the road layer is assumed as homogeneous medium. In the simulation, the random rough surface is illuminated by plane wave with incident angle of $\theta_i = 70°$ and $\phi_i = 0°$. The frequency in the simulation is 77 GHz, which is the most commonly used automotive radar's frequency band. The real part of the simulated tangential scattered E fields on the top of air layer is depicted in Figure 7.5. It shows that the wavefront of the scattered E fields is distorted by the random rough surfaces.

In order to obtain the scattered fields as a function of scattering angles, the near-field and far-field transformations can be applied [50]. Since the simulated E fields are in $x - y$ plane, denote the 2D inverse Fourier transform of the tangential scattered E fields $\boldsymbol{E}_{st}$ as $\boldsymbol{A}_s$:

$$\boldsymbol{A}_s\left(k_x, k_y, z_0\right) = e^{-ik_z z_0} \iint_{-\infty}^{\infty} \boldsymbol{E}_{st}\left(x, y, z_0\right) e^{-ik_x x - ik_y y} dxdy, \tag{7.11}$$

where k_x and k_y are the x and y components of the propagation constant of scattering wave, and it is determined by the scattering direction; z_0 is the height where the scattered E fields are collected. The electrical far fields then can be computed from $\boldsymbol{A}_s$ by [50]:

$$E_\theta\left(\theta, \phi, r\right) = ik_0 \frac{e^{ik_0 r}}{2\pi r}\left(A_x \cos\phi_s + A_y \sin\phi_s\right) \tag{7.12}$$

$$E_\phi\left(\theta, \phi, r\right) = ik_0 \frac{e^{ik_0 r}}{2\pi r} \cos\theta_s\left(A_y \cos\phi_s - A_x \sin\phi_s\right) \tag{7.13}$$

where θ_s and ϕ_s are the elevation and azimuth direction of observation point. A_x and A_y are the tangential component of $\boldsymbol{A}_s$. Because of the PBC, the scattered E fields from the rough surface at $(x+nL_x, y+mL_y, z)$ can be derived from $\boldsymbol{E}_s(x,y,z)$ by

$$\boldsymbol{E}_s(x+nL_x, y+mL_y, z) = \boldsymbol{E}_s(x,y,z)\, e^{i(nL_xk_{ix}+mL_yk_{iy})} \tag{7.14}$$

Then (7.11) can be rewritten as

$$\boldsymbol{A}_s(k_x,k_y,z_0) = e^{-ik_zz_0}\lim_{M,N\to\infty}\sum_{n=-N}^{N} e^{inL_x(k_{ix}-k_x)}\sum_{m=-M}^{M} e^{imL_y(k_{iy}-k_y)}\times \int_0^{L_x}\int_0^{L_y}\boldsymbol{E}_{st}(x,y,z_0)\, e^{ik_xx+ik_yy}dxdy \tag{7.15}$$

The summation term in (7.15) yields a sinc function if N or M is finite, and it becomes a δ function when N or $M\to\infty$:

$$\sum_{n=-N}^{N} e^{inL_x(k_{ix}-k_x)} = \begin{cases} 1+2N, & \text{if } k_x = k_{ix} + \frac{2\pi n}{L_x},\ n\in\mathbb{Z} \\ 0, & otherwise \end{cases} \tag{7.16}$$

Therefore, with the PBC, the scattered far fields can only have accurate solutions at discretized directions, which is also known as the Bragg angles. The specular reflection direction ($k_x = k_{ix}$, and $k_y = k_{iy}$) is always one of the Bragg angles no matter what values for L_x and L_y are chosen. In the radar application, however, in order to ensure the backscattering direction is one of the Bragg angles, the dimensions of the random rough surface have to satisfy the following condition:

$$L_x = \frac{\pi n}{k_{ix}} = \frac{\lambda n}{2\sin\theta_i\cos\phi_i}, n\in\mathbb{Z},\ L_y = \frac{\pi m}{k_{iy}} = \frac{\lambda m}{2\sin\theta_i\sin\phi_i}, m\in\mathbb{Z} \tag{7.17}$$

The integral of one cell in (7.15) can be evaluated numerically from the simulated scattered E fields from HFSS, and we can denote it as $\boldsymbol{A}_{t0}$:

$$\boldsymbol{A}_{t0} = e^{-ik_zz_0}\int_0^{L_x}\int_0^{L_y}\boldsymbol{E}_{st}(x,y,z_0)\, e^{ik_xx+ik_yy}dxdy \tag{7.18}$$

Substitute $\boldsymbol{A}_{t0}$ into (7.12) and (7.13), we can get the scattered far fields of one cell $\boldsymbol{E}^s$, and then the scattering coefficient can be derived from:

$$\langle\sigma_{pq}^0\rangle = lim_{r\to\infty}\frac{4\pi r^2}{L_xL_y}\frac{|\hat{p}\cdot\boldsymbol{E}^S|^2}{|\hat{q}\cdot\boldsymbol{E}^i|^2} \tag{7.19}$$

One example of the scattering coefficient of one random rough surface is shown in Figure 7.6. It shows the full polarimetric scattering coefficients as functions of Bragg angles for a rough surface illuminated by a plane wave with $\theta_i = 85°$. The dimension of the rough surface is about $5\times 5\lambda$, and its material is set as asphalt with dielectric constant of $3.18 + i0.1$. In the figure, V means the vertical polarization or TM mode, and H denotes the horizontal polarization or TE mode. HV refers the case where transmitting signal is V polarization, while the receiver is H polarization. It can be seen that the scattering coefficients for HV and VH polarization are almost

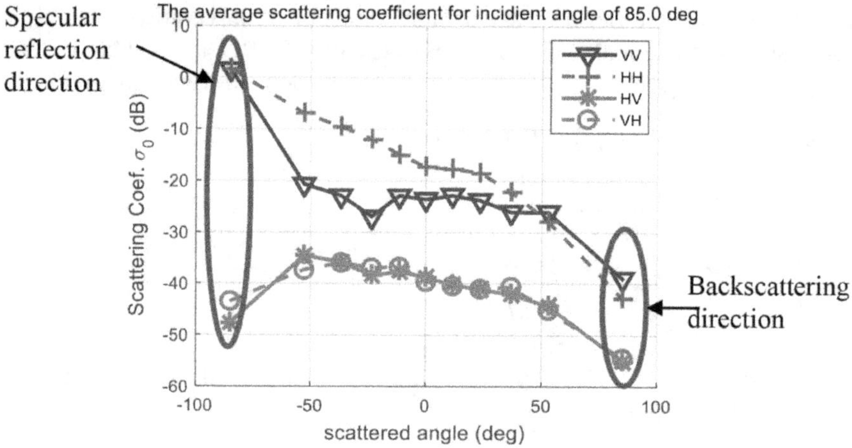

Figure 7.6 The scattering coefficients calculated from the simulated tangential scattered E fields as functions of Bragg angles for a rough surface with incident angle of 85°

identical for all Bragg angles, which follows the reciprocity of EM wave and indicates that the simulation achieves good accuracy.

7.3.2 Convergence analysis and statistics of the simulated results

Because of the randomness of the rough surface, the calculated scattering power from one sample also behaves randomly. Monte Carlo simulations are required to obtain the accurate scattering coefficient. To determine how many samples are needed for the scattering coefficients to reach convergence, 80 different random rough surfaces with almost the same statistical roughness ($s = 0.8$ mm, $l = 2$ mm) are generated and simulated. The rough surfaces are illuminated by the EM wave from 70°. The average backscattering scattering coefficients as functions of the number of samples are illustrated in Figure 7.7. It shows that the scattering coefficients for all polarizations will converge after 30 samples of simulation. Therefore, in the rest discussions, we choose to simulate about 40 realizations for the road surfaces with the same statistical roughness.

Theoretically, the larger of the dimensions of the rough surface, the better accuracy the simulation can achieve. However, the simulation time and memory usage will increase dramatically with the dimensions of the rough surface as well in the full-wave simulation. To find the optimal dimensions for the rough surface simulation, the rough surfaces with different dimensions but the same statistical roughness are generated and used in the backscattering simulation. Table 7.2 shows the simulation results and computational cost comparison for the rough surfaces with the same statistical roughness ($s = 0.5$ mm, $l = 1.2$ mm) and the same incident angle ($\theta_i = 70°$) but different dimensions from about 21 mm (5λ) to 31 mm (8λ). About 40 realizations are simulated for all three cases, and the data shown below

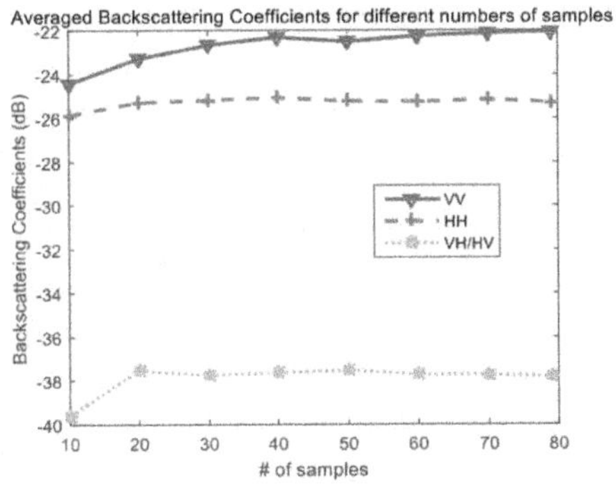

Figure 7.7 *The average backscattering coefficients as a function of the number of realizations of random rough surfaces with the same statistical features [34]*

are the averaged values. As can be seen, the scattering coefficients only have small differences (less than 1.5 dB) in all polarizations between the rough surfaces with the three dimensions. On the other hand, when the dimension of the rough surface increases from 21 to 31 mm, the simulation time and memory usage are increased by more than 300% and about 150%, respectively. Hence, by considering the tradeoff between accuracy and computational cost, around 5 × 5 λ is chosen as the dimensions of rough surfaces in the simulations.

Please note that even for the rough surface with 5 × 5λ, it still needs around eight hours to run in one CPU and more than 100 GB memory in one realization, and the required computational resources may be even more for rougher surfaces. In order to perform the simulation with such large number of computations, the high-performance computing platform is utilized and parallel computing with multiple CPUs is used to accelerate the simulations.

The monostatic scattering coefficients or backscattering coefficients exhibit randomness across different samples with the same statistical roughness. Characterizing the statistical features of the backscattering coefficients may help us better understand the scattering mechanism from random rough surfaces. To find the statistical behavior of the backscattering coefficients, the backscattering coefficients from different realizations are fit to some known statistical distributions, and it turns out

Table 7.2 *Comparison of the backscattering coefficients and computational cost between the rough surfaces with different dimensions [34]*

Dimensions (mm)	σ_{vv}^0 (dB)	σ_{vh}^0 (dB)	σ_{hh}^0 (dB)	CPU time (hours)	Memory (GB)
~21 × 21	–22.0	–37.9	–25.1	7.99	105
~25 × 25	–22.6	–38.6	–27.1	14.37	156
~31 × 31	–23.0	–39.4	–26.5	32.47	250

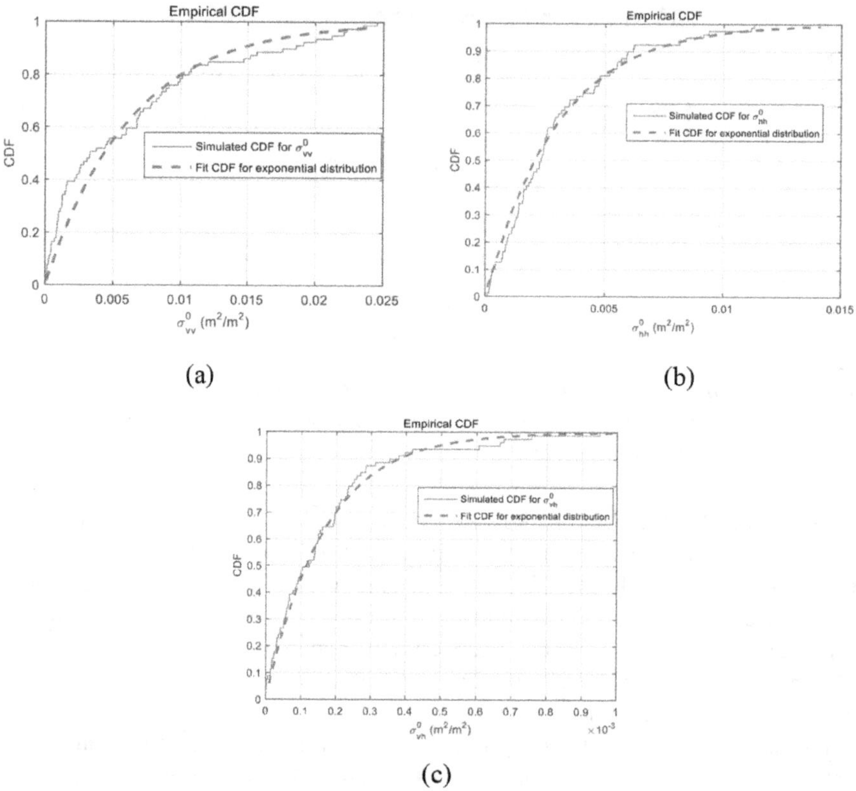

Figure 7.8 The CDF of the empirical backscattering coefficients and the fitted exponential distribution with (a) VV polarization, (b) HH polarization, and (c) VH/HV polarization [34]

that the exponential distribution can fit the backscattering coefficients very well. Figure 7.8 depicts the comparison between the empirical cumulative density function (CDF) of the backscattering coefficients and the CDF of the fitted exponential distribution for all polarizations. In this example, the RMS height is 0.8 mm, and the correlation length is about 1.8 mm. Good agreement can be observed between the empirical data and the fitted data.

The probability density function of the exponential distribution is given by:

$$f(x) = \begin{cases} e^{-x/\mu}/\mu & x \geq 0 \\ 0 & x < 0 \end{cases}. \tag{7.20}$$

The exponential distribution only has one parameter μ, which is the mean value of the random variable. The statistical distributions of the backscattering coefficients of road surfaces are potentially useful in the radar simulation and radar signal processing such as the Constant false alarm rate (CFAR) algorithm.

7.3.3 Reduced backscattering coefficients models of rough surface in mmWave band

In order to find the relation between the backscattering coefficients of the random rough surfaces and the incident angles, surface roughness and the dielectric constant of the road, massive simulations with different values for those parameters are performed. In particular, the simulated instances include the rough surfaces with the RMS height from 0.4 to 1 mm, and correlation length from 1 to 3 mm. The dielectric constant for road from 2 to 10, and the incidence angle ranges from 65° to 85°. For each combination of those parameters, about 40 realizations of random rough surfaces are generated and simulated, and each realization takes more than eight hours to finish. Note that the scattering coefficient is unitless, for generality, the RMS height and correlation length are normalized to the wavelength and represented by the unitless parameters *ks* and *kl*. *k* denotes the propagation constant in free space at 77 GHz. The dielectric constant is converted to the unitless reflection coefficient of flat surface with normal incidence ρ_0, which is defined as:

$$\rho_0 = \frac{\sqrt{\varepsilon_1} - \sqrt{\varepsilon_0}}{\sqrt{\varepsilon_1} + \sqrt{\varepsilon_0}}, \tag{7.21}$$

where ε_1 is the dielectric constant of the road, and ε_0 is the dielectric constant of air, which is roughly equal to 1.

To study how the backscattering coefficients change with different parameters, control variables are utilized. Figure 7.9 shows the examples of the polarimetric backscattering coefficients changing with the incident angles, while the rest parameters are kept the same. It turns out that the backscattering coefficients can be fit to the cosine functions of θ: $a \cos^b \theta$. Good fitting performance is shown in the figure. The fitting coefficients a and b will change when the other parameters such as *ks*, *kl*, and ε_r are different, therefore, a and b should be functions of *ks*, *kl*, and ε_r.

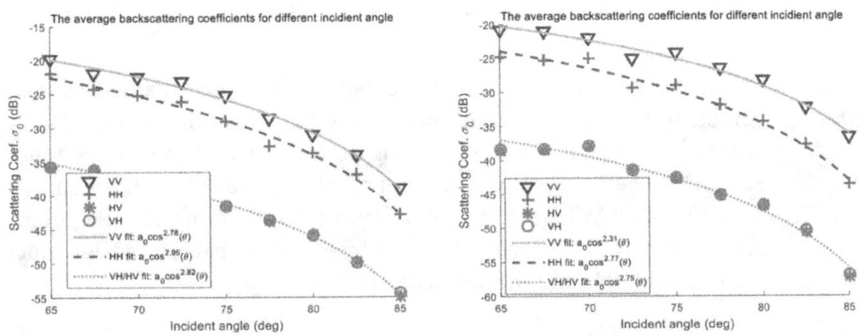

(a) (b)

Figure 7.9 The polarimetric backscattering coefficients as functions of incident angle for the rough surfaces with (a) ks = 1.29, kl = 3.23, $\varepsilon_r = 3.18 + i0.1$, *(b)* ks = 0.81, kl = 1.93, $\varepsilon_r = 3.18 + i0.1$ *[34]*

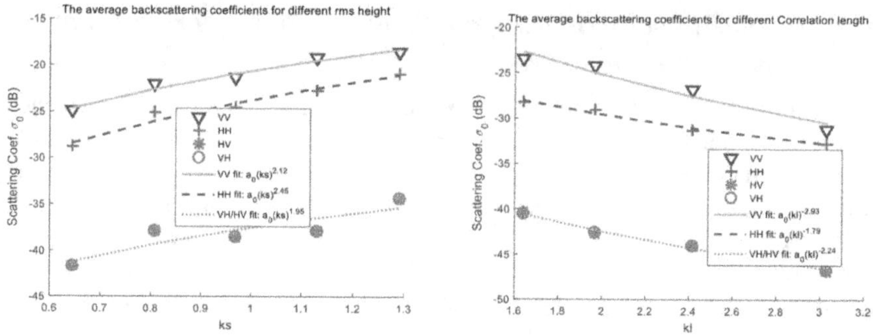

(a) (b)

Figure 7.10 The polarimetric backscattering coefficients as functions of (a) ks *and (b)* kl *of the rough surfaces with (a)* kl = 1.93, $\varepsilon_r = 3.18 + i0.1$, $\theta_i = 70°$ *and (b)* ks = 0.81, $\varepsilon_r = 3.18 + i0.1$, $\theta_i = 75°$ *[34]*

Similarly, the polarimetric backscattering coefficients can be fit as functions of *ks*, *kl*, and $|\rho_0|$. Figure 7.10 illustrates the backscattering coefficients vary with (1) *ks* and (2) *kl* of the random rough surfaces. It shows that the backscattering coefficients increase with the RMS height and decrease with the correlation length, which is expected as the surface becomes rougher with larger RMS height and smaller correlation length. The backscattering coefficients are fitted to the exponential functions of *ks* and *kl*, and as can be seen there are only small fitting errors.

Synthesize the observations of the backscattering coefficients as functions of all the parameters, and after some optimizations, the backscattering coefficients can be best modeled as functions with the following form:

$$\sigma_{pq}^0 = A\frac{\cos^{\alpha_1}\theta|\rho_0|^{\alpha_2}\left(ks\right)^{\alpha_3\cos\theta+\alpha_4}}{\left(kl\right)^{\alpha_5\cos\theta+\alpha_6}}\left(\frac{m^2}{m^2}\right),\quad p,\ q = h\ or\ v \tag{7.22}$$

where A, α_1 to α_6 are the fitting coefficients obtained by fitting all simulated backscattering coefficients with different parameters, and the values of them for different polarizations are given in Table 7.3. The fitting performance of the reduced model for the backscattering coefficients from random rough surface at 77 GHz is shown in Figure 7.11. The overall RMS error for VV polarization is 0.76 dB, for HH polarization is 0.69 dB, and for VH/HV polarization is 1.05 dB. It indicates good agreements between the reduced models and the simulated data.

7.3.4 Experiment results

To validate the reduced model based on full-wave simulations, a full polarimetric 77 GHz radar is utilized to measure the backscattering coefficients of road surfaces, and the measured data are compared with the data calculated by the reduced model. Water is very lossy at 77 GHz ($\varepsilon_r = 8.84 + i15.17$), and its skin depth can be calculated by the Debye model [51]:

Table 7.3 Coefficients of the reduced backscattering model for different polarizations [34]

Coefficient	σ^0_{vv}	σ^0_{hh}	σ^0_{vh}
A	10.77	2.73	0.927
α_1	2.67	3.15	2.96
α_2	1.86	1.24	2.61
α_3	5.40	1.38	−2.27
α_4	0.47	1.94	3.42
α_5	0.78	1.16	0.83
α_6	2.38	1.86	2.43

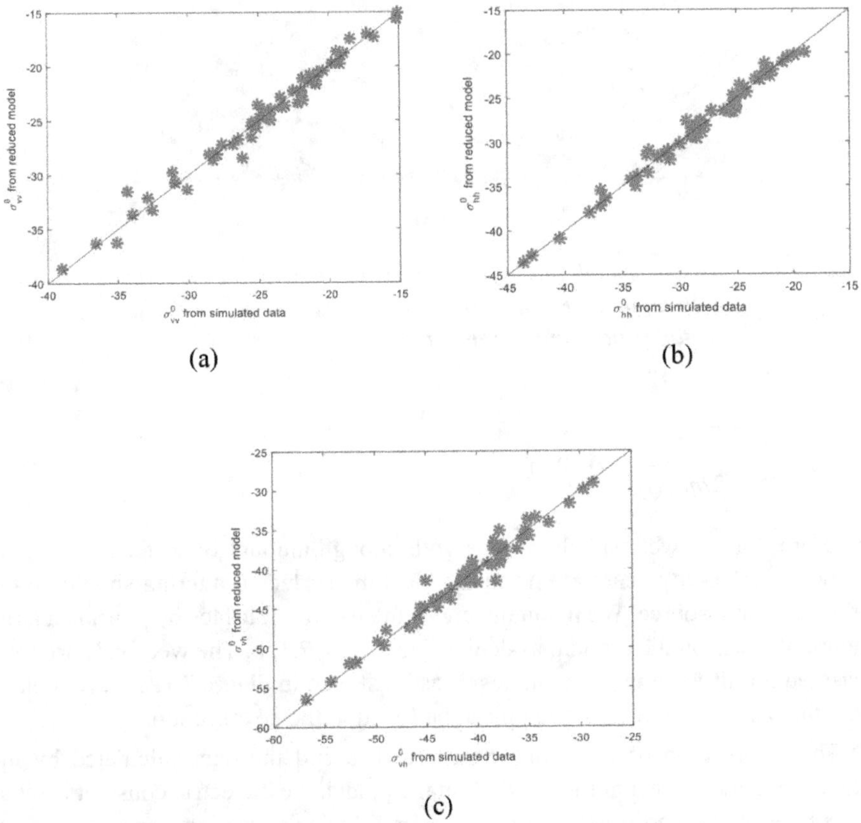

Figure 7.11 The fitting performance of the proposed reduced backscattering model vs. simulated backscattering coefficients with (a) VV polarization, (b) HH polarization, and (c) VH/HV polarization [33]

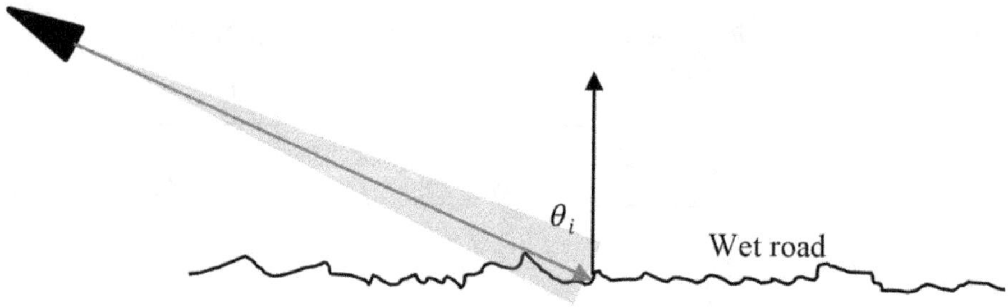

(a)

(b)

Figure 7.12 (a) The sketch of the radar measurement on the wet road surface with an incident angle θ_i, and (b) the photograph of the wet road surface under measurement

$$\delta = \frac{\lambda}{\pi}\sqrt{\frac{1}{2Im\left(\varepsilon_r\right)}} = 0.225 \text{ mm} \tag{7.23}$$

Therefore, for the wet asphalt surface with enough amount of water content, the EM wave can hardly penetrate its surface and the surface scattering should be the major scattering source. We manually create the wet road surface by pouring a large amount of water on the asphalt, as depicted in Figure 7.12b. The wet road surface is measured for different incident angles θ_i, as illustrated in Figure 7.12a. More details about the radar measurements setup can be found in the next section.

The comparison between the measured data and the data calculated by the reduced model is shown in Figure 7.13. In the model, the dielectric constant is set as $\varepsilon_r = 8.84 + i15.17$, and $s = 0.7$ mm, $l = 1.65$ mm, which describes the roughness of the wet asphalt surface. The measured data and the reduced model have good agreement, and the RMSE between the measured backscattering coefficients for VV, HH, and VH polarization are 1.3, 1.67, and 1.27 dB, respectively.

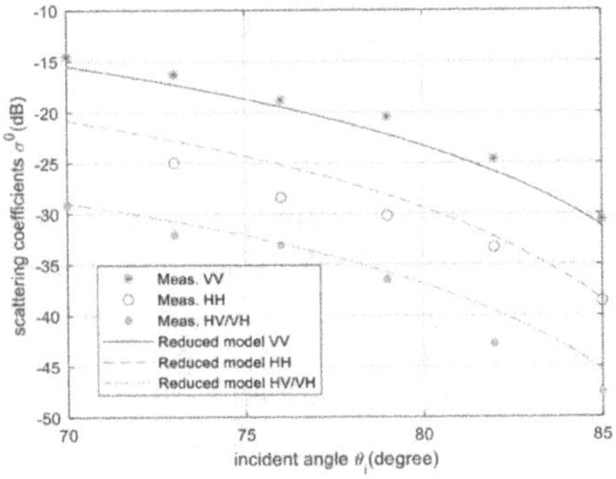

Figure 7.13 *The comparison between the measured backscattering coefficients and the prediction from the reduced model of wet asphalt surface*

7.4 Semi-empirical volumetric scattering model based on radiative transfer model

Dry roads such as dry asphalt or concrete are usually not lossy and allow the EM wave to penetrate its surface. The process of the volumetric scattering in the random medium can be roughly illustrated in Figure 7.14. The black circles represent the ingredient particles like rocks, sands, or air holes. Since the dimensions of the ingredient particles are comparable to the wavelength in mmWave band, the volumetric scattering may have strong contribution to the backscattering power [35]. As the permittivity, shapes, and the sizes of the ingredient particles are rather complex and randomized, it is very challenging to study the volumetric scattering from the random medium of road theoretically or numerically with full-wave methods. Instead, a semi-empirical model based on radiative transfer theory can be developed to model the polarimetric volumetric scattering coefficients at 94 GHz as a function of incident angle [29, 35]. In this section, the theoretical analysis will follow the derivation

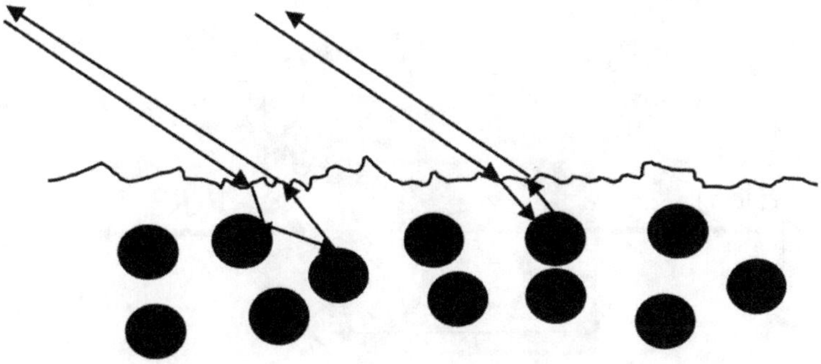

Figure 7.14 *The illustration of volumetric scattering wave in the backscattering direction*

in Reference [35] and the measurement data with 77 GHz radar [27] are utilized to fulfil the semi-empirical models.

7.4.1 Theoretical analysis of radiative transfer method

The radiative transfer method is widely applied in the volumetric scattering problems in remote sensing area [52–54]. Unlike many EM theoretical methods based on Maxell's equations, the radiative transfer method is based on energy conservation. The radiative transfer approach assumes that the particles are randomly distributed and no correlation between the fields scattered by different particles, which allows the incoherent addition of the scattered fields [39]. The key quantity in the radiative transfer equation is called the specific intensity $\boldsymbol{I}$, which is a 4×1 real vector that denotes the power density per unit solid angle, and is defined as [39]:

$$I = [I_v, I_h, U, V]^T = \frac{1}{\eta}\left[\langle |E_v|^2 \rangle, \langle |E_h|^2 \rangle, \langle 2\mathrm{Re}\,(E_v E_h^*) \rangle, \langle 2\mathrm{Im}\,(E_v E_h^*) \rangle\right]^T \quad (7.24)$$

where η denotes the intrinsic impedance of free space, E_v, E_h are the electric fields with vertical and horizontal polarization.

The vector radiative transfer equation describes how the specific intensity $\boldsymbol{I}$ in a given direction changes with the distance in the random medium, and is given by [35]:

$$\frac{d\boldsymbol{I}(\boldsymbol{R}, \hat{s})}{ds} = -\kappa \boldsymbol{I}(\boldsymbol{R}, \hat{s}) + \iint_{4\pi} \boldsymbol{P}(\hat{s}, \hat{s}')\boldsymbol{I}(\boldsymbol{P}, \hat{s}')d\Omega' \quad (7.25)$$

where κ denotes the extinction matrix that indicates the decay rate of the power of electromagnetic (EM) wave in the random medium, and it includes both the absorption and scattering; $\boldsymbol{P}(\hat{s}, \hat{s}')$ is a 4×4 phase matrix from the direction $\hat{s}'$ to the power flow direction $\hat{s}$. The process that described by the differential equation is depicted in Figure 7.15. The first term on the right-hand side $-\kappa \boldsymbol{I}(\boldsymbol{R}, \hat{s})$ describes the power decay of the EM wave propagation for a small distance, and the second term refers the process that the specific intensity in the same position but different directions $\hat{\boldsymbol{s}}'$ get scattered to the direction of interest $\hat{\boldsymbol{s}}$.

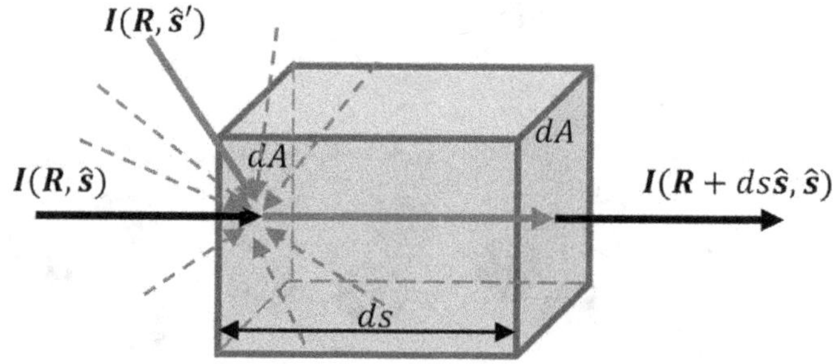

Figure 7.15 Propagation of the specific intensity in a differential volume of the random medium

The incident and scattered specific intensity on the road surface are given by [35]:

$$\boldsymbol{I}^i = \frac{1}{\eta}\left[\left\langle\left|E_v^i\right|^2\right\rangle, \left\langle\left|E_h^i\right|^2\right\rangle, \left\langle 2\mathrm{Re}\left(E_v^i E_h^{i*}\right)\right\rangle, \left\langle 2\mathrm{Im}\left(E_v^i E_h^{i*}\right)\right\rangle\right]^T \tag{7.26}$$

$$\boldsymbol{I}^s = \frac{r^2}{\eta A \cos\theta_s}\left[\left\langle\left|E_v^s\right|^2\right\rangle, \left\langle\left|E_h^s\right|^2\right\rangle, \left\langle 2\mathrm{Re}\left(E_v^s E_h^{s*}\right)\right\rangle, \left\langle 2\mathrm{Im}\left(E_v^s E_h^{s*}\right)\right\rangle\right]^T \tag{7.27}$$

where E_v^i denotes the incident E fields on the road surface, E_v^s denotes the scattered E fields on the receiver, A is the illumination area, r is the distance between radar and the road surface, and θ_s is the scattering direction. The backscattering coefficients can be expressed by the ratio of $\boldsymbol{I}^s$ and $\boldsymbol{I}^i$:

$$\sigma_{pq}^0 = \frac{4\pi\cos\theta_s I_p^s}{I_q^i}, \quad p, q = v \; or \; h \tag{7.28}$$

The differential equation (7.25) can be solved iteratively [39], and the first-order solution is given by [35]:

$$\boldsymbol{I}^s = \frac{1}{2\kappa}\boldsymbol{T}_{10}\boldsymbol{P}\boldsymbol{T}_{01}\boldsymbol{I}^i \tag{7.29}$$

where $\boldsymbol{T}_{01}$ is the 4×4 transmissivity matrix from air to the road medium, and $\boldsymbol{P}$ denotes the phase matrix. $\boldsymbol{T}_{01}$ is defined by [35]:

$$\boldsymbol{T}_{01} = \frac{n_1^2\eta_0\cos\theta_1}{n_0^2\eta_1\cos\theta_0}\begin{bmatrix} \left|t_{01}^v\right|^2 & 0 & 0 & 0 \\ 0 & \left|t_{01}^h\right|^2 & 0 & 0 \\ 0 & 0 & \mathrm{Re}\left(t_{01}^v t_{01}^{h*}\right) & -\mathrm{Im}\left(t_{01}^v t_{01}^{h*}\right) \\ 0 & 0 & \mathrm{Im}\left(t_{01}^v t_{01}^{h*}\right) & \mathrm{Re}\left(t_{01}^v t_{01}^{h*}\right) \end{bmatrix} \tag{7.30}$$

where n_0 and η_0 are the index of refraction and impedance of air, respectively, and n_1 and η_1 are those for the road medium. t_{01}^v denotes the transmission coefficients from air to the road with vertical polarization.

Note that (7.29) also holds for the scenario when the road is covered by a thin layer of homogeneous medium such as ice or water. Denote the thin layer medium as Medium 1, and the road as Medium 2, then we can just replace $\boldsymbol{T}_{10}$ and $\boldsymbol{T}_{01}$ by the three-layer transmissivity matrix $\boldsymbol{T}_{20}$ and $\boldsymbol{T}_{02}$ in (7.29).

The phase matrix in the backscattering direction for isotropic random media can be expressed as [35]:

$$\boldsymbol{P} = \begin{bmatrix} p_1 & p_2 & 0 & 0 \\ p_2 & p_1 & 0 & 0 \\ 0 & 0 & p_3 + p_2 & -p_4 \\ 0 & 0 & p_4 & p_3 - p_2 \end{bmatrix} \tag{7.31}$$

Finally, substituting (7.30) and (7.31) into (7.29), the backscattering coefficients can be calculated with (7.28) [35]:

$$\begin{pmatrix} \sigma_{vv}^0 \\ \sigma_{hh}^0 \\ \sigma_{vh}^0 \\ \sigma_{hv}^0 \end{pmatrix} = \frac{4\pi \cos\theta_0}{2\kappa} \begin{pmatrix} \left|t_{01}^v\right|^2 \left|t_{10}^v\right|^2 p_1 \\ \left|t_{01}^h\right|^2 \left|t_{10}^h\right|^2 p_1 \\ \left|t_{01}^h\right|^2 \left|t_{10}^v\right|^2 p_2 \\ \left|t_{01}^v\right|^2 \left|t_{10}^h\right|^2 p_2 \end{pmatrix} \tag{7.32}$$

In (7.32), the transmission coefficients for different polarizations can be calculated analytically with the estimated dielectric constant of road, and the rest variables p_1, p_2, and κ are very difficult to obtain theoretically. Therefore, the measurement data are utilized to determine their values.

7.4.2 Radiative transfer model with radar measurement data

The 77 GHz radar measurement data in [27] are utilized to obtain the values of $p_1/2\kappa$ and $p_2/2\kappa$ for several types of road surfaces. The measurements were taken by a full polarimetric radar, and the measurement setup is shown in Figure 7.16a. The radar is a continuous wave radar operated from 76.5 to 79.5 GHz with 3° beamwidth. It is placed on a turntable that can rotate in both elevation and azimuth directions. The height of the radar during the measurement is about 0.9 m. The road surfaces are measured with six different elevation incident angles θ_i from 70° to 85° with 3° step, and 21 different azimuth angles ϕ_i from −15° to 15° with 1.5° step. The measured data with different ϕ_i are considered as independent samples that are used in the backscattering coefficients' calculations.

The radar response of the road surface is measured in the frequency domain, and in order to separate the response of the road surface, inverse FFT is taken to obtain the time domain response, as shown in Figure 7.16b. Then the time domain signal is gated carefully to isolate the response from the illuminated area, and then

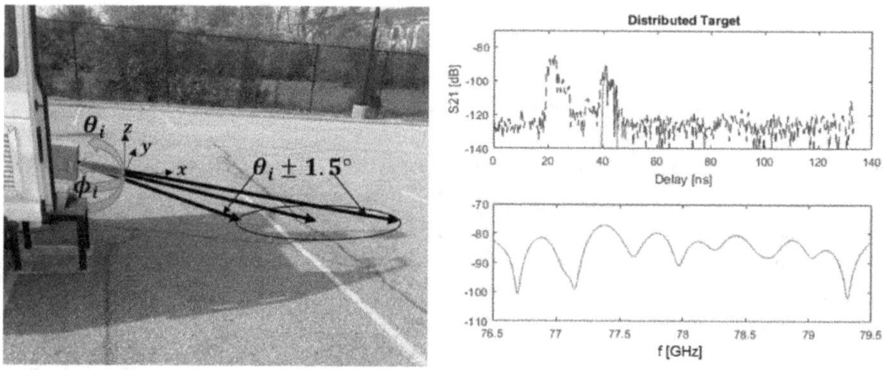

(a) (b)

Figure 7.16 (a) The road surface measurement setup with a 77 GHz polarimetric radar system, and (b) the time-domain response of one radar measurement and the FFT result of the gated signal for the road surface

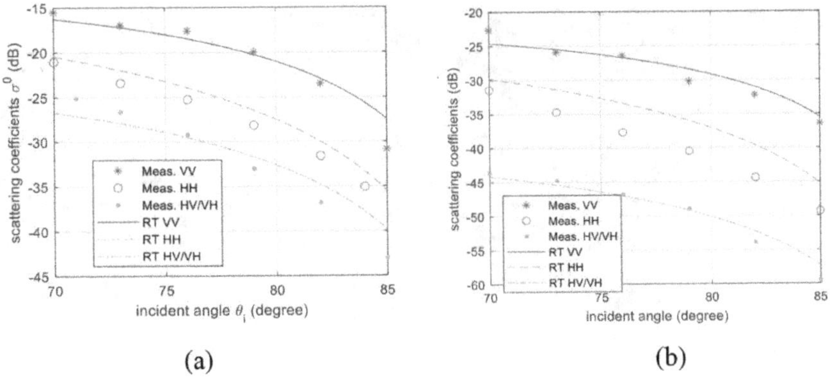

Figure 7.17 The comparison between the measured backscattering coefficients and the predictions by radiative transfer model for (a) asphalt and (b) concrete road surface

fast Fourier transform (FFT) is applied to the gated time domain signal to obtain the frequency response of the road surfaces. The frequency response of the road surfaces is calibrated to the RCS of the illuminated road surface by comparing with the frequency response of a known RCS targets such as a metallic sphere. In all, 15 different frequency response data with 200 MHz spacing are recorded and considered as independent samples. Recall that the measurements are repeated for 21 different azimuth angles ϕ_i; therefore, for each elevation incident angle θ_i, more than 300 samples are generated and used in the backscattering coefficients' evaluations with (7.9). The measured backscattering coefficients with different polarizations for asphalt and concrete surfaces are shown in Figure 7.17. The parameters $p_1/2\kappa$ and $p_2/2\kappa$ then are calculated with the measured data and are given in Table 7.4. The comparison of the measured data and the theoretical predictions by radiative transfer model are demonstrated in Figure 7.17. As can be seen, the theoretical predictions match the measured data well.

Table 7.4 Normalized phase matrix coefficients for different road surfaces

	Asphalt	**Concrete**
$p_1/2\kappa$	5.8e−3	8.3e−4
$p_2/2\kappa$	8.5e−4	1.67e−5

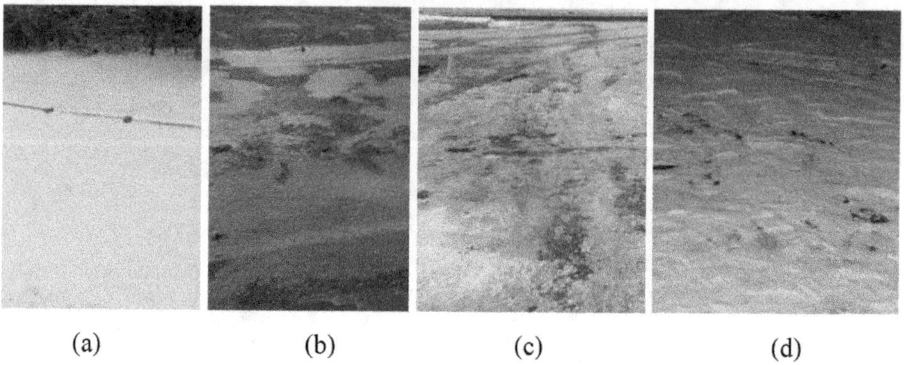

Figure 7.18 The photograph of the measured road surfaces covered by (a) thick snow, (b) thin ice, (c) slush, and (d) packed snow

7.5 Radar measurements for different road conditions at 77 GHz

Besides the wet road condition, icy or snowy road will also increase the risk of traffic accidents and therefore needs to be studied. To investigate whether automotive radar has the ability to identify different road conditions, the backscattering coefficients of different snowy or icy road conditions are measured by the 77 GHz polarimetric radar as well. The measured road conditions include the road surface covered by a thick snow layer (>3 cm thick), a thin ice layer, slush (mixture of snow and liquid water), and packed snow (around 1 cm thick), as demonstrated in Figure 7.18. Since the icy or snowy road surfaces may contain salt or dirt that is not pure water or ice, and the snow/ice layer exhibits random thickness and roughness, as illustrated in Figure 7.18, it is very difficult to analyze the mmWave scattering problem theoretically or numerically. Therefore, in this section, no EM theory or numerical methods are applied, and the analysis is solely based on the measurement data.

The backscattering coefficients with cross-polarization is usually 10 to 15 dB lower than those with co-polarization, as shown in Figures 7.13 and 7.17. In practical, the cross-pol response can easily be lower than the noise floor and may not be correctly detected by the automotive radars. Thereby, only the co-polarized backscattering coefficients are collected and compared in this study.

The vertical co-pol backscattering coefficients σ^0_{VV} as a function of incident angles for the asphalt road with different road conditions are depicted in Figure 7.19a. It shows that the values of σ^0_{VV} can be roughly categorized into three groups: (1) the asphalt road covered by thick snow, packed snow, and slush, (2) dry and wet asphalt, and (3) icy asphalt road surface. In general, the first group snow and slush road conditions have the highest value for σ^0_{VV}, and the third group icy asphalt shows the lowest σ^0_{VV}. For relatively small incident angle (about 70°), σ^0_{VV} of the second group is close to that of the first group, and in the near grazing direction (about 85°), σ^0_{VV} of the second group coincide with that of the third group. In this case, to achieve the

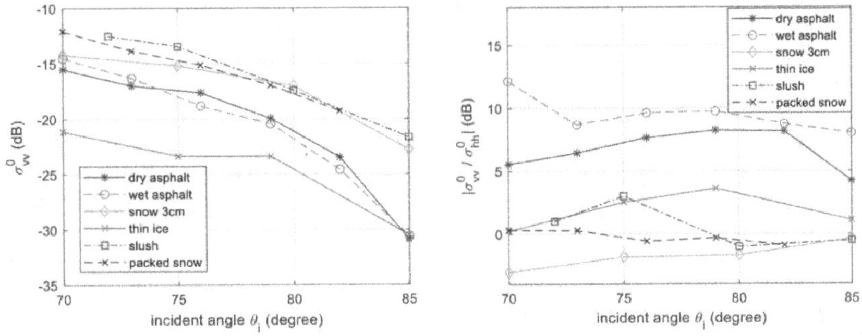

(a) (b)

Figure 7.19 *The comparison between the measured backscattering coefficients (a) σ_{vv}^0 and (b) $|\sigma_{vv}^0/\sigma_{hh}^0|$ for the asphalt with different road conditions at 77 GHz*

Table 7.5 *Average co-pol backscattering coefficients ratio for different road conditions at 77 GHz*

Road conditions	Dry asphalt	Wet asphalt	Thick snow	Thin ice	Slush	Packed snow
$\lvert\sigma_{vv}^0/\sigma_{hh}^0\rvert$(dB)	6.69	9.49	–1.73	1.85	0.61	–0.24

best differentiability among the three groups, the incident angle to the rough surface is recommended to be between 75° and 80° if only one radar with fixed incident angle is used.

Figure 7.19b demonstrates the ratio between the co-pol backscattering coefficients with vertical and horizontal polarizations $|\sigma_{vv}^0/\sigma_{hh}^0|$ for different road conditions. It shows that the wet asphalt has the highest value of $|\sigma_{vv}^0/\sigma_{hh}^0|$ for all angles, and the second highest value of $|\sigma_{vv}^0/\sigma_{hh}^0|$ belongs to the dry asphalt, the rest road conditions have relatively low difference for the backscattering coefficients between vertical and horizontal polarizations. The $|\sigma_{vv}^0/\sigma_{hh}^0|$ in dB averaged for all incident angles for different road conditions is given in Table 7.5. Recall that if only σ_{vv}^0 is examined, the wet and dry asphalt can hardly be separated, but when the horizontal polarization data are taken into consideration, the parameter $|\sigma_{vv}^0/\sigma_{hh}^0|$ can be utilized to provide more differentiability between the wet and dry road. The thick snow, packed snow, and slush road conditions, however, are still not able to be separated with the horizontal polarized data, as the $|\sigma_{vv}^0/\sigma_{hh}^0|$ values are quite close for them. Nevertheless, these three road conditions may bring the similar risk for the driving safety and can be classified as the snowy road condition.

The strategy of identify the road conditions with radar backscattering coefficients can be simply described as: (1) evaluate the backscattering coefficients for

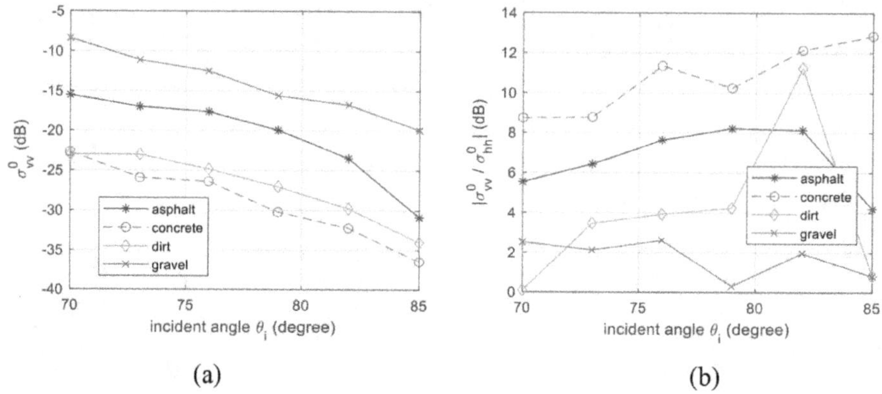

Figure 7.20 The comparison between the measured backscattering coefficients (a) σ_{vv}^0 and (b) $|\sigma_{vv}^0/\sigma_{hh}^0|$ for different road types

both vertical and horizontal polarizations from the radar data; (2) check the σ_{vv}^0 value and categorize it into one of the three groups: snowy, dry/wet, and icy; and (3) If the road is classified as the second group dry/wet, check the value of $|\sigma_{vv}^0/\sigma_{hh}^0|$ to see if it is dry or wet.

Other than the road conditions, the backscattering coefficients σ_{vv}^0 and σ_{hh}^0 depend on the surface roughness and material property of road as well. The measured σ_{vv}^0 and the ratio between v-pol and h-pol $|\sigma_{vv}^0/\sigma_{hh}^0|$ for dry asphalt, concrete, dirt roads, and gravel roads are depicted in Figure 7.20. It shows that the dirt roads and concrete roads have significantly lower σ_{vv}^0 values than that of dry asphalt, and gravel roads, on the other hand, have stronger σ_{vv}^0 compared to that of dry asphalt. Notice that the backscattering coefficients σ_{vv}^0 and $|\sigma_{vv}^0/\sigma_{hh}^0|$ of gravel roads are comparable to those of the snowy asphalt road and the backscattering coefficients of dirt/concrete roads are similar to those for icy asphalt, which indicates that the dry gravel road may be mistaken as snowy road solely with radar data. Therefore, other information such as camera data or the GPS data together with the prior knowledge of the road type should be utilized to assess the road conditions.

7.6 Conclusion

Correct recognition of the road conditions is critical for driving safety. Identifying the road conditions with radar data is attractive since first radar can detect the road in front of the vehicle to provide some reaction time, and second, the mmWave radar is a robust sensor that can work under inclement weather conditions.

The radar signals scattering from the road surface have two contributors: surface scattering and volumetric scattering. In this chapter, the surface scattering at 77 GHz is studied with the full-wave simulation approaches, and the backscattering coefficients are modeled as empirical functions of the surface roughness, dielectric constant as well as the incident angle. The volumetric scattering is modeled with the radiative transfer theory and a semi-empirical model has been developed for the dry asphalt and concrete road. Moreover, the measured backscattering coefficients for different road conditions and road types are presented and analyzed, which shows that the polarimetric backscattering coefficients are promising to be applied to identify the dry, wet, icy and snowy road conditions.

There are still some limitations and the need for significant future work before radar data can be utilized for road recognition. First, the detectable range of the road condition is limited by the height of automotive radar, for example, even for a 2 m height radar with incident angle of 80°, its detectable range is only about 11 m. Second, as the backscattering power from the road is very weak, high gain antennas and large transmitting power are required, but they are usually limited by the radar design and not available for most automotive radars on the market. The measured road conditions presented in this chapter may not represent all possible cases that the vehicles may experience, for example, the road condition with heavy rain is not considered. In order to generate a comprehensive and robust road recognition model, more measurements for all kinds of road conditions are desirable.

References

[1] Federal Highway Administration *Snow and ice [online]*. 2020. Available from https://ops.fhwa.dot.gov/weather/weather_events/snow_ice.htm

[2] Federal Highway Administration *Rain and flooding [online]*. 2020. Available from https://ops.fhwa.dot.gov/weather/weather_events/rain_flooding.htm

[3] Andrey J. 'Long-term trends in weather-related crash risks'. *Journal of Transport Geography*. 2010, vol. 18(2), pp. 247–58.

[4] Brodsky H., Hakkert A.S., Shalom Hakkert A. 'Risk of a road accident in rainy weather'. *Accident Analysis & Prevention*. 1988, vol. 20(3), pp. 161–76.

[5] Pisan P.A., Goodwin L.C., Rossetti M.A. 'US highway crashes in adverse road weather conditions'. *24th Conference on International Interactive Information and Processing Systems for Meteorology, Oceanography and Hydrology*; New Orleans, LA, 2008.

[6] Persson B.N.J., Tartaglino U., Albohr O., Tosatti E. 'Rubber friction on wet and dry road surfaces: the sealing effect'. *Physical Review B*. 2005, vol. 71(3), p. 035428, .

[7] Ghandour R., Victorino A., Doumiati M., Charara A. 'Tire/road friction coefficient estimation applied to road safety'. *18th Mediterranean Conference on Control and Automation, MED'10*; IEEE, 2010. pp. 1485–90.

[8] Han I., Luan B., Hsieh F. 'Development of autonomous emergency braking control autonomous emergency braking control system based on road friction'. *IEEE International Conference on Automation Science and Engineering (CASE)*; IEEE, 2014. pp. 933–37.

[9] Singh K.B., Ali Arat M., Taheri S. 'An intelligent tire based tire-road friction estimation technique and adaptive wheel slip controller for antilock brake

system'. *Journal of Dynamic Systems, Measurement, and Control*. 2013, vol. 135(3).

[10] Wellstead P.E., Pettit N.B.O.L. 'Analysis and redesign of an antilock brake system controller'. *IEE Proceedings – Control Theory and Applications*. 1997, vol. 144(5), pp. 413–26.

[11] Castillo Aguilar J.J., Cabrera Carrillo J.A., Guerra Fernández A.J., Carabias Acosta E. 'Robust road condition detection system using in-vehicle standard sensors'. *Sensors*. 2015, vol. 15(12), pp. 32056–78.

[12] Singh K.B., Ali Arat M., Taheri S. 'An intelligent tire based tire-road friction estimation technique and adaptive wheel slip controller for antilock brake system'. *Journal of Dynamic Systems, Measurement, and Control*. 2013, vol. 135(3), pp. 1–26.

[13] Edelmann J., Gobbi M., Mastinu G., Ploechl M., Previati G. 'Friction estimation at tire-ground contact'. *SAE International Journal of Commercial Vehicles*. 2015, vol. 8(1), pp. 182–88.

[14] Eichhorn U., Roth J. *Prediction and monitoring of tyre/road friction; XXIV FISITA*. London: Congress; 1992. pp. 7–11.

[15] Umeno T., Ono E., Asano K., *et al*. '*Estimation of Tire-Road Friction Using Tire Vibration Model*'. *SAE Technical paper*. 2002.

[16] Gustafsson F. *Slip-based Estimation of Tire-Road Friction*. Linköping University; 1995.

[17] Lee C., Hedrick K., Yi K. 'Real-time slip-based estimation of maximum tire–road friction coefficient'. *IEEE/ASME Transactions on Mechatronics*. 2004, vol. 9(2), pp. 454–58.

[18] Ahn C., Peng H., Tseng H.E. 'Robust estimation of road friction coefficient using lateral and longitudinal vehicle dynamics'. *Vehicle System Dynamics*. 2012, vol. 50(6), pp. 961–85.

[19] Andersson M., Bruzelius F., Casselgren J., *et al*. *Road Friction Estimation*. Trollhättan, Sweden: Saab Automobile AB; 2007.

[20] Sebastian G., Vattem T., Lukic L., Burgy C., Schumann T. 'RangeWeatherNet for lidar-only weather and road condition classification'. *IEEE Intelligent Vehicles Symposium (IV)*; Nagoya, Japan, IEEE, 2021. pp. 777–84.

[21] Jokela M., Kutila M., Le L. 'Road condition monitoring system based on a stereo camera'. *IEEE 5th International Conference on Intelligent Computer Communication and Processing*; Cluj-Napoca, Romania, IEEE, 2009. pp. 423–28.

[22] Jonsson P. 'Classification of road conditions: from camera images and weather data'. *IEEE International Conference on Computational Intelligence for Measurement Systems and Applications (CIMSA) Proceedings*; Ottawa, ON, Canada, IEEE, 2011. pp. 1–6.

[23] Kawai S., Takeuchi K., Shibata K., Horita Y. 'A method to distinguish road surface conditions for car-mounted camera images at night-time'. *12th International Conference on ITS Telecommunications*; Taipei, Taiwan, IEEE, 2012. pp. 668–72.

[24] Roychowdhury S., Zhao M., Wallin A., Ohlsson N., Jonasson M. 'Machine learning models for road surface and friction estimation using front-camera images'. *International Joint Conference on Neural Networks (IJCNN)*; Rio de Janeiro, IEEE, 2018. pp. 1–8.

[25] Shibata K., Takeuch K., Kawai S., Horita Y. 'Detection of road surface conditions in winter using road surveillance cameras at daytime, night-time and twilight'. *International Journal of Computer Science and Network Security*. 2014, vol. 14(11), p. 21.

[26] Viikari V.V., Varpula T., Kantanen M. 'Road-condition recognition using 24-GHz automotive radar'. *IEEE Transactions on Intelligent Transportation Systems*. 2009, vol. 10(4), pp. 639–48.

[27] Giallorenzo M., Cai X., Nashashibi A., Sarabandi K. 'Radar backscatter measurements of road surfaces at 77 ghz'. *IEEE International Symposium on Antennas and Propagation & USNC/URSI National Radio Science Meeting*; Boston, MA, IEEE, 2018. pp. 2421–22.

[28] Asuzu P., Thompson C. "Road condition identification from millimeter-wave radar backscatter measurements"'. *IEEE Radar Conference (RadarConf18)*; Oklahoma City, OK, 2018. pp. 0012–16.

[29] Li E.S., Sarabandi K. 'Low grazing incidence millimeter-wave scattering models and measurements for various road surfaces'. *IEEE Transactions on Antennas and Propagation*. 1999, vol. 47(5), pp. 851–61.

[30] Sarabandi K., Li E.S. 'Characterization of optimum polarization for multiple target discrimination using genetic algorithms'. IEEE Trans Antennas Propag. 1997, vol. 45(12), pp. 1810–17.

[31] Tsang L., Kong J.A., Ding K.-H. *Scattering of Electromagnetic Waves: Theories and Applications*. Vol. 27. New York: John Wiley & Sons; 2004.

[32] Ogilvy J.A., Foster J.R. 'Rough surfaces: Gaussian or exponential statistics?' *Journal of Physics D: Applied Physics*. 1989, vol. 22(9), pp. 1243–51.

[33] Tsang L., Kong J.A., Ding K.-H., Ao C.O. *Scattering of Electromagnetic Waves: Numerical Simulations*. Vol. 25. New York: John Wiley & Sons; 2004.

[34] Cai X. 2020. 'Autonomous Vehicles: Mmw Radar Backscattering Modeling of Traffic Environment, Vehicular Communication Modeling, and Antenna Designs'. Ann Arbor, PhD. Diss

[35] Sarabandi K., Li E.S., Nashashibi A. 'Modeling and measurements of scattering from road surfaces at millimeter-wave frequencies'. *IEEE Trans Antennas Propag*. 1997, vol. 45(11), pp. 1679–1688.

[36] Kim Y.R. *Modeling of asphalt concrete*. New York: McGraw-Hill Education; 2008.

[37] Chang C.-M., Chen J.-S., Wu T.-B. 'Dielectric modeling of asphalt mixtures and relationship with density'. *Journal of Transportation Engineering*. 2011, vol. 137(2), pp. 104–11.

[38] Polívka J. 'Measuring the permittivity of concrete at millimeter waves'. *International Journal of Infrared and Millimeter Waves*. 1996, vol. 17(10), pp. 1673–83.

[39] Ulaby F.T., Long D.G., Blackwell W.J, *et al. Microwave radar and radiometric remote sensing*. Vol. 4. Ann Arbor, MI: University of Michigan Press; 2014.

[40] Liebe H.J., Hufford G.A., Manabe T. 'A model for the complex permittivity of water at frequencies below 1 THz'. *International Journal of Infrared and Millimeter Waves*. 1991, vol. 12(7), pp. 659–75.

[41] Sihvola A. 'Mixing rules with complex dielectric coefficients'. *Subsurface Sensing Technologies and Applications*. 2000, vol. 1(4), pp. 393–415.

[42] Jiang J.H., Wu D.L. 'Ice and water permittivities for millimeter and submillimeter remote sensing applications'. *Atmospheric Science Letters*. 2004, vol. 5(7), pp. 146–51.

[43] Rice S.O. 'Reflection of electromagnetic waves from slightly rough surfaces'. *Communications on Pure and Applied Mathematics*. 1951, vol. 4(2-3), pp. 351–78.

[44] Nieto-Vesperinas M. 'Depolarization of electromagnetic waves scattered from slightly rough random surfaces: a study by means of the extinction theorem'. *Journal of the Optical Society of America*. 1982, vol. 72(5), pp. 539–47.

[45] Beckmann P., Spizzichino A. *The Scattering of Electromagnetic Waves from Rough Surfaces*. Pergamon, New York: Norwood; 1987.

[46] Elfouhaily T.M., Guérin C.-A. 'A critical survey of approximate scattering wave theories from random rough surfaces'. *Waves in Random Media*. 2004, vol. 14(4) R1–40.

[47] Ye H., Jin Y.-Q. 'Parameterization of the tapered incident wave for numerical simulation of electromagnetic scattering from rough surface'. *IEEE Transactions on Antennas and Propagation*. 2005, vol. 53(3), pp. 1234–37.

[48] Pak K., Johnson J., Tsang L., Chan C.H. 'Backscattering enhancement of electromagnetic waves from two-dimensional perfectly conducting random rough surfaces based on Monte Carlo simulations'. *Journal of the Optical Society of America A*. 1995, vol. 12(11), pp. 2491–99.

[49] Kuang L., Jin Y.-Q. 'Bistatic scattering from a three-dimensional object over a randomly rough surface using the FDTD algorithm'. *IEEE Transactions on Antennas and Propagation*. 2007, vol. 55(8), pp. 2302–12.

[50] Wu J. 2016. 'Radar Sub-surface Sensing for Mapping the Extent of Hydraulic Fractures and for Monitoring Lake Ice and Design of Some Novel Antennas'.Ann Arbor, PhD Diss

[51] Vahidpour M., Sarabandi K. 'Millimeter-wave doppler spectrum of walking bodies'. *IEEE Transaction on Geoscience Remote Sensing*. 2012, vol. 50(7), pp. 2866–79.

[52] Leung Tsang., Ishimaru A. 'Radiative wave equations for vector electromagnetic propagation in dense nontenuous media'. *Journal of Electromagnetic Waves and Applications*. 1987, vol. 1(1), pp. 59–72.

[53] Ulaby F.T., Sarabandi K., Mcdonald K., Whitt M., Dobson M.C. 'Michigan microwave canopy scattering model'. *International Journal of Remote Sensing*. 1990, vol. 11(7), pp. 1223–53.

[54] Chang W., Tan S., Lemmetyinen J., Tsang L., Xu X., Yueh S.H. 'Dense media radiative transfer applied to SnowScat and SnowSAR'. *IEEE Journal of Selected Topics in Applied Earth Observations and Remote Sensing*. 2014, vol. 7(9), pp. 3811–25.

Chapter 8

Radar-based gesture sensing

Lina Ma[1,2] *and Changzhan Gu*[1,2]

This chapter presents a basic introduction to short-range radar and types of gesture radar. The chapter is mainly divided into three parts, including continuous wave (CW) radar, frequency-modulated CW (FMCW) radar, and in the Internet of Things (IoT) era several application examples of short-range radar-based motion sensing in gesture recognition. The simple theory and signal processing algorithm of the radar is also briefly introduced. With the advantage of the small form factor, low power consumption, and low cost, we can envision the ubiquitous applications of short-range radar technology in the coming IoT era. This chapter introduces their potential application in automotive, mobile and wearables, smart home hardware, augmented reality/virtual reality, gaming consoles, metaverse, healthcare, and many other exciting applications and use cases.

8.1 Introduction

With the advent of IoT smart devices, such as smartphones, thermostats, smart refrigerators, TVs, and game consoles, the world has developed into a new era, where artificial intelligence (AI) is on the rise, and human–computer interaction (HCI) is developing toward directions of more natural, intuitive, and more efficient [1–3].

In the past few years, the keyboard and the mouse, which are seen as the most commonly used input devices, have been widely used in the functionality of computing devices for interacting with or controlling. As technology advances, the input devices and techniques are also updated iteratively as well as possess new freedom and flexibility. For example, different from the traditional wired mouse, the wireless mouse is free from the shackles of wires. Thanks to the wireless nature, more freedom is provided for the users in the mobility and placement of the mouse. Note that the new kind of input devices like wireless mouse still requires users to physically

[1]MoE Key Lab of Artificial Intelligence, AI Institute, Shanghai Jiao Tong University, Shanghai, Minhang District, China
[2]MoE Key Lab of Design and EMC of High-Speed Electronic Systems, Department of Electronic Engineering, Shanghai Jiao Tong University, Shanghai, Minhang District, China

operate, like hold, touch, click, or move. Furthermore, users rely heavily on input devices. If the input devices are missing or malfunctioning, the users cannot do anything but wait for new input devices. Therefore, HCI independent of physical devices gives users additional freedom [4].

Thanks to the advancements of cloud computing and improved accuracy of speech recognition, digital voice assistants are increasingly used everywhere in our daily lives and increasingly appear on various occasions, such as home environments, workplaces, automotive, entertainment venues, etc. Voice control and gesture control are natural and further advanced ways to interact with the physical world. Suppose we just entered the house on a cold night, we might say something like "Xiao Ai, turn on the light and air conditioner in the master bedroom" to the voice interaction devices. When we play man–machine games in an entertainment venue such as doubles tennis, users prefer a combination of speech and gesture over speech or gesture alone to control the game interface for increased immersion [5].

Hands, as the most primitive natural tool, have been used to interact with the physical world by expressing gestures for decades [6]. Due to the flexibility of the hand function, the hand gesture could achieve fluid, effective, and rapid manipulation to enter simple and accessible commands to computing devices [7]. However, it is not simple for humans to capture the fine movement of the hand and fingers in free space. Optical techniques, including multicamera [8], structured light [9], and time of flight [10], have gradually become the research of interest and drawn great attention in recent years. Computer vision technology is a traditional and common way to obtain images that can be further analyzed to identify hand gestures. Furthermore, such advanced technologies even may be integrated into consumer hardware devices, such as gaming consoles [11] and smartphones [12]. Unfortunately, they are still limited from latency and technical constraints, such as resolving the overlapped fingers and requiring instrumentation of the hands, as well as being out of order in darkness. Moreover, due to the impenetrability of the camera, the implementation of this sensing method is limited [4]. It should be noted that no input technology is universally optimal for all computing devices and users. The choice of input techniques not only depends on the computing device but is also based on the specific demand constraints, which include requirements of size, power consumption, sensing precision, bill of material cost, ease of integration into the computing device, sensitivity to light and environmental noise, and many other considerations.

Compared with conventional optical-based sensors and camera-based solutions, microwave radar systems have their own natural advantages. First, thanks to the working principle of radar that transmits radio frequency (RF) signals to the target and the target's movement can be extracted by analyzing the backscattering reflections, the microwave radar system is a contactless solution that brings good convenience for users [13]. Second, microwave radar systems do not depend on light, which greatly get rid of the limitation of light sensitivity [14]. Third, microwave radar systems can penetrate obstacles well [15]. More importantly, radar sensors are more sensitive to small movements.

For emerging applications in wearable, mobile, and especially automotive, it is a promising and attractive form of HCI. Considering a potential operational scenario

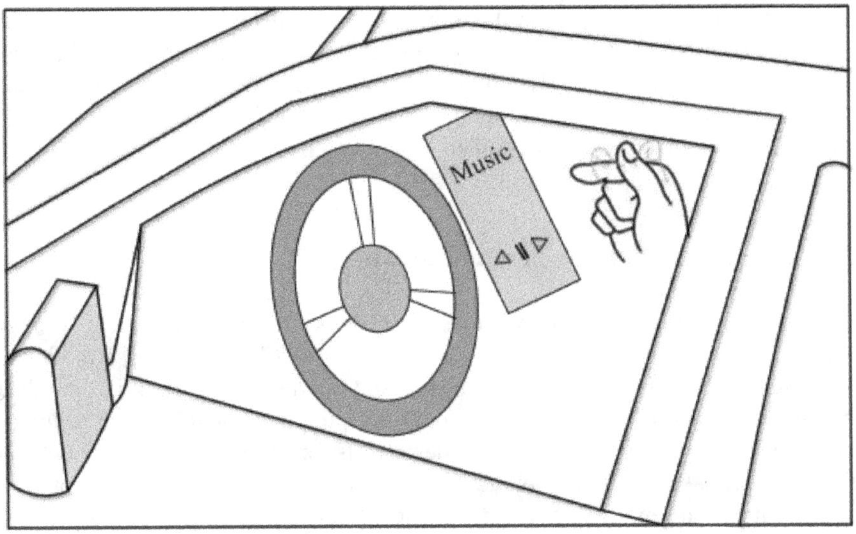

Figure 8.1 *Potential operative scenarios in automotive*

in automotive, shown in Figure 8.1, by recognizing different gestures, drivers can perform intelligent interactions with the car, such as voice, volume adjustment, recommended route switching, radio selection, etc. During the past few years, breakthroughs have been made to wirelessly and precisely detect hand gestures by using radar. This chapter reviews the latest advancements in motion-sensing technologies using radar, with an emphasis on the applications of gesture sensing for HCI.

8.2 Fundamentals of short-range radar

Radar-based RF sensing has been a mature field of engineering and applied sciences since the first radar systems were developed in the 1930s [16]. Mainstream radar research is mainly focused on long-range applications, such as the detection and tracking of large moving objects at distances of hundreds of meters or even kilometers. So-called long-range radars involve in air defense systems [17], atmospheric radars [18], oceanographic radars [19], and geophysical monitoring systems. Over the past decades, due to the applications in collision avoidance and autonomous driving systems, automotive radar has become an interesting area in both academia and industry [20, 21]. The working range of the automotive radar is tens or hundreds of meters, so it can be classified as medium-range radar. On the other hand, short-range radar refers to distances of several meters or even less than a meter.

In the 1970s, short-range radar technology was first used in healthcare applications, such as obtaining very small movements of target objects of human vital signs such as human respiration and heartbeat [22]. Short-range radars possess a very wide range of emerging applications, including wireless healthcare [23, 24], medical [25, 26], gesture sensing [27, 28], and IoT [29, 30]. Due to the short-range nature of radar and modern consumer applications, well-established hardware systems and computational methods for long- and medium-range radars may not be applicable to short-range applications directly where radar must be small enough to fit compact consumer hardware that runs on limited battery power and computing resources and tracks dynamic objects, such as human hands or chest walls, with submillimeter accuracy.

8.3 Radar systems for gesture sensing

According to signal modulation schemes, the radar systems can be generally classified into three categories: impulse, single-tone CW, and FMCW or phase-modulated CW (PMCW). Due to the features of broadband nature and the limited time delay resulting from the short backscattering round trip, impulse radar is not appealing in short-range applications. The short time delay calls for an extremely fast switch, and the impulse signal needs to be obtained by a high-speed analog-to-digital converter (ADC), which is often a power-hungry device. On the other hand, CW or FMCW/PMCW radars allow for lower power operation and a higher level of chip integration, making them more attractive for mobile, wearable, automotive, gaming consoles, and other consumer hardware applications. Several examples of radar systems for gesture determination are discussed in the following.

8.3.1 CW radar

8.3.1.1 Basic theory

Thanks to the advantages of a simple structure, high sensitivity, narrow spectral occupancy, and a high level of chip integration, CW radar sensors have a broad application prospect in various scenarios.

Figure 8.2 shows the block diagram of a common CW radar. As shown in Figure 8.2, CW radar detects the movements of the target by transmitting a single-tone electromagnetic wave to the target and receives the reflected signal, which modulates the motion information of the target. Generally speaking, a single-tone signal is generated by a voltage-controlled oscillator. Then, the signal goes through balun 1 and is divided into two channels. One is used as a transmitting (Tx) channel, and the other is used as a mixer channel by balun 2 for receiving (Rx). The signal for

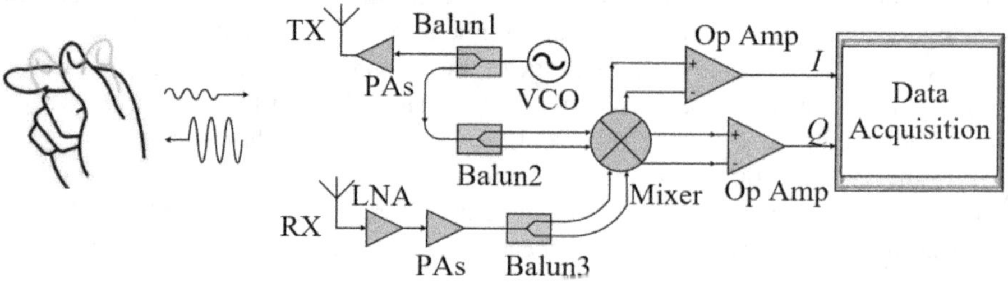

Figure 8.2 Common configuration of CW radar

Tx is amplified by a gain block, which usually consists of one or two power amplifiers (PAs), which depend on the transmit power needed and are radiated through the Tx antenna. After receiving the differential signals, which are amplified from the received signal by a gain block consisting of a low-noise amplifier (LNA) and/or PAs and converted by balun 3, two differential signals are mixed through the quadrature mixer. The outputs are represented as a pair of differential in-phase (I) and a pair of differential quadrature-phase (Q) baseband signals. With further intermediate frequency (IF) amplification, the signals are converted into digital signals and are processed by the microcontroller. After proper processing, the motion trajectory of movement signs can be obtained. The theory is detailed as follows.

The transmitted signal $T(t)$ is modeled as [31]:

$$T(t) = \exp\left(j\left(2\pi f_c t + \varphi_0\right)\right) \tag{8.1}$$

where f_c is the frequency of $T(t)$, φ_0 is the initial phase.

The reflected signal $R(t)$ is

$$R(t) = \rho T(t - \Delta t) \tag{8.2}$$

where ρ is the propagation loss and Δt is the time interval of the signal from the TX antenna to the RX antenna. $\Delta t = \frac{2x(t)}{c}$, where $x(t)$ is the distance of the target.

After mixed, the IF signal is obtained:

$$\begin{aligned} S(t) = T(t)R^*(t) &= \rho\exp\left(j\left(2\pi f_c\Delta t\right)\right) = \rho\exp j\left(2\pi f_c\tfrac{2x(t)}{c}\right) \\ &= \rho\exp j\left(\frac{4\pi\Delta x(t)}{\lambda_c} + \frac{4\pi d_0}{\lambda_c}\right) \\ &= I(t) + jQ(t) \overset{def}{=} \rho e^{j\Phi(t)} \end{aligned} \tag{8.3}$$

where $\lambda_c = c/f_c$ is the wavelength of the electromagnetic wave, $x(t) = \Delta x(t) + d_0$, $\Delta x(t)$ is the motion of the target at around distance d_0.

The motion of the target can be extracted from the phase $\Phi(t)$:

$$x(t) = \frac{c\phi(t)}{4\pi f_c} = \frac{\lambda_c\Phi(t)}{4\pi} \tag{8.4}$$

Note that, $\Phi(t)$ is a periodic function whose cycle is 2π. Therefore, only the vibration of $x(t)$, i.e., $\Delta x(t)$ can be extracted. The speed of the target is as follows:

$$v = \frac{dx(t)}{dt} = \frac{\frac{c}{4\pi f_c}d\Phi(t)}{dt} = \frac{cf_d}{2f_c},\ v \ll c \tag{8.5}$$

where $f_d = \frac{1}{2\pi}\frac{d\Phi(t)}{dt}$ is the Doppler shift.

8.3.1.2 Phase demodulation algorithm

As discussed in section 8.3.1.1, the motion of the target can be extracted from phase $\Phi(t)$. Thus, the key to extracting the motion is to demodulate $\Phi(t)$. Many demodulation algorithms have been proposed. The classic algorithm is arctangent demodulation algorithm [31–33], as shown in (8.6).

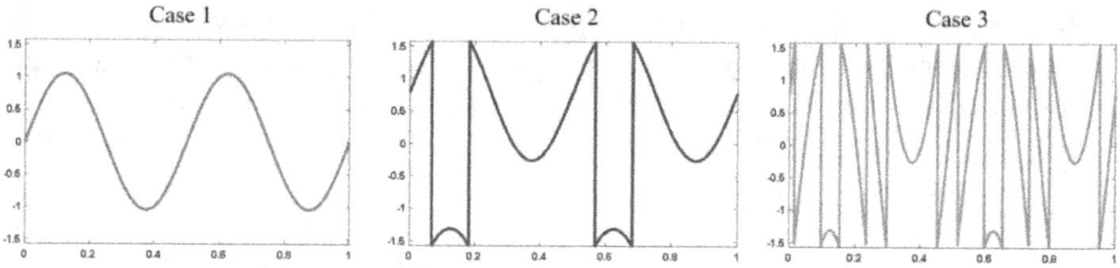

Figure 8.3 Three cases of arctangent demodulation results

$$\Phi_{\text{atan}}(t) = arctan\left(\frac{Q(t)}{I(t)}\right) \tag{8.6}$$

However, the value of the arctangent function is restricted to $(-\pi/2, \pi/2)$. Figure 8.3 shows three cases of arctangent demodulation. The vibration amplitude of $\Phi(t)$ is smaller than π in case 1 and case 2. The transition occurs when the direct current (DC) component is close to $-\pi/2$ or $\pi/2$. When the vibration amplitude of $\Phi(t)$ is larger than π, as shown in case 3, the transitions are inevitable. Compensations of π or $-\pi$ need to be done at the transition point. However, automatic compensation is difficult to implement when the vibration amplitude becomes larger and noise exists [34].

Another phase demodulation algorithm is the differentiate and cross multiply (DACM) algorithm [25]. Unlike the arctangent demodulation algorithm, the DACM algorithm obtains the result by the derivation of the arctangent function [35]:

$$\omega(\text{t}) = \frac{\text{d}\Phi_{\text{atan}}(t)}{\text{d}t} = \frac{I(t)Q'(t)-I'(t)Q(t)}{I(t)^2+Q(t)^2} \tag{8.7}$$

where $Q'(t)$ and $I'(t)$ are the time derivatives of $Q(t)$ and $I(t)$, respectively. $\omega(\text{t})$ is also the differential of $\Phi(t)$. Therefore, $\Phi(t)$ can be reconstructed by the integration of $\omega(t)$. In digital domain, the $\Phi[n]$ could be reconstructed by Reference [35] as follows:

$$\Phi[n] = \sum_{k=2}^{n} \frac{I[k]\{Q[k]-Q[k-1]\}-\{I[k]-I[k-1]\}Q[k]}{I^2[k]+Q^2[k]} \tag{8.8}$$

where the differentiation is approximated by a forward difference, and the integration is replaced with an accumulation. Note that $\omega(t)$ is a single-valued function, thereby, the codomain restriction does not exist in this algorithm.

Recently, a modified DACM algorithm is reported [36], which has simplified expression but much improved performance for high-linear motion detection. Different from the traditional DACM algorithm shown in (8.7), the modified DACM algorithm does not include arctangent function. The baseband I/Q signal shown in Figure 8.2 could be modeled as follows:

$$I(t) = \cos\left[\frac{4\pi\Delta x(t)}{\lambda_c} + \frac{4\pi d_0}{\lambda_c}\right] \tag{8.9}$$

$$Q(t) = \sin\left[\frac{4\pi\Delta x(t)}{\lambda_c} + \frac{4\pi d_0}{\lambda_c}\right] \tag{8.10}$$

First, apply differentiation to (8.9), (8.10):

$$I'(t) = -\sin\left[\frac{4\pi\Delta x(t)}{\lambda_c} + \frac{4\pi d_0}{\lambda_c}\right] \cdot 4\pi\Delta x'(t)/\lambda_c \tag{8.11}$$

$$Q'(t) = \cos\left[\frac{4\pi\Delta x(t)}{\lambda_c} + \frac{4\pi d_0}{\lambda_c}\right] \cdot 4\pi\Delta x'(t)/\lambda_c \tag{8.12}$$

Second, since $\cos^2\Phi + \sin^2\Phi = 1$, $\Delta x'(t)$ can be extracted from (8.11) and (8.12):

$$\Delta x'(t) = \frac{\lambda}{4\pi}\left[I(t)\,Q'(t) - I'(t)\,Q(t)\right] \tag{8.13}$$

As shown above, the differential form of the target motion does not include arctangent function. The motion trajectory $\Delta x(t)$ can be obtained by applying an integration function to $\Delta x'(t)$. In digital domain, the motion trajectory can be obtained by

$$x[n] = \frac{\lambda}{4\pi}\sum_{k=2}^{n} I[k-1]\,Q[k] - I[k]\,Q[k-1] \tag{8.14}$$

It should be noted that the relationships between $\omega(t)$, $\Phi[n]$ and $\Delta x'(t)$, $x[n]$ are $\Delta x'(t) = \omega(t)\,\lambda/4\pi$ and $x[n] = \Phi[n]\,\lambda/4\pi$, respectively. Comparing (8.8) and (8.14), the proposed algorithm does not include $I^2[k] + Q^2[k]$, which decreases the signal-to-noise ratio requirements of the signal and thereby improves the accuracy of the demodulation.

8.3.1.3 Advancements in CW radar

Over the past few decades, efforts have been made to minimize CW radar sensors for integration, including board-level and chip-level system integration [37]. However, only a few people use CW radar for gesture recognition [38] and studied the feasibility of implementing basic functions such as mouse movement and clicking based on CW radar. Due to the broad radiation patterns of the Tx/Rx antenna, the incident CW is scattered not only by the objects of hands and fingers, but also by the torso, forearm, and background clusters. Thus, the radar-received signal is a complex combination of Doppler phase shifts from multipoint motion. Precise extraction of motion patterns or signatures of gestures is important for reliable practical applications.

In recent research, Rx uses a homodyne direct conversion architecture to be cost-effective. To eliminate the DC offset problem and mitigate the effects of flicker noise in direct conversion architectures [39], a symmetric subcarrier is modulated into the transmitted CW. Since human motion is generally slow, which is about in the Hertz range, bandpass sampling is used to digitize the down-converted subcarriers using a low-speed, low-cost ADC. An arcsin algorithm is developed to demodulate I/Q signals to recover phase information, and a new motion-imaging algorithm is proposed to reconstruct two-dimensional (2D) motion based on this. Compared with frequency-domain micro-Doppler methods, the proposed remote gesture sensing is based on linear motion reconstruction and is able to identify explicit features of gesture motion [38].

The 2D motion-imaging model is shown in Figure 8.4(a). It is assumed that the subject hand is limited to moving in the *xy* plane, with an instantaneous location $(x(t), y(t), 0)$. The transmitting antenna, T, located at (x_T, y_T, z_T), has a distance of d_T

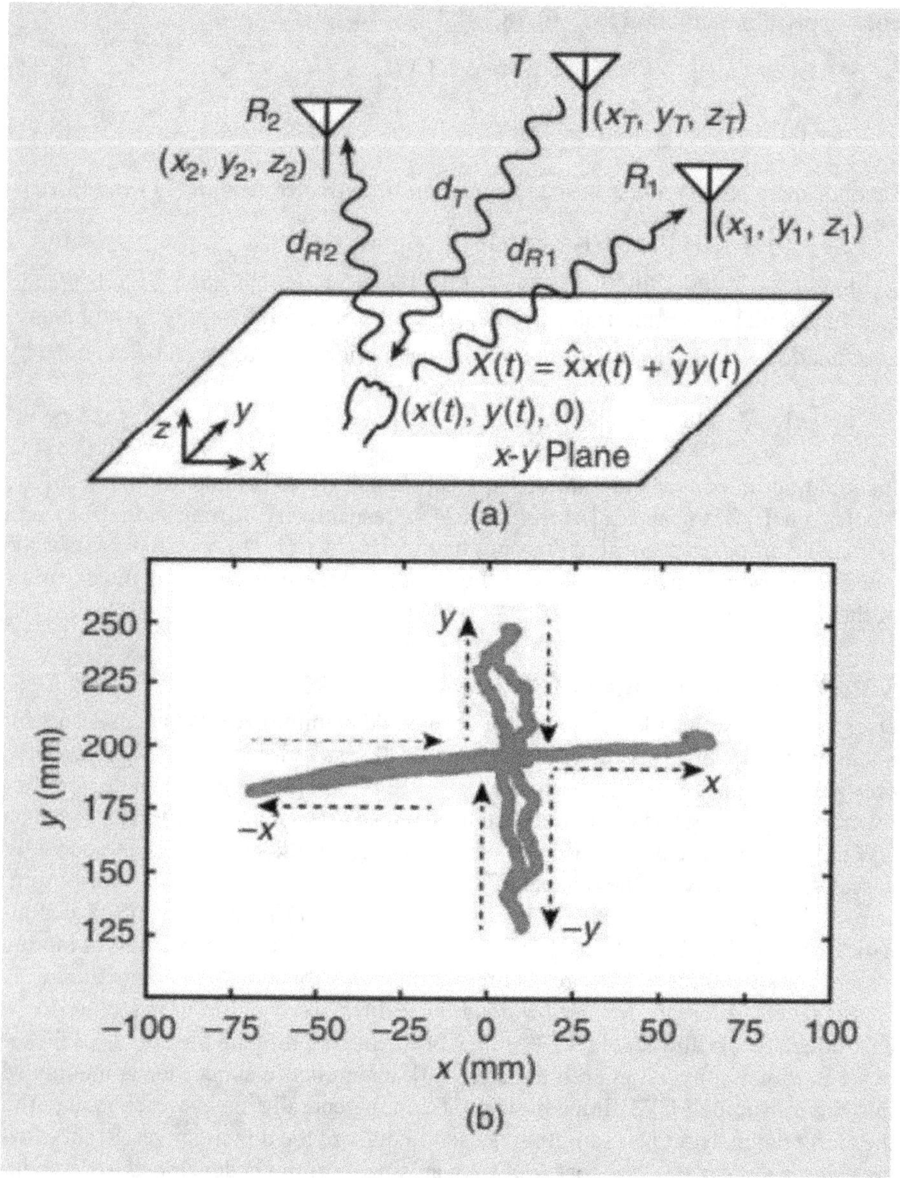

Figure 8.4 *A 5.8-GHz CW radar sensing the hand gesture that mimics a function of mouse control: (a) a 2D motion-imaging model and (b) a reconstructed motion pattern of a moving hand in the* xy *coordinate [38]*

to the hand. Likewise, the two receiving antennas, $R_1\left(x_1, y_1, z_1\right)$ and $R_2\left(x_2, y_2, z_2\right)$, are separated with distances of d_{R_1} and $d_{R_2}$to the hand. In radar sensing, the total phase delay is accumulated from the round-trip distance between the transmitting antenna and the receiving antennas. In this case, the round-trip distances are $\mathrm{d}_1 = \mathrm{d}_T + \mathrm{d}_{R_1}$ and $\mathrm{d}_2 = \mathrm{d}_T + \mathrm{d}_{R_2}$ for the two receiving antennas, respectively. Based on the model, the location in the *xy* plane can be derived as follows:

$$x\left(t\right) = \frac{\mathrm{d}_1\mathrm{d}_2\left(\mathrm{d}_1 - \mathrm{d}_2\right) + \mathrm{d}_1 x_2^2 - \mathrm{d}_2 x_1^2}{2\left(\mathrm{d}_1 x_2 - \mathrm{d}_2 x_1\right)} \tag{8.15}$$

$$y\left(t\right) = \sqrt{\left(\mathrm{d}_1 + x_1\right)\left(\mathrm{d}_1 - x_1\right)\left(\mathrm{d}_2 + x_2\right)\left(\mathrm{d}_2 - x_2\right)\left(\mathrm{d}_1 - \mathrm{d}_2 + x_1 - x_2\right)} \tag{8.16}$$

In the experiment, the subject was required to sit in front of a table and move one hand on a 10 cm × 10 cm region on the table. The reconstructed trajectory of the hand motion is shown in Figure 8.4(b), which indicates the hand moving in the *x* and *y* directions. It is experimentally demonstrated that the radar sensor can also successfully identify various click actions [38].

8.4 FMCW radar

8.4.1 Basic theory

The composition of the FMCW radar system mainly includes the following: sawtooth wave signal generator, transceiver module, digital-to-analog converter, digital signal processor, and controller. The signal is generated by the sawtooth wave signal generator, passed through the frequency multiplier and the PA in turn, and transmitted by the transmitting antenna. The electromagnetic wave is reflected by the object and received by the receiving antenna. The received signal is mixed with the transmitted signal after passing through an LNA to obtain an IF signal. Finally, the IF signal is filtered and ADC sampled sequentially, and then input to the back end for further signal processing. The block diagram of a typical FMCW radar system is shown in Figure 8.5.

For FMCW radar, a variety of modulations are possible. The transmitter frequency can slew up and down as follows: sine wave, sawtooth wave, triangle wave, square wave, etc.

Sawtooth modulation is the most used in FMCW radars, so the following analysis is based on the sawtooth wave.

The sawtooth wave of FMCW radar is a frequency linear modulation method. The frequency of electromagnetic waves changes linearly with time.

Denote T_c as the chirp repetition period, B is the bandwidth of the chirp, $S = B/T_c$ is the slope of the frequency modulation. The mathematical expression for the transmitted signal within one frequency ramp interval is as follows:

$$s_{\mathrm{Tx}}\left(t\right) = \exp\left(\mathrm{j}\left(2\pi f_c t + \pi \mathrm{S}t^2 + \phi\right)\right) \tag{8.17}$$

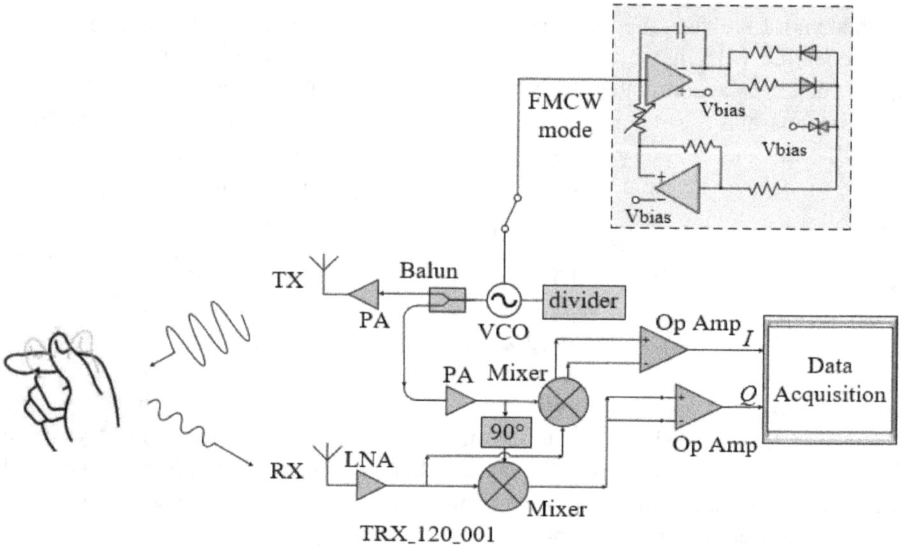

Figure 8.5 Block diagram of a typical FMCW radar system

where f_c is the center frequency of the frequency ramp, ϕ is the initial phase residual, $t \in \left[-T_c/2,\ \ T_c/2\right]$. Suppose that there is a reflection point, and its distance from the radar as a function of time is $R\left(\tau\right)$. Assuming that the movement of the scattering point is relatively slow, thus $R\left(\tau\right)$ can be regarded as a constant within a certain period. This is a "stop-and-go" hypothesis, which is extremely common when dealing with slow moving targets. Therefore, for the scattering point located at $R\left(\tau\right)$, the echo signal received by the FMCW radar is a function of the time delay Δt and a certain amplitude attenuation σ of the transmitted signal. Among them, the time delay is as follows:

$$\Delta t = \frac{2R\left(\tau\right)}{c} \tag{8.18}$$

Therefore, the echo signal can be expressed as follows:

$$\begin{aligned} s_{\mathrm{Rx}}\left(t\right) &= \sigma s_{\mathrm{Tx}}\left(\mathrm{t} - \Delta t\right) \\ &= \sigma \cdot \exp\left(\mathrm{j}\left(2\pi f_c\left(t - \Delta t\right) + \pi \mathrm{S}\left(t - \Delta t\right)^2 + \phi\right)\right) \end{aligned} \tag{8.19}$$

According to Figure 8.5, the received signal is mixed with the transmitted signal. After that, the resulting mixed signal is low-pass filtered and the IF signal is obtained:

$$\begin{aligned} s_{\mathrm{IF}}\left(t\right) &= f_{LPF}\left\{s_{\mathrm{Tx}}\left(t\right) \cdot s_{\mathrm{Rx}}\left(t\right)\right\} \\ &= \sigma \cdot \exp\left(\mathrm{j}\left(\frac{4\pi SR(\tau)t}{c} + \frac{4\pi f_c R(\tau)}{c} - \frac{4\pi SR^2(\tau)}{c^2}\right)\right) \end{aligned} \tag{8.20}$$

It can be seen from (8.18–8.20) that the obtained IF signal is a sinusoidal motion with frequency f_{IF}:

$$f_{IF} = \frac{2SR(\tau)}{c} \tag{8.21}$$

8.4.2 Advancements in FMCW radar

As far as we know, project Soli is the world's first radar-based sensing system, which is designed for gesture sensing meeting the requirements of modern HCI [39]. The architecture of its sensing system is based on an FMCW, and the center operating frequency is 60 GHz. The radar sensor chip consists of a four-channel Rx and a two-channel Tx [40], which enable a high-range resolution of ~2 cm. By combining and leveraging micro-Doppler and time-domain analyses, the sensing system captures complex hand movements with millimeter accuracy. Thanks to the high integration and system minimization which are more suitable for millimeter-wave operation, it is essential for Soli to be incorporated into mobile, wearable, and IoT devices. Since the size of antenna is small enough, antenna-in-package techniques can be employed to integrate the metal redistribution layer in an embedded wafer-level ball grid array (eWLB) package.

A block diagram of the millimeter-wave 60-GHz radar transceiver is shown in Figure 8.6 [40]. The radar chip includes four fully differential Rx chains and two fully differential Tx chains. Each Tx chain is realized using a differential cascode stage, which offers a relatively high gain as well as a relatively high output impedance and facilitates output impedance matching. The Tx output is matched to a 100-Ω differential load impedance. The Tx output power could be adjusted using a six-bit digital-to-analog converter that controls the current level flowing through the cascode [40]. Differential topology is utilized in the Tx to reduce Tx-to-Rx coupling by limiting the leakage via the common ground on the chip. An active RF distribution network is used to feed a single local oscillator (LO) to all Tx and Rx channels. It consists of a three-way Wilkinson power divider that feeds the three LO buffer stages, each of which is integrated with a balun for single-to-differential conversion. The Rx chain includes the differential LO buffer stage that drives the double-balanced mixer cell. Unlike the conventional Gilbert cell, the mixer cell's common emitter stage was removed from the design to improve linearity. The stage of the LNA was omitted in the design to increase the overall linearity of the Rx stage [40]. It is expected that a 9-dB single-sideband noise figure would be seen at the IF output of the Rx chain.

The linearity in an FMCW radar system is crucial and could sometimes be more important than noise because the linear chirping is directly related to the sensing performance [41]. Both hardware and software phase-locked loop systems are enabled on a chip via a programmable frequency divider with two division ratios. The integrated voltage-controlled oscillator is based on the push–push architecture [42] and feeds an LO to the distribution network as well as to the divider chain. It has a wide tuning range to cover the entire 57–64-GHz band, with a measured phase noise lower than –80 dBc/Hz at a 100-KHz offset.

To achieve a higher SNR at the radar Rx output, the Tx employed a differential antenna configuration so as to maintain the same linear polarization as the Rx

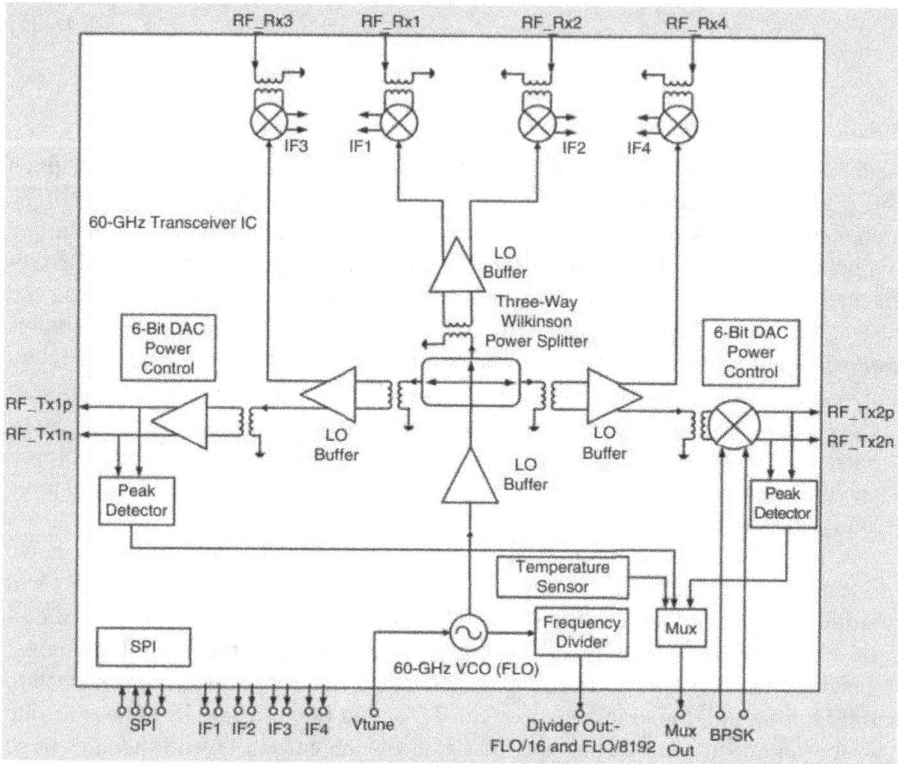

Figure 8.6 A block diagram of the millimeter-wave 60-GHz radar transceiver [40]. VCO, voltage-controlled oscillator; SPI, serial peripheral interface; BPSK, binary phase shift keying; Mux, multiplexer.

antennas. Tx-to-Rx coupling is an important factor in radar sensing because it may generate a large DC offset at the mixer output, which challenges the subsequent IF or baseband amplifiers [43, 44]. Moreover, in an FMCW architecture, it may generate low-frequency clutter that challenges the identification of objects in proximity to the radar sensor [45]. To reduce Tx-to-Rx coupling, the ground plane with additional eWLB balls and a package layout was introduced to maximize the distance and improve isolation between the antennas. The achieved isolation between the Tx and Rx antennas is more than 40 dB. However, the coupling path from Tx to Rx is not limited to the antennas but also includes the internal on-chip coupling between Tx and Rx circuit chains, which sometimes even dominate. Special circuit design methodologies, such as feedback compensation, may be introduced to mitigate the mutual coupling at the on-chip circuit level [28].

8.5 Radar in the IoT era

Except for gesture sensing, the short-range radar will find various other applications in the coming IoT era. For example, our daily activities at home could be actively monitored by the radar without disturbing our habits and regular lifestyle. Various vital signs of humans and pets such as heartbeat and respiration, sudden events (such as falling), walking, and moving trajectories could be detected. In automotive, driver's blink detection could be collected by wearable glasses using a short-range millimeter-wave radar sensing. The collected data may be uploaded seamlessly to the cloud or other remote computing terminals. With the help of AI technologies, statistics, classification, and analysis of data could help us learn more about physical conditions ourselves, as well as form a better habit or improve the quality of our lives in fruitful ways.

With the rise of the IoT concept, researchers have begun to investigate using radar technology for various IoT applications. For example, by training and classifying different kinds of materials, the RadarCat researchers demonstrated that they can recognize different objects in real time [46]. Suppose there is a physical object dictionary, a painting, and a photo-editing application on the floor, the radar system could predict accurately which object is placed on it, such as copper versus steel and milk versus water. B-reel demonstrated the feasibility of performing a custom gesture to serve as a password, which can be used to create an innovative gesture-based security system [47]. As can be seen, many doors are opened when radar technology is scaled down into a small form factor and made available to the technology community in general. In automotive, researches on driver fatigue detection are also constantly followed up. By putting short-range radar sensing on wearable glasses, a microwave Doppler sensor has been developed to monitor a driver's fatigue level on the basis of the eye blink duration and its frequency. Moreover, with emerging concepts such as metaverse, radar-based motion sensing will have broader application prospects.

8.6 Conclusion

This chapter summarizes the basic introduction to short-range radar and types of gesture radar. On that basis, several application examples of short-range radar in gesture sensing are introduced, including CW radar, FMCW radar, and in the IoT era. With the advantage of the small form factor, low power consumption, and low cost, we can envision the ubiquitous applications of short-range radar technology in the coming IoT era, including automotive, mobile and wearables, smart home hardware, augmented reality/virtual reality, gaming consoles, metaverse, healthcare, and many other exciting applications and use cases. It has been demonstrated that radar is truly a viable, powerful, and attractive technology to enhance user interaction.

References

[1] Alexe B., Deselaers T., Ferrari V. 'Measuring the objectness of image windows'. *IEEE Transactions on Pattern Analysis and Machine Intelligence.* 2012, vol. 34(11), pp. 2189–202.

[2] Sarwar M., Soomro T.R. 'Impact of smartphones on society'. *European Journal of Scientific Research.* 2013, vol. 98(2), pp. 216–26.

[3] He K., Zhang X., Ren S., Sun J. 'Deep residual learning for image recognition'. *2016 IEEE Conference Computer Vision and Pattern Recognition (CVPR)*; Las Vegas, NV, 2016. pp. 770–78.

[4] Gu C., Lien J. 'Concurrent detection of absolute distance and relative movement for sensing action gestures',U.S. Patent 20180157330A1.

[5] Hauptmann A.G. 'Speech and gestures for graphic image manipulation'. *ACM SIGCHI Bulletin.* 1989, vol. 20(SI), pp. 241–5.

[6] Mitra S., Acharya T. 'Gesture recognition: a survey'. *IEEE Transactions on Systems, Man and Cybernetics, Part C.* 2007, vol. 37(3) 311–24.

[7] Zhai S., Milgram P., Buxton W. 'The influence of muscle groups on performance of multiple degree-of-freedom input'. *Proceedings of the SIGCHI Conference on Human Factors in Computing Systems*; Vancouver, BC, 1996. pp. 308–15.

[8] Lu H., Zhang Y., Li Y., *et al.* 'Depth MAP reconstruction for underwater Kinect camera using Inpainting and local image mode filtering'. *IEEE Access.* 2017, vol. 5, pp. 7115–22.

[9] Zhang Y., Xiong Z., Wu F. 'Hybrid structured light for scalable depth sensing'. *Proc. 2012 19th IEEE International Conference on Image Processing*; Orlando, FL, 2012. pp. 17–20.

[10] Kilpelä A., Pennala R., Kostamovaara J. 'Precise pulsed time-of-flight laser range finder for industrial distance measurements'. *Review of Scientific Instruments.* 2001, vol. 72(4) 2197–202.

[11] Han J., Shao L., Xu D., Shotton J. 'Enhanced computer vision with microsoft kinect sensor: a review'. *IEEE Transactions on Cybernetics.* 2013, vol. 43(5), pp. 1318–34.

[12] Cardinal D. ExtremeTech. Available from 2017 [Accessed Oct 2018].

[13] Watson-Watt R. 'Radar in war and in peace'. *Nature.* 1945, vol. 156(3959), pp. 319–24.

[14] Wang F.-K., Horng T.-S. 'Seeing through walls with a self-injection-locked radar to detect hidden people'. *IEEE/MTT-S International Microwave Symposium Digest*; Montreal, QC, 2012. pp. 1–3.

[15] Liu X., Leung H., Lampropoulos G.A. 'Effect of wall parameters on ultra-wideband synthetic aperture through-the-wall radar imaging'. *IEEE Transactions on Aerospace and Electronic Systems.* 2012, vol. 48(4), pp. 3435–49.

[16] Watson-Watt R.. ' 'Radar in war and in peace''. *Nature.* 1945, vol. 156(3959), pp. 319–24.

[17] Adrian O. 'Future surface radar technology: from air defence to air and missile defence'. *2007 IEEE Radar Conference*; Waltham, MA, 2007. pp. 49–54.

[18] Isom B., Cheong B.L., Yeary M., *et al.* 'The Atmospheric Imaging Radar (AIR) for high resolution observations of severe weather'. *Proc. IEEE Radar Conf.*; Kansas City, MO, USA, 23-27 May; 2011–627–32.

[19] Williams P.D.L. 'Civil marine radar – a review and a way ahead'. *Journal of Navigation*. 1998, vol. 51(3), pp. 394–403.

[20] Murad M. 'Requirements for next generation automotive radars'. *IEEE Radar Conference (RadarCon13)*; Ottawa, ON, 2013. pp. 1–6.

[21] Hasch J., Topak E., Schnabel R., Zwick T., Weigel R., Waldschmidt C. 'Millimeter-wave technology for automotive radar sensors in the 77 ghz frequency band'. *IEEE Transactions on Microwave Theory and Techniques*. 2012, vol. 60(3), pp. 845–60.

[22] Lin J.C. 'Noninvasive microwave measurement of respiration'. *Proceedings of the IEEE*; 1975. pp. 1530–1530.

[23] Lin F., Zhuang Y., Song C., *et al.* 'SleepSense: a noncontact and cost-effective sleep monitoring system'. *IEEE Transactions on Biomedical Circuits and Systems*. 2017, vol. 11(1), pp. 189–202.

[24] Huang M.-C., Liu J.J., Xu W., Gu C., Li C., Sarrafzadeh M. 'A self-calibrating radar sensor system for measuring vital signs'. *IEEE Transactions on Biomedical Circuits and Systems*. 2016, vol. 10(2), pp. 352–63.

[25] Chen K.M., Huang Y., Zhang J., Norman A. 'Microwave life-detection systems for searching human subjects under earthquake rubble or behind barrier'. *IEEE Transactions on Biomedical Engineering*. 2000, vol. 47(1), pp. 105–14.

[26] Lubecke V.M., Boric-Lubecke O., Host-Madsen A., Fathy A.E. 'Through-the-wall radar life detection and monitoring'. *IEEE/MTT-S International Microwave Symposium*; Honolulu, HI, 2007. pp. 769–72.

[27] Gu C., Li R., Jiang S.B., Li C. 'A multi-radar wireless system for respiratory gating and accurate tumor tracking in lung cancer radiotherapy'. *Annual International Conference of the IEEE Engineering in Medicine and Biology Society*; Boston, MA, 2011. pp. 417–20.

[28] Gu C., Li R., Zhang H, *et al.* 'Accurate respiration measurement using DC-coupled continuous-wave radar sensor for motion-adaptive cancer radiotherapy'. *IEEE Transactions on Bio-Medical Engineering*. 2012, vol. 59(11), pp. 3117–23.

[29] Wang G., Gu C., Inoue T., Li C. 'A hybrid FMCW-Interferometry radar for indoor precise positioning and versatile life activity monitoring'. *IEEE Transactions on Microwave Theory and Techniques*. 2014, vol. 62(11), pp. 2812–22.

[30] Peng Z., Ran L., Li C. 'A K -band portable FMCW radar with beamforming array for short-range localization and vital-doppler targets discrimination'. *IEEE Transactions on Microwave Theory and Techniques*. 2017, vol. 65(9), pp. 3443–52.

[31] Wang J., Wang X., Chen L., Huangfu J., Li C., Ran L. 'Noncontact distance and amplitude-independent vibration measurement based on an extended DACM algorithm'. *IEEE Transactions on Instrumentation and Measurement*. 2014, vol. 63(1), pp. 145–53.

[32] Kim S., Nguyen C. 'On the development of a multifunction millimeter-wave sensor for displacement sensing and low-velocity measurement'. *IEEE Transactions on Microwave Theory and Techniques*. 2004, vol. 52(11), pp. 2503–12.

[33] Park B.-K., Boric-Lubecke O., Lubecke V.M. 'Arctangent demodulation with DC offset compensation in quadrature Doppler radar receiver systems'. *IEEE Transactions on Microwave Theory and Techniques*. 2007, vol. 55(5), pp. 1073–9.

[34] Itoh K. 'Analysis of the phase unwrapping algorithm'. *Applied Optics*. 1982, vol. 21(14),2470.

[35] Schadt F., Mohr F., Holzer M. 'Application of Kalman filters as a tool for phase and frequency demodulation of IQ signals'. *2008 IEEE Region 8 International Conference on Computational Technologies in Electrical and Electronics Engineering*; Novosibirsk, Russia, 21-25 Jul; 2008. pp. 421–4.

[36] Xu W., Gu C., Mao J. -da 'Noncontact high-linear motion sensing based on A modified differentiate and cross-multiply algorithm'. *IEEE/MTT-S International Microwave Symposium (IMS)*; Los Angeles, CA, 2020.

[37] Gu C., Li C. in Mosta F.I., Boric-Lubecke O., Lin J. (eds.) *Physiological radar sensor chip development, in Medical and Biological Microwave Sensors and Systems*. Cambridge, U.K: Cambridge Univ. Press; 2017.

[38] Fan T., Ma C., Gu Z., *et al.* 'Wireless hand gesture recognition based on continuous-wave Doppler radar sensors'. *IEEE Transactions on Microwave Theory and Techniques*. 2016, vol. 64(11), pp. 4012–20.

[39] Lien J. 'Soli: ubiquitous gesture sensing with millimeter wave radar'. *ACM Transactions on Graphics*. 2016, vol. 35(4), pp. 1–19.

[40] Nasr I., Jungmaier R., Baheti A, *et al.* 'A highly integrated 60 ghz 6-channel transceiver with antenna in package for smart sensing and short-range communications'. *IEEE Journal of Solid-State Circuits*. 2016, vol. 51(9), pp. 2066–76.

[41] Jiang S.B. 'Radiotherapy of mobile tumors'. *Seminars in Radiation Oncology*. 2006, vol. 16(4), pp. 239–48.

[42] Chakraborty A., Trotta S., Wuertele J., Weigel R. 'A D-band transceiver front-end for broadband applications in a 0.35 μm SiGe bipolar technology'. *2014 IEEE Radio Frequency Integrated Circuits Symposium*; Tampa, FL, USA, 01-03 Jun; 2014. pp. 405–8.

[43] Gu C., Li C., Lin J., Long J., Huangfu J., Ran L. 'Instrumentbased noncontact Doppler radar vital sign detection system using heterodyne digital quadrature demodulation architecture'. *IEEE transactions on instrumentation and measurement*. 2010, vol. 59(6), pp. 1580–8.

[44] Gu C., Li C. 'From tumor targeting to speech monitoring: accurate respiratory monitoring using medical continuous-wave radar sensors'. *IEEE microwave magazine*. 2014, vol. 15(4), pp. 66–76.

[45] Wang G., Muñoz-Ferreras J.-M., Gu C., Li C., Gómez García R. 'Application of linear-frequency-modulated continuous-wave (LFMCW) radars for tracking of vital signs'. *IEEE Transactions on Microwave Theory and Techniques*. 2014, vol. 62(6), pp. 1387–99.

[46] Yeo H.-S., Flamich G., Schrempf P., Harris-Birtill D., Quigley A. 'RadarCat: radar categorization for input & interaction'. *Proceedings of the 29th Annual Symposium on User Interface Software and Technology*; Tokyo, Japan, 2016. pp. 833–41.

[47] b-reel [online] Available from https://www.b-reel.com

Chapter 9

In-cabin vital sign monitoring

Fu-Kang Wang[1] *and Pin-Hsun Juan*[1]

9.1 Introduction

As automobile craftsmanship evolves, automobiles now include intelligent technology, environmentally friendly manufacturing process, luxurious interior, and comfort, other than basic transportation needs. These additions provide much better driving and passenger experiences. Nowadays, the number of vehicles per capita is significantly increased, safety, however, should always be the priority for every car, and the definition of safety covers drivers, passengers, freight, pedestrians, and other vehicles.

9.1.1 Driving safety: emerging threats

In addition to common causes of traffic accidents such as drunk driving or not following traffic rules, new threats are emerging every day. Whether the causes of accidents are sleep deprivation, long commutes, or long-distance driving, drowsy drivers show poor performances in attention, response, and decision-making time. According to the survey conducted by the National Sleep Foundation, 40% of the drivers had fallen asleep while driving. Statistics from the National Highway Traffic Safety Administration shows that, in the United States, 100,000 traffic collisions were caused by sleep-deprived drivers. These incidents caused 71,000 injuries and 1,550 casualties, including 109 billion dollars of losses [1]. Investigators could not confirm which of the fatal incidents were caused by sleep-deprived drivers because they might not admit they were asleep or were unaware of their fatigue. In most occurrences, no exact evidence could be found. However, it is a consensus from the academia and public sectors that the actual number may far exceed everyone's expectations.

The increase in the senior population rate is attributed to the improvement of economical, medical, and hygienic conditions in developed countries. But the

[1]Department of Electrical Engineering, National Sun Yat-Sen University, Kaohsiung City, Taiwan

rapid changes in lifestyles have induced cardiovascular diseases from which the elderly suffers. It is a haunted nightmare for all urban citizens because of the wasted time spent in stagnant traffic jams, frequent traffic accidents and road maintenance, annoying honks, polluted air, and ever-deteriorating traffic. These arouse emotions, and emotions are further turned into another form that is called "road rage," dangerous behavior attributed to drivers' tension and anger [2]. Under a mentally and physically unstable condition, a driver might be a threat to public safety.

Even with a pleasant air temperature at 22°C (72 °F), the temperature inside a vehicle would increase 1.5°C (3.2 °F) every 5 min, and after 60 min, regardless of the position of vehicle windows (closed or open), the interior temperature rises to an astonishing temperature of 47°C (117 °F) [3]. To a child seated in a child safety seat, such temperature can be fatal, as the thermoregulation of a child is not fully developed and the increase in body temperature of a child is 3 to 5 times faster compared to adults. Exposed to a temperature of 40°C (104 °F), the fastest time record of death caused by heat stroke is only 15 min. In 2018, 52 precious lives perished in the United States because either parents negligently left sleeping children or children locked themselves in vehicles [4]. Similar tragedies might happen again and again if the effects of climate change and distracted lifestyles escalate.

9.1.2 Prevention methods and their contradictions

From 2022, the European Union mandates that an advanced driver assistance system (ADAS) must be installed on all new passenger vehicles to lower global traffic deaths which are more than 1.3 million people per year all over the world. The regulation reminds us that road safety is a basic requirement for every human being. To encounter the emerging threat to driving safety, global scientists and engineers join hands to develop better and more accurate sensors, which have been incorporated into technologies developed by leading automobile manufacturers including ATTENTION ASSIST by Mercedes-Benz, Driver Alert Control (DAC) by Volvo, and Smart Senior by BMW. Current technologies can be divided into three categories to detect driver fatigue. The first category is the vehicle-based method, which makes decisions based on the vehicle driving status such as lane departure, speed patterns, and operations of the steering wheel and pedal [5]. However, these systems only provide warnings and interventions for high-risk driving behavior that has happened, but to effectively reduce the possibilities of fatal accidents, an alternative way is to implement preventive measures.

The second category is behavior monitoring [6]. When the driver is tired, the explicit expressions are frequent yawns and blinks, eventually closed eyes and head nods. Next, the driver would wake up suddenly and stretch himself/herself. This cycle repeats itself again and again, and such expressions and behavior can be seen when people are in pain or full of emotions such as anger, depression, and fidgety. Car seats have direct contact with people in cars and can receive various motions created by drivers and passengers. Behavior monitoring requires multiple sensing techniques, including image capturing, pressure detection, and acceleration monitoring. However, monitoring results may vary for different subjects who experience different physical and mental conditions. Subjects may also change in appearance or position according to different environments, times, and cultures, and have to be independently monitored in the long term for data collection and analysis. Actually, the pressure sensor cannot differentiate an entity between a human being, an animal, and an ordinary object, and furthermore, when a child is asleep and motionless, the accelerometer would detect nothing.

Most biotechnology research relates the physiological and mental condition of subjects to vital signs. For example, respiration rates (RRs) are lowered as driving time increases [7]; drivers with irregular heart rates (HRs) may cause more traffic accidents [8]; in clinical diagnosis, HR variability is widely attributed to the diagnosis of cardiovascular diseases, dysautonomia, emotions, and stress [9]; and cardiopulmonary activity is used to differentiate an object from a sleeping human. In other words, vital signs (including breath and HR) used for fatigue prevention, physical/mental monitoring, and organism existence detection are now the third category for driving safety.

Among current vital sign sensing technologies, electrocardiography (ECG) is undoubtedly the most mature and widely applied technology. Conductive gel is applied to the ECG electrodes, which are then attached to the skin around the heart, to acquire polarized and depolarized signals in neurons of the atrium and ventricle. The signals are then transmitted through an amplifier, matched, and processed to show information that allows us to learn critical clinical information [10]. In a similar way, electroencephalography and electrooculogram can detect the activities of the brain and eyeballs, respectively, to determine the status of the driver. Conductive gel, however, may dry and foul, and therefore undermine the results, and as a solution for these flaws, dry electrodes are new products proposed by researchers. Flaws of dry electrodes are high impedance between the electrode and the skin, as well as inconsistent signals during measurement. These flaws attribute to the product's low penetration rate [11]. When pressure is applied to a specific solid material, a voltage difference can be detected between the front and back of the stressed material. The voltage difference attributes to a process called the direct piezoelectric effect, which converts mechanical energy into electricity. By adjusting the thickness of each later, the piezoelectric sensor located at the chest or wrist can be used to measure the subject's breath, heartbeat, and pulse [12]. Usually, the piezoelectric material would be secured by a belt or straps to avoid noise caused by frictional electricity. The use of piezoelectric material allows us to reduce the size of products and get rid of the entangled wires once and for all. However, the ECG and products using piezoelectric materials both require direct contact with the skin making them inapplicable for users with wounds or allergies in a long-term monitoring scenario.

A MEMS-based accelerometer measures the variations of resistance or capacitance caused by the shape or movement of an inertial measurement device to measure acceleration readings. An MEMS-based accelerometer is smaller in size and costs lower. Today it has become a standard equipment in smart handheld devices and its applications are blooming. In 2011, Kwon *et al.* used a three-axis accelerometer embedded in an iPhone to monitor subjects' HRs [13]. In Reference [14], the impedance of a well-designed resonator would change with the distance to the skin, and the RF proximity sensor makes use of this effect to combine a resonator with an active circuit to create an oscillator, which is worn at the chest of the subject. The transmission frequency is shifted according to the cardiopulmonary activity, and the demodulated output signal can be used to learn breath and HR [14]. These two methods do not require direct contact with the skin. The only requirement is a flexible belt to secure the sensor. In all, the above wearable technologies [10–14] do not provide optimal comfort and cause inconvenience to customers.

Considering user experience and portability, the remote sensing method with a range of over dozens of centimeters is obviously the better choice when it comes to vital sign monitoring. The latest techniques adopt light, sound, or electromagnetic (EM) waves to achieve the intended purpose. We will provide a brief introduction to the different techniques. Camera-based photoplethysmography (PPG) receives light reflected from people's faces to determine the intensity of hemoglobin (Hgb) in blood. Hgb would absorb and reflect different wavelengths of light to achieve the purpose of remote vital sign sensing [15]. However, PPG requires steady light emission, and non-transparent objects cannot block the passage between the camera and skin, and also vital signs of users who wear sunglasses or hats would not be acquired. In addition, weather, sunlight, skin colors, birthmarks, and tattoos all affect sensing accuracy. In 2019, Xu *et al.* proposed a novel technique that used a smartphone's speaker to generate a 20-kHz soundwave and used the same phone's microphone to collect reflected signals in the environment. By analyzing the frequency distribution of the reflected wave, i.e., energy spectrum density, we can remotely measure the rise and fall of a driver's chest caused by breathing [16]. It should be noted that vehicle noises and harmonics would affect sensing accuracy, and high-frequency soundwaves can be unpleasant for infants as well as pets. Furthermore, visual images and sound in vehicles would be recorded at all times by these two techniques [15, 16]. However, continuous recording may increase computing complexity and even involves privacy and information security concerns.

Widely applied in wireless communication, EM wave has lower attenuation when travelling through the air, can penetrate nonmental medium, and is reliable in various environments. As the transmission of EM waves might be blocked or affected if subjects move around in the test field, scholars have attempted to perform non-contact vital sign sensing using the received signal strength [17] and channel state information [18] of Wi-Fi. The technique requires two connected wireless communication devices, one of which transmits and the other receives. With a complex hardware structure, the system-to-noise ratio (SNR) in such a bistatic system needs to be improved. Methods for improvement include, e.g., maintaining the line-of-sight (LOS) path between the two devices [17], using plentiful sensor nodes with corresponding routing protocols [19], or combining sensing information of multiple subcarriers [18] or multiple frequency channels [19]. Moreover, a sophisticated digital signal processing technology is strongly recommended to relieve the potentially heavy computing load, considering the constructive and destructive interferences caused by vehicle part movements and multipath signals, as well as the interference caused by passenger movements and vital signs. Other than that, how can we be sure which passenger is showing a vital sign anomaly? In section 9.2, you will learn one classic application of EM wave, radar, and its pros and cons in terms of vital sign sensing.

9.2 In-vehicle radar-based vital sign monitoring

The radar system has been a highly sensitive military confidentiality since WWII, and its development, with the technologies of electronic circuits and computers, has also played an important role in industrial and civilian fields. In the 1970s, the radar system was first applied in noncontact vital sign sensing [20]. This technology can detect vital signs of people who are bleeding, burned, or wearing heavy clothes. In addition, comfort can be increased during the sensing period because no gadgets are needed to be worn. It is a high-profile technology in terms of its advantages of applicability and convenience. In this chapter, we will introduce the development of vital sign sensing radar, as well as its challenge to the in-vehicle scenario.

9.2.1 Continuous-wave radar

Figure 9.1 illustrates a conventional continuous-wave (CW) radar for noncontact vital sign monitoring. The signal generator outputs two radio-frequency (RF) sinusoidal signals, one of which is sent through the transmit antenna (TX Ant.) to the environment as $T(t)$, and the other of which is fed into the IQ mixers. Signals received by the receive antenna contain two major components: $D(t)$, with phase modulation information, is a Doppler signal that is reflected from a moving object in relation to the radar; $C(t)$ is a clutter signal that includes reflected signals of stationary objects in the environment, TX-RX coupling, and signal leaks caused by hardware imperfection. $C(t)$, mathematically speaking, is equivalent to an attenuated and time-delayed $T(t)$. $D(t)$ and $C(t)$ are received by the receive antenna (RX Ant.), amplified with a low-noise amplifier (LNA), and sent to the other input of the IQ mixers. Multiplied by each other and filtered, the two input signals of IQ mixers can lead to output IQ signals $I(t)$ and $Q(t)$. Subsequent signal processing produces the Doppler information resulting from relative movement between the subject and the

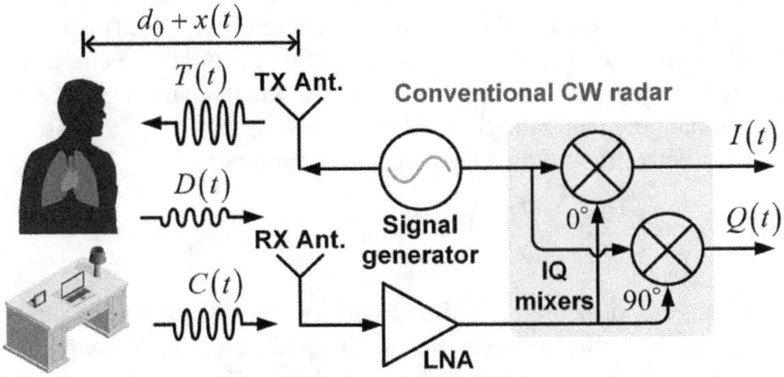

Figure 9.1 Conventional CW radar for noncontact vital sign monitoring

radar in the environment. It should be noted that a signal generator usually adopts the phase-locked loop (PLL) system, but in this experiment, the unlocked voltage-controlled oscillator (VCO) can be used to replace the PLL and greatly reduce the system complexity and circuit cost. This is because under a short sensing range (<5 m), range correlation can suppress phase noise to avoid the covering of Doppler information [21].

According to Reference [21], the output signals can be presented, respectively, as

$$I(t) = DC_I + A_{CW} \cos\left(\frac{4\pi \cdot (d_0 + x(t))}{\lambda_T}\right) \tag{9.1}$$

and

$$Q(t) = DC_Q - A_{CW} \sin\left(\frac{4\pi \cdot (d_0 + x(t))}{\lambda_T}\right) \tag{9.2}$$

where the DC offsets, DC_I and DC_Q, of two channels I and Q are the mixture of $T(t)$ and $C(t)$. Considering the mismatch of the IQ mixers, the two numbers may not be the same. A_{CW}, λ_T, d_0, and $x(t)$, which are on the right sides of (9.1) and (9.2), are, respectively, output amplitude, operating wavelength, initial distance between the radar and the subject, and movements caused by breath, heartbeat, and random body movements (RBMs). It should be noted that, if

$$4\pi \cdot d_0/\lambda_T = n \cdot \pi,\ n \in N \tag{9.3}$$

is true, the amplitude of $I(t)$ will be minimized, and if $x(t)$ is much smaller than the operating wavelength, the frequency of $I(t)$ will be twice of $x(t)$ [22]. This is the null point, which indicates the worst position of sensing effectiveness between the radar and the subject. And $Q(t)$ shows a maximum wave fluctuation and a minimum distortion. This is called the optimum point, the best position of sensing effectiveness between the radar and the subject. When

$$4\pi \cdot d_0/\lambda_T = (n+0.5) \cdot \pi,\ n \in N \tag{9.4}$$

is true, now $I(t)$ is at the optimum point, whereas $Q(t)$ is at the null point. According to (9.3) and (9.4), the null and optimum point would change alternatively, the spacing between the two points is 1/8 of the operating wavelength, and the spacing between adjacent null points and between adjacent optimum points is $\lambda_T/4$.

Null points might affect vital sign readings, and to avoid that, generally, the CW radar uses IQ mixers to generate output IQ signals, which go through DC offset calibration and arctangent demodulation (AD) to produce the Doppler phase shift [23] caused by the relative motion between the subject and radar:

$$B(t) = \tan^{-1}\left(\frac{DC_Q - Q(t)}{I(t) - DC_I}\right) = \frac{4\pi}{\lambda_T}\left(d_0 + x(t)\right) \tag{9.5}$$

The real challenge is to accurately estimate and compensate for the DC offset caused by clutter signals. Referring to Figure 9.2 and in an ideal world, the output IQ signals can be plotted as an arc in Figure 9.2's blue solid line. This arc is a piece of an ideal circle with a radius of A_{CW}, with origin as the center. Because of the DC offset, the

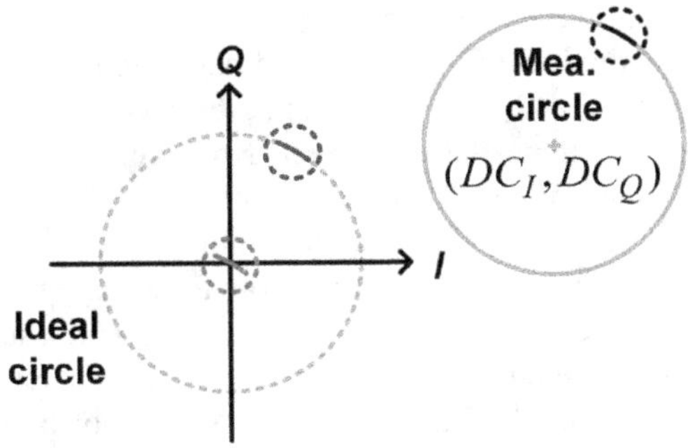

Figure 9.2 Constellation diagram of output IQ signals of a CW radar

actual measurement result is the black solid line, and the center is shifted to (DC_I, DC_Q). Considering (9.1) and (9.2), the DC components in the output IQ signals contain not only DC_I and DC_Q that should be compensated but also DC information resulting from the distance between the subject and radar. If a capacitor is used to compensate all DC components, the arc would move to the origin, as indicated by the red solid line in Figure 9.2. This would cause a serious demodulation distortion and the accurate frequency of vital signs could hardly be determined [23].

To calibrate DC offsets, a hardware solution is to send the signal generated by the signal generator to a clutter compensation circuit composed of a phase shifter and a variable gain amplifier. The compensation circuit generates a compensation signal, the amplitude of which is the same as the clutter signal but with a reversed phase. Next, the compensation signal, passing through a coupler or a power combiner, is combined with the received signal to remove clutter components [24]. However, this structure does not include an algorithmic mechanism to accurately determine the compensation signal. The mainstream method is using digital signal processing to calibrate DC offset. The Levenberg–Marquardt (LM) algorithm [25] is used to continuously adjust unknown A_{CW}, DC_I, and DC_Q, with trigonometric functions, to yield

$$(I(t) - DC_I)^2 + (Q(t) - DC_Q)^2 = A_{CW}^2 \tag{9.6}$$

where the calibrated IQ signals would match the ideal circle in Figure 9.2. During the computing process, the gradient of residual error would perform an iterative calculation until the error is small enough to be accepted. Different from the LM method, compressed sensing (CS) is a technique focusing on the sparsity of the output IQ signal's trajectory rather than absolute error. The programing of CS is much more complex, and some of the signals would be ignored to avoid convergence [26] where signal quality is poor. The above two calibration methods require a massive amount of iterative calculation to yield the calibrated results, and although accuracy and tolerance for input signal quality are high, calculation efficiency is low.

Among non-iterative methods, the algebraic method chooses several points on the IQ trajectory and measures parameters such as the center of the circle and radius to perform DC offset calibration [27]. If the length of the arc of the IQ trajectory is shorter, it means that a small displacement occurs between the test object and the radar, and the error rate would rise drastically. In 2007, Park proposed the linear demodulation (LD) method. He first turned and paralleled the measurement trajectory to the imaginary axis, and the center was at the real axis. Next, he adopted the heuristic estimator to calculate the DC offset. Nonetheless, the sensing error rate increased with the displacement of the test object. In 2017, Gao combined the LD method with the radius calibration technique to improve the sensing error [28]. First, the test object displacement was increased to obtain a bigger Doppler shift and a longer IQ polar coordinate trajectory [29]. The mixer's amplitude and imbalanced phase were calibrated and then applied to a small displacement scenario to reduce the errors. If the displacement of the test object could not be adjusted, the radar antenna would be moved to increase the Doppler shift for calibration. The system complexity is, however, higher as additional mechanical devices are required. The above non-iterative methods [27–29] may provide superior computing efficiency, but the accuracy may be inconsistent as the displacement of the object varies and the system's SNR is more demanding. In general, the above DC offset calibration methods [25–29] have a hard time balancing between accuracy, computing efficiency, and reliability. It should be noted that an exaggerating DC offset may cause serious distortion because of the saturation in both the baseband amplifier and the analog-to-digital converter (ADC).

As shown in (9.5), the Doppler shift is inversely proportional to the operating wavelength. Thus, the most instinctive way to improve the sensitivity of the CW radar is to increase the transmit frequency. However, when using millimeter wave to conduct vital sign sensing, a larger displacement would cause serious nonlinear distortion, such as high-order harmonics and intermodulation. If necessary, the distortion can be processed in time domain [29] and frequency domain [30] to clearly measure the subject's cardiopulmonary activity, thereby reducing the computing efficiency. In addition, transmission loss is much higher for millimeter waves, the use of which is therefore not applicable for sensing subjects wearing heavy clothes. The frequency of vital signs may be so close to DC to be affected by flicker noise. In 2010, Mostafanezhad was the first to propose a coherent low intermediate frequency (IF) radar to improve system sensitivity [31]. This radar consisted of an IF oscillator, which elevated the frequency of the output IQ signals, and an RF oscillator responsible for frequency adjustment. The output IQ signals were sampled and processed using digital quadrature demodulation to obtain the Doppler information. Compared to direct conversion radar, this method increases the SNR by 7 dB. In addition, as the IF carrier from the IF oscillator was sampled for demodulation, a range correlation was used to eliminate the residual noises of the two oscillators [32], and a high sampling rate ADC, which avoided aliasing, was used to improve the SNR by 15 dB. Also, an additional RF filter was required to filter the spurious signals under this single-sideband architecture [31, 32]. Recently, a double-sideband architecture [33] that does not require an RF filter was proposed. The working principle of this architecture is that the IF carrier and output IQ signals are added, respectively, to generate two envelopes. After AD, the Doppler phase is obtained. Not only the demands for ADC performances are greatly reduced but also the improvement of residual noise for SNR has remained at 15 dB. It should be noted that only by properly adjusting the amplitude of the IF carrier to avoid folded envelopes can the system provide correct sensing results.

Air and blood will repeatedly flow in and out of a contracting and relaxing lung and heart when the subject breathes and his/her heart beats. The movement of the chest can be observed to calculate the RR and HR. Clinically speaking, ECG readings are measurements of the voltage signals that control the heart, while radar readings of the displacement variations of skin surfaces caused by cardiopulmonary activities. In general, chest movements caused by breaths and heartbeats in adults can be measured by millimeters and micrometers. This explains why numerous researchers have delved into the research for increasing sensitivity. Meanwhile, the radar is extremely sensitive to random movements such as coughing, itching, and shaking.

Empirical mode decomposition (EMD) [34], wavelet transform [35], matched filter [36], and polynomial fitting [37] are different techniques that aim to remove the Doppler shift caused by body movements. The cyclostationary method [38] is also employed to investigate cycles in raw signals. However, similar time-domain algorithms [34–38] can only process signals with specific premises, while frequency-domain algorithms discover sporadic body movements and focus on stable vital sign signals with the multiplication of spectrum results in different time periods [39], joint time-frequency analysis (JTFA) [40], and generalized Warblet transform [41] that combines iterative calculations into signal parameters. However, these methods [39–41] have never been applied in any in-vehicle environments where vibration exists continuously, and accurate and real-time information cannot be acquired in such an urgent scenario. In Reference [42], Lee applied the multiple signal classification (MUSIC) algorithm to enhance frequency resolution and discovered heartbeat signals that were hidden and buried in a spectral result flooded with vehicle vibration signals and harmonics. One problem of MUSIC is that the number of tones in the spectrum has to be pre-defined for further calculation, and another is the difficulty to differentiate the correct vital sign frequency from a complex spectrum.

In 2018, Li proposed an architecture that eliminated RBMs during the sensing period [43]. As indicated in Figure 9.3, two CW radars were, respectively, placed in front of and in the back of the subject, and the sensing results were plotted in the yellow and green rectangle. Cardiopulmonary activities of the subject were omnidirectional, and when the subject moved to CW radar 1, he/she would move away from CW radar 2 at the same time or vice versa. After the demodulation phases from the radar outputs were added together, the effects of RBMs were cancelled, and vital sign information was enhanced as shown in the blue rectangle. Many devices satisfying various demands were derived from this concept. For example, two radars were placed on the same side of the subject to detect the chest and abdomen [44]; four CW radars were used to cancel two-dimensional RBMs [45]; an active tag with frequency-shifting ability was placed at a fixed position to stabilize the handheld CW radar [46]; the fundamental-and-harmonic dual-frequency CW radar was used to sense the subject and a stationary wall, respectively, to eliminate the shaking of radar [47]. The above-mentioned post-baseband processing methods [43–47] require operating frequency change or polarization diversity techniques to increase

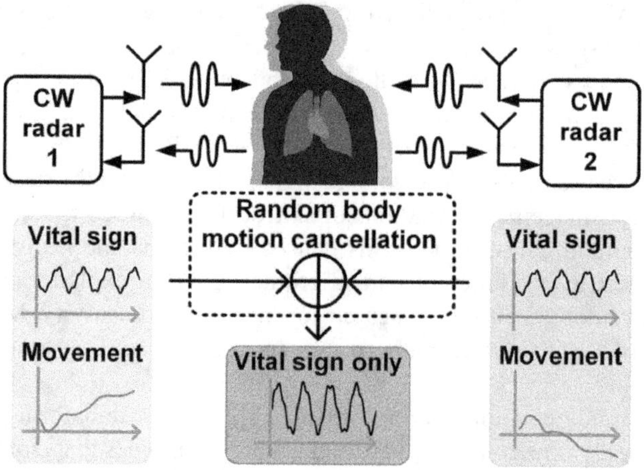

Figure 9.3 CW radar setup for cancelling the RBM of the subject

the isolation between the two echoes to avoid interferences. In addition, excessively fluctuating signals caused by body movements and clutter-caused DC offset may result in baseband circuit saturation, and the cancelling range is limited to millimeters [43–47]. An alternative way is to combine an accelerometer to reduce shaking [48] or an additional camera to cancel body movements with a handheld radar [49]. Such a hybrid method not only causes system complexity and cost to spike, but it is also limited to a restrained scenario. Overall, the above methods [34–49] are too limited to apply in in-vehicle environments and require more investments from researchers.

9.2.2 Frequency-modulated continuous-wave radar

Besides the noncontact vital sign sensing [50], the CW radar has been popular in emerging applications including animal husbandry, structural safety, machinery inspection, and posture recognition due to its excellent sensitivity and compact architecture [51]. However, an unmodulated CW radar, uncapable of providing distance information from the test object, is limited in applications. Sections 9.2.2 and 9.2.3 introduce two structures with ranging capability: frequency-modulated continuous-wave (FMCW) radar and impulse-radio ultra-wideband (IR-UWB) radar. Besides inferior Doppler sensitivity, FMCW and IR-UWB radars would still face other common defects such as range resolution and hardware complexity.

The architecture of an FMCW radar is basically the same as the conventional CW radar, as shown in Figure 9.1. The major difference is that the output signal of the signal generator is a frequency-modulated signal rather than a sinusoidal signal. The operation diagram of the FMCW radar is indicated in Figure 9.4. When a triangular wave with a cycle of T_s is used for frequency modulation, the black line shows the frequency versus time graph of the transmit signal $T(t)$, and the sweep bandwidth is B. The echo signal reflected from the test object would be received by

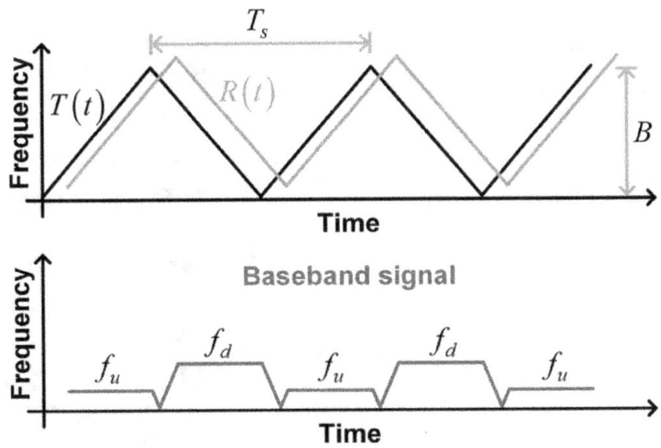

Figure 9.4 Operation of an FMCW radar with triangular chirp modulation

the radar and represented as $R(t)$, plotted by the green line. The sweep bandwidth of $R(t)$ is also B, but a frequency shift from $T(t)$ occurs because of the Doppler effect caused by the relative motion to the radar. The frequency of the baseband signal is down-converted and demodulated by a mixer, and the frequency versus time graph is shown in Figure 9.4 and plotted by the blue line. During the time when frequency ascends and descends, the stable frequencies of the baseband signal are f_u and f_d, respectively. The following equations

$$\begin{cases} R_s = T_s \cdot (f_u + f_d) / (8B) \\ V_s = c \cdot (f_u - f_d) / (4f_c) \end{cases} \tag{9.7}$$

give us the distance R_s and relative speed V_s between the test object and the radar [52]. c and f_c in (9.7) are, respectively, the speed of light and center frequency. If V_s is smaller, a longer T_s is required to lower the frequency resolution in order to tell the difference between f_u and f_d, but on the contrary, T_s needs to be much smaller than the vibration frequency of the test object to provide the adequate time resolution. This is trade-off scientists need to make for vital sign sensing. Moreover, if multiple moving objects exist in the environment, using a wrong stable frequency in the calculation process would result in nonexistent test objects (ghost target). This problem may be solved by applying multiple chirps [53] but may cause another problem: the increase in complexity of the system and calculation. Moreover, the combination of chips remains undecided because it is not possible to learn the number of objects in advance.

Figure 9.5 shows us another signal processing method of the FMCW radar: two-dimensional fast Fourier transform (2-D FFT). The left side of Figure 9.5 is a structure that uses ramp signals for frequency modulation and the radar sends a signal to sweep frequencies. The sweep cycle and bandwidth are, respectively, T_s and B. Then a mixer is used to down-convert and demodulate the frequencies, and the corresponding baseband signal for every chirp is indicated by the blue line in Figure 9.5.

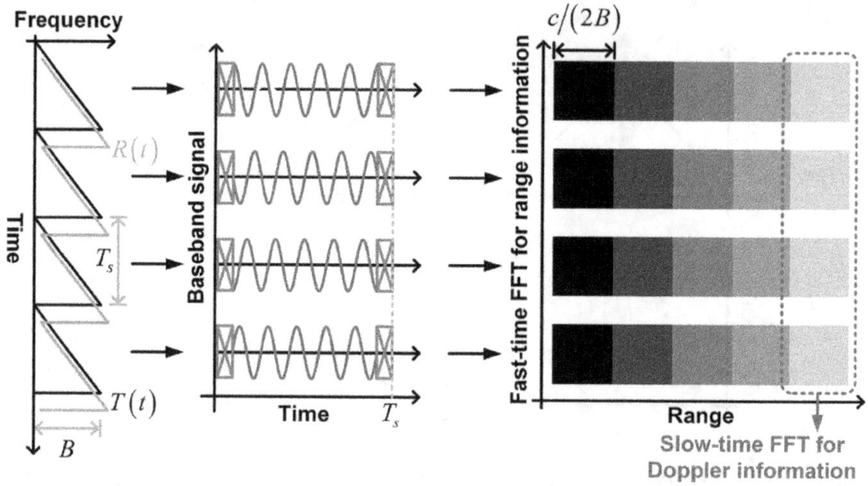

Figure 9.5 2-D FFT for range-Doppler detection of FMCW radar

After disregarding the imperfect sectors caused by $R(t)$ time delay and margin of the ramp signal, the time span is slightly smaller than T_s. Every segment of the baseband signal is processed with the FFT, and the spectrum is shown on the right side of Figure 9.5. Each tone represents a test object, and the center frequency of a tone can be converted to provide the distance from the test object to the radar. It should be emphasized that the minimum spacing between tones is known as the range resolution, $c/(2B)$. Within this range, any test object more than one cannot be recognized. Amplitude and phase information is recorded at every fixed frequency (range) in every spectrum. The recorded information is processed by the FFT again to provide the Doppler information at the specific location, and the time span is determined by the observed frequency range. The time spans of the input signals in the two FFT conversions are immensely different (units are 1,000 times different, milliseconds and seconds, when applying to noncontact vital sign sensing), and therefore they are named fast-time and slow-time FFTs.

FMCW radar is highly anticipated because of its ability to provide distance and Doppler information, as well as noncontact vital sign sensing for subjects at different locations. Technology bottlenecks still need to be overcome such as flicker noise caused by active circuits. The power spectral density of flicker noise is inversely proportional to frequency and usually would cover the target component in the baseband signal. An intuitive solution is to increase the delay inside the circuit or lower sweep cycle T_s to move the target beat frequency outside the noise interval. If we enlarge the frequency difference between the transmit and receive signals, the coherence between them is also lowered, and the invalidity of the range correlation mechanism leads to a higher residual phase noise. Additionally, this method reduces the maximum coherent processing interval (CPI) to influence the frequency resolution and SNR of the slow-time FFT spectrum. To improve Doppler sensitivity, another method is to apply a hybrid FMCW–CW architecture signal generator, which alternatively outputs sinusoidal and frequency-modulated signals to yield subjects' vital sign information and distance information. The application range is limited if multiple subjects exist since pairing of the subjects and the sensing information requires the subjects to be situated in different azimuths and the use of antennas with beam-forming capability [54].

Another critical bottleneck is the range resolution limited by sweep linearity and bandwidth. Previous references indicate that, with the predistortion technique, compensation can be made if the frequency variation curve of the signal generator is measured in advance [55]; or, by approximating the baseband signal to high order polynomial models, nonlinear characteristics can be calibrated by nonuniform sampling [56]. The effects of the above two methods can be influenced by environments or approximate methods. Currently, the principal focus of the design of signal generators is to improve the range resolution. Techniques such as delta–sigma modulation [52], offset architecture [57], and direct digital synthesizer [58] are implemented in the design of FMCW generators to achieve the goals of low-phase noise, wide tuning range, fast settling time, and precise frequency control. But the complexity, power consumption, and cost would rise as compared with the CW radar. If the sweep bandwidth or the number of antenna units is confined, a common signal processing technique called the super-resolution algorithm [59] is applied in 2-D FMCW imaging systems. MUSIC and estimation of signal parameters using the rotational invariance technique utilize the eigen and singular value decomposition to differentiate signals and white noise subspaces, in order to improve the spectral and direction-of-arrival estimation resolution. However, besides the heavy computation load and ghost target issue, the performance of such methods is affected by several factors [60]. In short, the efficiency and robustness of super-resolution algorithms remain a challenge that needs to be resolved.

9.2.3 IR-UWB radar

The development of FMCW radar is restrained by operating bandwidth, severe interruption to other wireless communication products due to continuous broadcasting of EM waves, and the lack of a dedicated frequency band. For example, when an FMCW radar operates at 5.725–5.875 GHz, which is the industrial, scientific, and medical (ISM) band, theoretically speaking its range resolution is 1 m and does not satisfy the demand for the in-vehicle scenario. If the range resolution is expected to be smaller than 0.5 m, an ISM frequency band of 61–61.5 GHz or an automotive radar band of 76–81 GHz is needed. However, the costs of millimeter-wave components and packaging testing remain obstinately high. As a matter of fact, the IR-UWB radar, as depicted on the left of Figure 9.6, is the best solution for this problem. This radar transmits a periodic RF pulse signal for sensing. By adjusting the waveform and cycle T_s of the transmit signal, the regulation for UWB spectral mask, an equivalent isotropically radiated power is smaller than –41.3 dBm/MHz within the frequency range of 3.1–10.6 GHz, can be fulfilled and the range resolution $c/(2B)$ can be down to 0.02 m. The middle part of Figure 9.6 shows the baseband signal after echoes are received by the radar. The time span of the transmit pulse

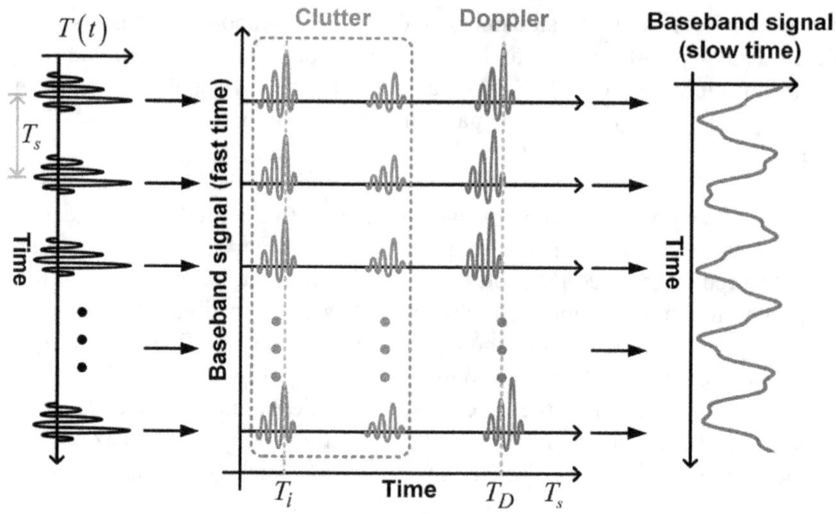

Figure 9.6 Sensing principle of an IR-UWB radar

is T_s, and the longest time is needed to calculate the fast-time baseband signals in each section. The pulse waveforms of the baseband signal represent different objects in the environment, and the time when waveforms appear tells us the distance, $R_i = cT_i/2$, between the objects and the radar. The fluctuating waveforms caused by the clutter signal are plotted in the red line and remain in a fixed time and position. If the object shows relative motions to the radar, the waveform position would change as shown by the blue line. The Doppler information of a moving object can be found by observing the amplitude or phase of a fast-time baseband signal at a specific time (i.e., T_D). The slow-time baseband signal, which combines multiple sectors, is indicated on the right of Figure 9.6.

Figure 9.7 introduces the hardware structure of an IR-UWB radar. The RF sinusoidal signal would be, respectively, sent to the transmitter that generates transmit

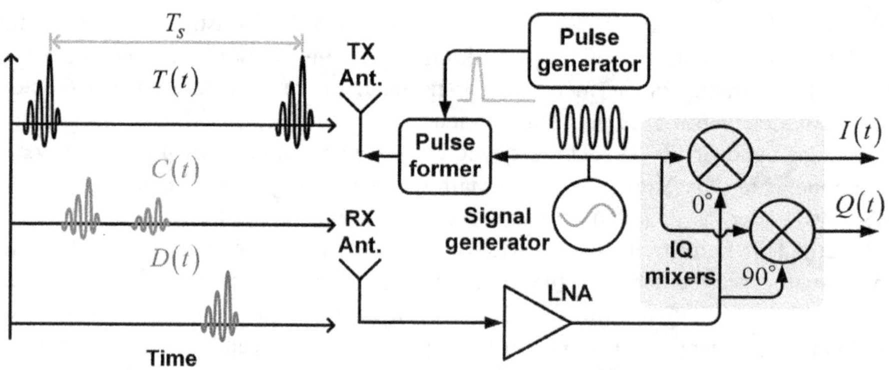

Figure 9.7 Block diagram of an IR-UWB radar

signal and the IQ mixers as a local oscillating signal. Compared to the CW and FMCW radar, the IR-UWB radar requires a pulse generator, which comprises a variable delay circuit and a digital logic circuit, to provide a baseband pulse wave with a width of hundreds of picoseconds [61]. Next, a pulse former is used to convert the input RF sinusoidal signal and baseband pulse signal to the RF pulse signal, which is then transmitted through the UWB antenna. The echo is received by the antenna, amplified by the LNA, and mixed with the sinusoidal signal. The mixer outputs a baseband signal, which is then sampled and digitally processed. The pulse former can be created with several methods: 1. Mixer [62], using RF sinusoidal signal as the RF signal and baseband pulse wave as the LO signal; 2. Switch [63], using RF sinusoidal signal as the input signal and baseband pulse wave to control whether the RF sinusoidal signal can pass through the switch; and 3. Switched-mode power amplifier (PA) [64], using RF sinusoidal signal as the input signal, baseband pulse wave to control the amplifying rate. The above methods [62–64] can all generate RF pulse wave signals for the purpose of sensing. An RF pulse wave is characterized by broadband and high peak-to-average power ratio, and the corresponding active components of a transceiver are generally more complex in design.

An IR-UWB radar only transmits a signal in a very short time, and the energy is dispersed in a wide bandwidth and is appealing to military applications because the radar wave is difficult to be detected or intercepted [51]. But this radar shows dissatisfactory performance in SNR, due to its smaller echo power, and poor sensing sensitivity in minor chest movement. The usual way of elevating pulse width, duty cycle, or pulse repetition frequency (PRF) is obviously impracticable in the UWB standard under the regulations for spectral mask. The usual solutions at present include using logarithmic, AD, or complex signal demodulation (CSD) to acquire phase variation of the baseband pulse signal caused by chest movements [65, 66]. The average-removed output IQ signals, $I'(t)$ and $Q'(t)$, through CSD are represented as complex numbers, which are $I'(t) + jQ'(t)$, on which the complex Fourier transform is performed to acquire vital signs. The complex method is more resistive to DC offset; thus, the precise DC offset calibration can be eliminated before CSD [43]. Additionally, with the state-space method to further process the demodulation information from AD and CSD, the SNR of heartbeat signals is increased and the measurement results are more accurate [66]. However, the above methods may still result in covered heartbeat signals due to harmonics or intermodulation components. Combining the CW radar with the pulse radar would alternatively transmit RF pulse and sinusoidal signals for sensing, but such a hybrid pulse-CW radar [67] is still bound by the lack of applicable scenario as the hybrid FMCW-CW radar [54] is. In Reference [68], Wang *et al.* placed the transmitter and the receiver at the front and back of the subject when performing penetration sensing. The SNR measured was superior. However, to lower the propagation loss, the range of operating frequency of this system was 224–836 MHz, unsuitable for the spectrum allocation in the UWB system.

As shown in Figure 9.6, the pulse waveform in the baseband signal of the IR-UWB radar is the reflection of objects in the environment. Turning on the receiver only at a specific time can reduce the effect of clutters. The IR-UWB radar, with its astonishing range resolution as low as 2 cm, fits the in-vehicle scenario where sensing of people at different locations and with different postures up to 32 [69] is required. Using this radar allows us to know whether the driver is using a cell-phone when driving [70], whether children are left in cars suffering heat injuries, or whether the driver is deficient in attention and is prone to accidents. Moreover, the slow- and fast-time method is beneficial for in-vehicle vital sign sensing [70–74] by avoiding vibrations of the vehicles and influences caused by RBMs of drivers and passengers. In Reference [70], a specific location of the target was first determined. Considering the periodic cycle of the signals produced by breaths and heartbeats, the slow-time baseband signal was separated into several segments of sub-signals, each of which was fitted using sinusoidal signals. This process continued until the recreated signal showed a maximum correlation. Yang conducted a JTFA on slow-time baseband signals of a longer sliding window length (30 s) to provide a stable breath information [71]. Comparing the sensing results of 16 locations in the car, the best location was found where the radar was set at the rear view mirror and when an LOS transmission path was established at the subject's chest. To avoid interference, Schires adopted a near-field antenna with an IR-UWB radar and placed the radar at the back of the subject for sensing [72]. To acquire accurate RR and HR information, Schires needed to compare two sets of vital signs from different sources. One source used a cross correlation for different fast-time windows; another used the slow-time baseband signals that went through DC offset calibration, AD, filtering, and frequency-domain processing.

When using the EMD method, the average of upper and lower envelopes is deducted in the recursion of the raw signal, and the difference signal satisfies the regulations regarding local maxima/minima, zero crossing, and envelopes in the intrinsic mode function (IMF). This process is repeated until no new IMF can be acquired. The EMD is completed to acquire the instantaneous frequency of unstable or nonlinear signals, and vital signs can still be obtained even if the subject's body is moving randomly [34]. This method, however, is sensitive to noise, and during the recursive process, the number of IMFs cannot be decided. The load of calculation is therefore increased, and real-time information cannot be provided. In 2014, Dragomiretskiy and Zosso proposed the variational mode decomposition (VMD) [75]. The core assumption is the limited bandwidth of the individual modes. The major concepts include Wiener filtering, Hilbert transform, frequency mixing, and heterodyne demodulation. Not only a non-recursive signal decomposition process increases calculation efficiency, but also the VMD achieves popularity due to its excellent robustness and can work with other algorithms for the IR-UWB radar's vital sign sensing. Yang calculated the correlation of adjacent slow-time baseband signals to further determine and deduct the superposed signal caused by the multipath in the calculations of the VMD [73]. Or, as Reference [74] did, a smooth filter and a loopback filter were used to lower the effect of noise and clutter, after the subject's position was discovered and a VMD calculation was performed. These two methods can both effectively reduce the effect of body movements and increase accuracy. Nonetheless, as a vehicle of subject experiences a considerable relative displacement (higher than the range resolution), it is still challenging for the technology of the current IR-UWB radar [70–74] to continuously track positions and obtain vital signs.

During the operation of the IR-UWB radar, the transmitter and the receiver can only be opened at a specific time. Considering the confined space inside a vehicle, the synchronizing mechanism for switching must be as precise as a nanosecond. Therefore, a high sampling rate ADC is required to sample the baseband signal's pulse waveform for subsequent signal processing. In Reference [72], the ADC's sampling rate was higher than 40 GHz, leading to a demanding requirement for the overall speed, memory, and noise of the circuit. In Reference [76], a solution of hardware architecture was proposed to generate two FM pulse signals, one of which was a transmit signal and the other was a reference signal. The PRFs were f_p and $f_p-\Delta f$. After the reference signal and the distance between the test object and the radar were mixed together, it could be found that the fast-time window length was the inverse of Δf, and the requirement of sampling rate of the ADC is effectively reduced. In the experiment, the fifth Gaussian derivative with a standard deviation of 51 ps and full-width at half-maximum of 180 ps was chosen. When f_p and Δf were chosen to be 200 MHz and 25 Hz, the ADC's sampling time was 30 μs, and the range resolution was up to 0.56 mm, it was possible to detect minor displacement of the chest caused by breaths and heartbeats. However, this system requires an additional FM pulse signal generator, and a maximum sensing range limited to 0.75 m. And the processing of baseband signals can be more complex when multiple subjects exist. Overall, the precise synchronizing mechanism as well as the state-of-the-art hardware components results in the pertinaciously high cost of the IR-UWB radar and limits its popularity among industrial and civilian fields.

9.3 Potential technology: self-injection-locked radar

When a conventional radar system is applied to detecting the vital signs of a subject wearing heavy clothes inside a vehicle, the system may suffer the problem of ineffective sensing due to the severe interferences of multiple paths and clutters, random movements of vehicles or human bodies, and the existence of multiple subjects. Therefore, a novel architecture, named self-injection-locked (SIL) radar, that shows enormous potential was invented. The following section would briefly introduce the working principle and developmental status of this radar with relevant references for interested readers to find technical specifications and math deduction.

9.3.1 Sensing principle and clutter immunity

The theoretical model of the SIL technology has been used to simulate how the PLL compromises communication quality because of the PA and the leaked signal of the VCO in wireless communication systems [77, 78]. This technology was first applied in the radar system in 2010 [79], and since then, it has attained distinction in the applications of cardiopulmonary monitoring [80]. Figure 9.8 displays a conventional SIL radar architecture, the high-frequency oscillating signals of which are generated by the SIL oscillator (SILO). The oscillating signals are then separated

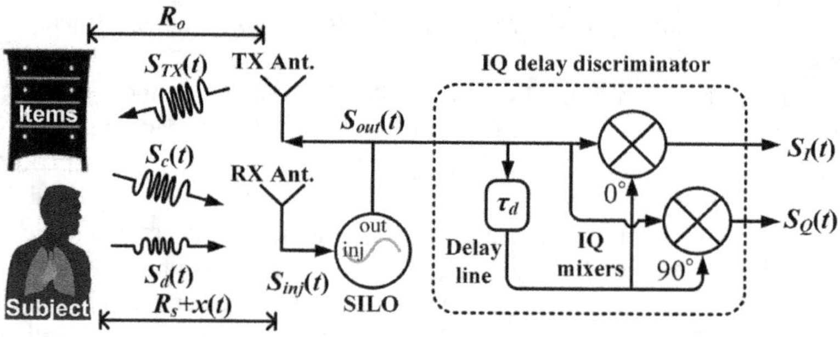

Figure 9.8 Block diagram of the conventional SIL radar

into two channels, one connected to a transmit antenna that transmits EM waves to air and the other connected to an IQ delay discriminator that performs noncoherent frequency demodulation to acquire baseband signals. It is noteworthy that the echo signal is received by the receive antenna and sent to the SILO, rather than the IQ mixers, to form an SIL loop. Benefitted from its excellent sensitivity, stationary clutter cancelling capability, and low system complexity, the SIL radar has received much attention and continued to attract researchers who delved into related research.

Assume that an inherent oscillation signal $S_{osc}(t)$ of the SILO has a constant amplitude E_{osc} and an instantaneous frequency $\omega_{osc}(t)$. The injection signals, $S_c(t)$ and $S_d(t)$, are, respectively, a clutter signal reflected from a stationary object and the Doppler signal from the subject, and the constant amplitudes of them are, respectively, $E_{inj,c}$ and $E_{inj,d}$. The phase difference of $S_c(t)$ and $S_d(t)$ from $S_{osc}(t)$ is, respectively, α_c and $\alpha_d(t)$. When the SILO is operating in SIL state, the output signal of the oscillator can be represented as $S_{out}(t)$, the constant amplitude as E_{out}, and the instantaneous frequency as $\omega_{out}(t)$. Assuming that the injection signal $S_{inj}(t)$ is much weaker than the inherent oscillation signal, based on Adler's analysis [81] and previous research [82, 83], the SILO's instantaneous output frequency can be represented as

$$\omega_{out}(t) = \omega_{osc} - \omega_{LR,c}\sin\alpha_c - \omega_{LR,d}\sin\alpha_d(t) \tag{9.8}$$

where

$$\begin{cases} \alpha_c = \dfrac{2\omega_{osc}}{c}R_o \\ \alpha_d(t) = \dfrac{2\omega_{osc}}{c}\left(R_s + x(t)\right) \end{cases} \tag{9.9}$$

where R_o and R_s are the distances of the SIL radar from the stationary object and subject, and $x(t)$ is the displacement fluctuation of the subject's chest. $\omega_{LR,c}$ and $\omega_{LR,d}$ are, respectively, the frequency locking ranges of the clutter signal and the Doppler signal and can be represented as

$$\begin{cases} \omega_{LR,c} = \left(\omega_{osc} \times E_{inj,c}\right)/2Q_{\text{tank}} \times E_{osc} \\ \omega_{LR,d} = \left(\omega_{osc} \times E_{inj,d}\right)/2Q_{\text{tank}} \times E_{osc} \end{cases} \tag{9.10}$$

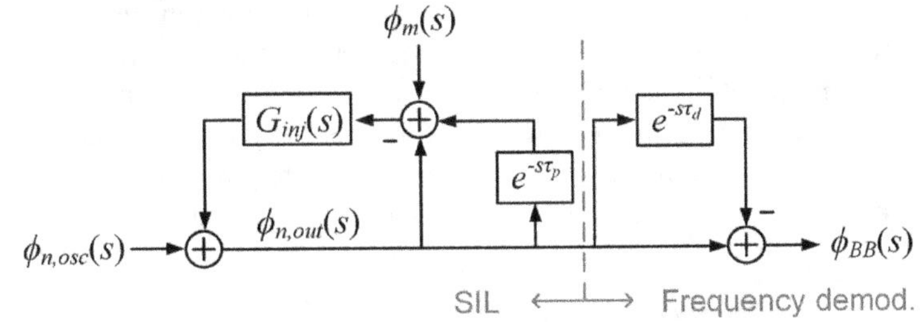

Figure 9.9 Frequency-domain of the SIL radar shown in Figure 9.8

Q_{tank} is the quality factor of the LC tank circuit in an oscillator. As indicated in (9.1), the output frequency of the SILO may be affected by the injection signal and the maximum frequency shift equals the locking range. The frequency deviation caused by the injection phase is represented as the phase-to-frequency modulation. It should be noted that if the injection signal is too strong and has a specific phase difference with $S_{osc}(t)$, the SILO may be unstable and show a drastic and irregular variation in its output frequency [84]. After the output signal of the IQ delay discriminator is arctangent demodulated, baseband signal $S_{BB}(t)$ can be represented as

$$S_{BB}(t) = \arctan\left(-S_Q(t)/S_I(t)\right) = S_{dc} + S_{ac}(t) \tag{9.11}$$

where

$$\begin{cases} S_{dc} \approx \pi \mathrm{Mod}\left(\omega_{osc}\tau_d, \pi\right) - \omega_{LR,c}\tau_d \sin\left(\dfrac{2\omega_{osc}}{c}R_o\right) + \omega_{LR,d}\tau_d \sin\left(\dfrac{2\omega_{osc}}{c}R_s\right) \\ S_{ac}(t) \approx \dfrac{2\omega_{osc}\omega_{LR,d}\tau_d}{c} \cos\left(\dfrac{2\omega_{osc}}{c}R_s\right) \cdot x(t) \end{cases} \tag{9.12}$$

The subject's vital sign can be obtained by $S_{ac}(t)$, where τ_d is the time delay in the frequency discriminator. The longer the τ_d, the larger the amplitude of AC signals. The DC offset S_{dc} can be compensated by selecting an appropriate operating frequency ω_{osc} or adding an additional phase shifter to adjust the injection phase [85].

Figure 9.9 shows the equivalent model of the SIL radar in the frequency domain without considering clutter signals. $\phi_{n,osc}(s)$ is the instantaneous phase noise of the SILO, and $\phi_m(s)$ is directed by the Doppler signal. The relation of the output instantaneous phase, according to [86], can be represented as

$$\phi_{n,out}(s) = H_m(s) \cdot \phi_m(s) + H_o(s) \cdot \phi_{n,osc}(s) \tag{9.13}$$

$$\begin{cases} H_m(s) = \dfrac{\omega_{LR}\cos\alpha_{d,0}}{s + \omega_{LR}\cos\alpha_{d,0} \cdot \left(1 - e^{-s\tau_P}\right)} \approx \dfrac{\omega_{LR}\cos\alpha_{d,0}}{s\left(1 + \omega_{LR}\cos\alpha_{d,0} \cdot \tau_P\right)} \\ H_o(s) = \dfrac{s}{s + \omega_{LR}\cos\alpha_{d,0} \cdot \left(1 - e^{-s\tau_P}\right)} \approx \dfrac{1}{1 + \omega_{LR}\cos\alpha_{d,0} \cdot \tau_P} \end{cases} \tag{9.14}$$

where $\alpha_{d,0}$ is the constant phase of the injection signal and is equal to $2\omega_{osc}R_s/c$. τ_P is the time delay caused by the SIL loop. In (9.13) and (9.14), it can be found that the decrease of the modulation frequency would increase $H_m(s)$, and this phenomenon is suitable for amplifying low-frequency modulation information such as vital signs. On the contrary, $H_o(s)$ shows an attenuation that increases with τ_p. With a far enough range, this second phenomenon can be used to suppress the phase noise. The equivalent architecture of the IQ delay discriminator in Figure 9.8 is represented on the right side of Figure 9.9. When $s\tau_d$<<1 is satisfied, the frequency-domain expression of the Laplace-transformed baseband signal can be approximated as

$$\begin{aligned}\phi_{BB}(s) &\approx \phi_{n,out}(s)\cdot s\tau_d\\ &= \frac{\omega_{LR}\cos\alpha_{d,0}\cdot\tau_d}{1+\omega_{LR}\cos\alpha_{d,0}\cdot\tau_p}\cdot\phi_m(s)+\frac{s\tau_d}{1+\omega_{LR}\cos\alpha_{d,0}\cdot\tau_p}\cdot\phi_{n,osc}(s)\end{aligned} \tag{9.15}$$

According to (9.14), $G_{inj}(s)$ in Figure 9.9 is actually $\omega_{LR}\cos\alpha_{d,0}/s$. Considering the effects of integral and differential caused by the vital sign information of the SIL mechanism and delay discriminator, the SIL radar can be equivalent to a first-order delta–sigma modulator, the noise shaping of which can elevate the system's SNR [86]. Comparing the SNR ratio before and after the delta-sigma process, the SNR gain is

$$G_{SNR}(s) = \left|\frac{\omega_{LR}\cos\alpha_{d,0}}{s}\right|^2 \tag{9.16}$$

According to (9.10) and (9.16), the SNR gain increases with the square of ω_{osc}. The Doppler phase shift is directly proportional to ω_{osc}, and if ω_{osc} doubles, the system's SNR would be increased by 16 times, while the SNR gain of conventional CW radar is four times. This also shows that the SIL radar is more significant in the SNR gain brought by frequency increase. Benefitted from the locking range and a vital sign frequency usually smaller than 10 Hz, the SNR gain in actual practice may reach 70 dB [86]. In consideration of the frequency shift caused by the clutter signal, the SNR gain affected by the clutter can be obtained as

$$G_{SNR,c}(s) = \left|\frac{\omega_{LR,d}\cos\left(\alpha_{d,0}\left(1-\omega_{LR,c}\cdot\sin\alpha_c/\omega_{osc}\right)\right)}{s}\right|^2 \tag{9.17}$$

Comparing (9.16) and (9.17), we learned that the maximum values of the two are identical, meaning that the clutter does not affect SNR's optimal performance.

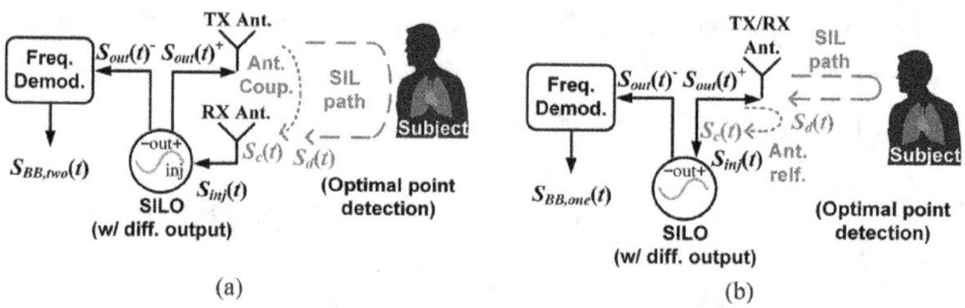

Figure 9.10 Experimental setups for vital sign detection. (a) Two-antenna SIL radar with effect of antenna coupling clutter. (b) One antenna SIL radar with effect of antenna reflection clutter.

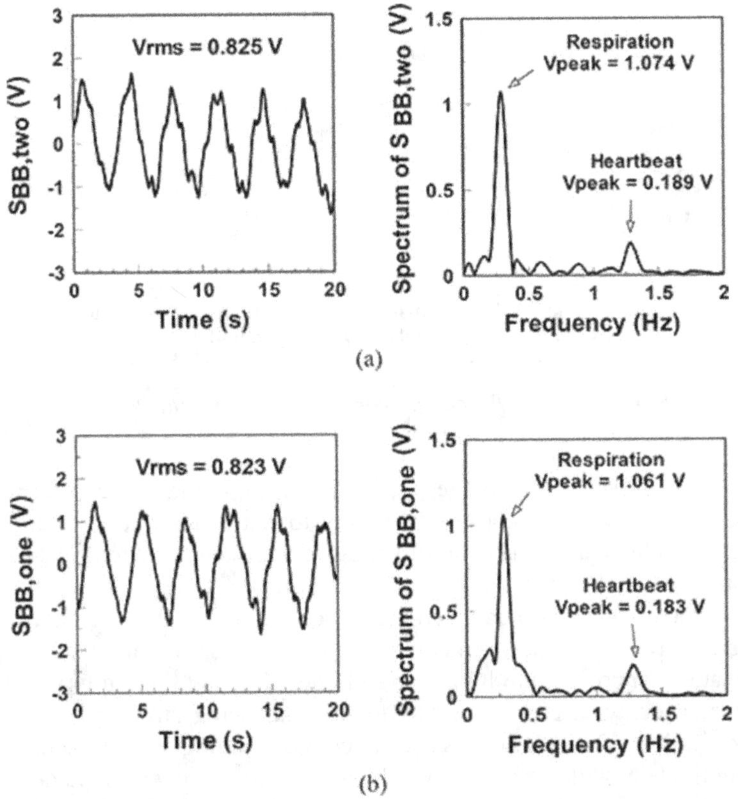

Figure 9.11 Measured results when using the experimental setup in Figure 9.10. (a) Two-antenna SIL radar. (b) Single-antenna SIL radar

Figure 9.10a,b are, respectively, two-antenna and single-antenna SIL radar architectures for noncontact vital sign monitoring. Different from the conventional CW radar that requires a circulator or coupler with high isolation to separate the transmit and receive signals, single-antenna SIL radar can be connected to the antenna through the output port of SILO, which does not include a buffer circuit, and the received echo signal would be injected to the SILO to operate in the SIL state [87]. Without considering the environmental factors, clutters of the two-antenna architecture are caused by the coupling path between the TX/RX antennas, while the clutters of the single-antenna architecture are caused by the reflected signals of the antenna. From the previous study [88], the transmission coefficient between Tx and Rx antennas in Figure 9.10a is around 0.03. As for the single-antenna SIL radars in Figure 9.10b, the antenna reflection coefficient is around 0.2. Figure 9.11 demonstrates the vital-sign results in the experiment in Figure 9.10. The root-mean-square (rms) values of the time-domain waveform in Figure 9.11a,b are 0.825 and 0.823 V, respectively. The output spectra are Fourier-transformed, and the peak tones associated with respiration are 1.074 and 1.061 V, respectively. Those associated with

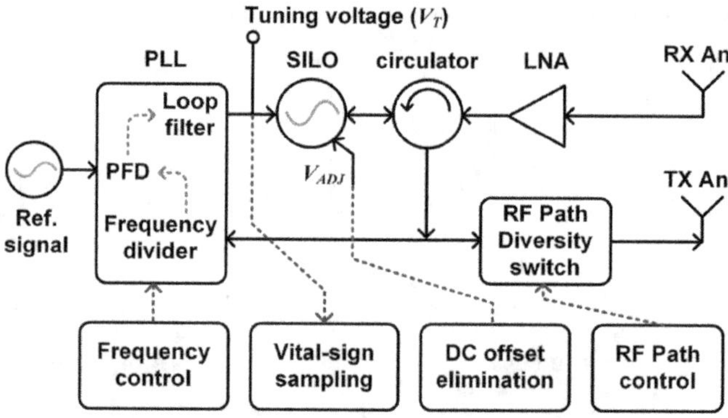

Figure 9.12 *Block diagram of the PSIL radar system*

heartbeat are 0.189 and 0.183 V, respectively. Based on the above experiment and theory, the SIL radar shows strong immunity to stationary clutter. Therefore, the two-antenna architecture, which aims to mitigate the antenna clutter, can be replaced by the single-antenna one to reduce system complexity.

The output frequency-modulated signals of the SILO should be accompanied by a non-coherent frequency demodulator. Figure 9.8 shows a common IQ delay discriminator composing one IQ mixer and one RF delay line. In terms of circuit integration, a delay discriminator has trouble balancing size, cost, and loss. In Reference [89], a differentiator was used to convert the frequency-modulated signal to the amplitude modulated signal, which passed through an envelope detector and a baseband amplifier to provide a demodulated signal. The increase in operating frequency would help reduce the size of the microstrip differentiator. However, gain flatness should be considered to avoid future distortion.

9.3.2 EM interference and nonlinear distortion

As seen in (9.1), the output frequency of a conventional SIL radar is modulated by the injection signal, and the center frequency cannot be determined due to the clutters and will cause electromagnetic (EM) interference (EMI) to other wireless devices. *Wu et al.* proposed phase- and self-injection-locked (PSIL) radar [90], the architecture of which is portrayed in Figure 9.12. The core element of the PSIL is composed of a SILO with two voltage tuning ports that are connected, respectively, to the PLL and the DC offset cancellation circuit. The oscillator is affected by the PLL as well as the SIL. If the output frequency of the SILO varies according to the subject's chest movement, the PLL would automatically modulate frequency tuning voltage V_t to cancel the frequency shift caused by the SIL, maintain the output frequency, and solve the EMI issue. In addition, vital sign information can be retrieved from V_t. If the system is troubled by the voltage shift of the demodulated signal caused by clutters and hardware imperfections, a DC offset cancellation circuit can cancel the voltage shift by providing an inverse voltage. Additionally, from (9.16), we found that the injection phase $\alpha_{\mathcal{T}0}$ will cause null detection point at every odd multiple of

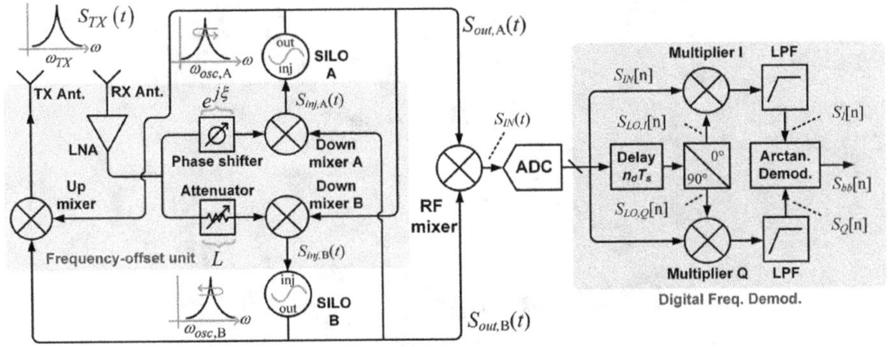

Figure 9.13 *Block diagram of FOSIL radar architecture with digital frequency demodulation*

90°, and therefore we can solve this issue by switching the RF path to provide a 90° phase difference.

The frequency-offset SIL (FOSIL) radar was proposed in 2020 [91]. Shown on the left side of Figure 9.13, a FOSIL radar combines two separate SIL radars, the two of which are distinguished by putting A and B in the subscripts. The output signals of the two SILOs were mixed by the up-conversion mixer and radiated into the air through an antenna. The RX antenna received echo signals, which were amplified and separated into two channels. One channel passed through the phase shifter and was connected to the down-conversion mixer A as the RF signal. The LO signal of the down-conversion mixer A was the output signal of SILO B. The down-converted signal $S_{inj,A}(t)$ was the injection signal of SILO A and was injected into SILO A to establish an SIL state. Likewise, the other channel passed through a VA and a down-conversion mixer, and the output signal $S_{inj,B}(t)$ was the injection signal of SILO B. According to (9.1), the frequency shift was related to the locking range and injection phase of the oscillator. In this architecture, the transmission coefficients of the variable phase shifter and the VA were, respectively, represented as $e^{j\xi}$ and L. With the help of the following condition

$$\begin{cases} L = \dfrac{Q_B}{Q_A} \times \dfrac{\omega_{osc,A}}{\omega_{osc,B}} \times \dfrac{E_{osc,B}}{E_{osc,A}} \\ \xi = \pi \end{cases} \tag{9.18}$$

the frequency shift of the two SILOs opposes each other during the sensing period. Thus, the output frequency of the FOSIL radar can be fixed to solve the EMI issue.

This architecture can be coupled with a digital non-coherent frequency demodulation technique [92], as shown on the right of Figure 9.13. The output signals of SILO A and B are represented as the RF and LO signals of the RF mixer, the output signal of which can be presented as a discrete signal $S_{IN}[n]$ after sampled by the ADC. After being processed by time delay and Hilbert transform, the orthogonal signals $S_{LO,I}[n]$ and $S_{LO,Q}[n]$ are multiplied by $S_{IN}[n]$. The output signal would go through a low-pass filter and AD to acquire baseband signal $S_{bb}[n]$, the amplitude of

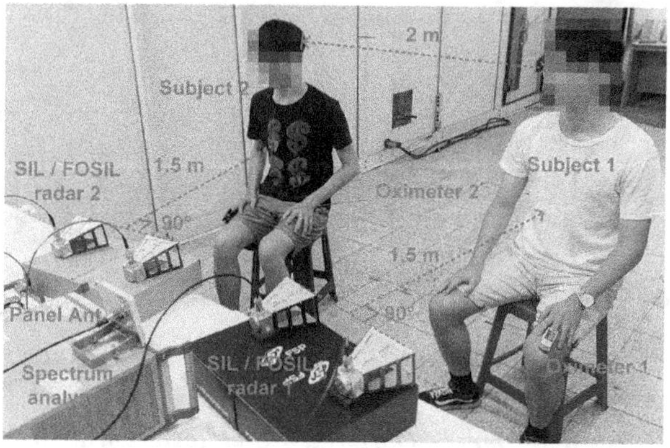

Figure 9.14 *Photograph of the environment in the two-radar experiment for noncontact vital sign monitoring of two seated subjects using SIL and FOSIL radars*

which is twice as that of the baseband signal of a conventional SIL radar. In addition, due to the low-IF architecture, the influence of flicker noise can be avoided and the physical delay line can be removed, thereby significantly enhancing the degree of integration. Since frequency demodulation is processed through software, this architecture does not require additional DC offset calibration and therefore significantly reduces computation load. In Reference [92], a 2.9-GHz SIL radar and a 5.8-GHz FOSIL radar with a front-end design composing two 2.9-GHz radars were used. The SNR of the FOSIL radar was 28 dB better than the SIL radar. Figure 9.14 shows an experimental setup for vital sign sensing involving a two-radar architecture. The difference of the center frequency in the two radars was 2 MHz, and the antennas were slightly inclined to avoid interruption. An extra antenna was used to connect with a spectrum analyzer. The wireless spectra in the environment are indicated by Figure 9.15, where the gray dotted lines, gray dots, and black solid

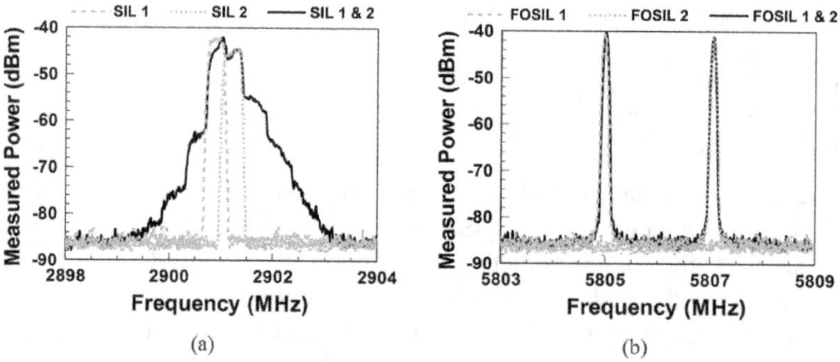

Figure 9.15 *Measured spectra of wireless propagation signal when using the experimental setup in Figure 9.14. (a) Conventional SIL radar. (b) FOSIL radar.*

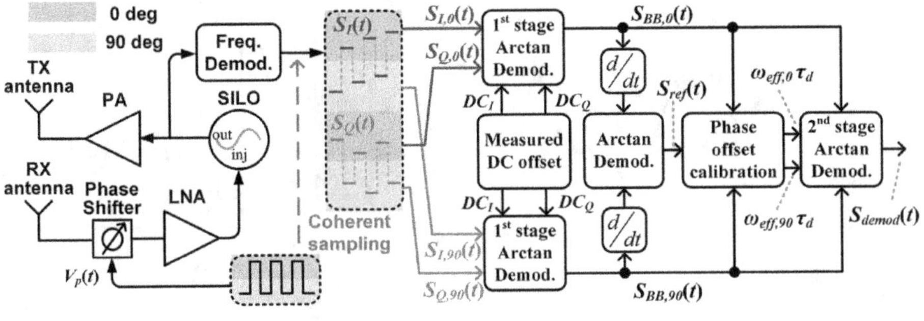

Figure 9.16 Block diagram of the proposed QSIL radar and two-stage AD

lines are, respectively, the representative measurements when radar A was turned on, radar B was turned on, or both radars were turned on. When the TX frequency of the SIL radar coincided with the frequency locking range of another radar, as shown in Figure 9.15a, a mutual injection locking state would appear to cause severe interruption and generates numerous redundancies in the spectra. Both the SIL radars could not correctly monitor the vital signs of the subjects in the front. However, the measurements when using the FOSIL architecture are shown in Figure 9.15b. The output frequencies of the two radars remained constant and showed no difference in frequencies when radars were operated individually. In addition, the measured results of breaths and heartbeats matched perfectly with the reference signals, proving that the FOSIL radar is applicable to different kinds of multiple-users scenarios.

The SIL mechanism applies nonlinear frequency modulation to the oscillator. If the test object shifts more than 0.1 of the operating wavelength, the demodulated signal would be distorted and thus limits the applications of a conventional SIL radar. This issue was proposed by the quadrature SIL (QSIL) [93], the architecture of which is shown in Figure 9.16. A voltage control phase shifter that provided a phase delay of 0°/90° was added to the injection port of the SILO. The blue and tan blocks, respectively, denoted the control signals and IQ signals under different phase delays. The signals were coherently sampled in the backend, and four channels of baseband signals in the SILO were represented as $S_{I,0}(t)$, $S_{I,90}(t)$, $S_{Q,0}(t)$, and $S_{Q,90}(t)$. Next, a two-stage AD was proceeded as depicted on the right side of Figure 9.16. Under the same phase delay, deducting the phase shift in the IQ signals caused by hardware imperfection, the signals were paired before an AD was conducted to generate $S_{BB,0}(t)$ and $S_{BB,90}(t)$. $S_{\mathrm{demod}}(t)$, which represented the Doppler shift of the moving object, was obtained after the phase offset of the baseband signal caused by clutters was deducted before an arctangent

It should be noted that the causes of the DC offset and phase offset in a QSIL radar are different. The former can be acquired by using an IQ delay discriminator to demodulate an FM signal with different center frequencies. The latter can be deducted using a derivative, and the experiment will be described in the following paragraphs. In Reference [94], a 10×15 cm^2 metal plate moved in a triangular wave trajectory with a constant speed of 0.05 Hz and a maximum displacement of

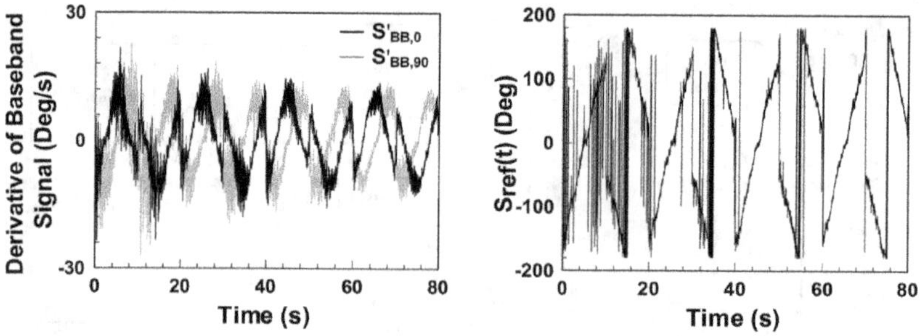

(a) (b)

Figure 9.17 Signals for calibrating the unwanted phase offset of baseband signals. (a) Derivative of baseband signals. (b) Reference signal.

5 cm, which was 1.5 m away from a 2.4-GHz QSIL radar prototype. Differentiation was applied to baseband signals with different phase delays. The results of $S'_{BB,0}(t)$ and $S'_{BB,90}(t)$ are shown in Figure 9.17a, where high-frequency noises resulted in unsmooth waveforms. When the system's SNR was high, an AD was performed on $S'_{BB,0}(t)$ and $S'_{BB,90}(t)$, and the result $S_{ref}(t)$ was the demodulated signal $S_{demod}(t)$. However, derivatives increased high-frequency noises especially when the test object's instantaneous speed was zero. The discontinuous wave fluctuations in Figure 9.17b were caused by $S_{ref}(t)$ in Figure 9.17a. Accordingly, $S_{ref}(t)$ was used as a reference signal for the calibration of phase shift. Two constant phases α_{d1} and α_{d2} in $S_{ref}(t)$ were used as reference values, and a threshold θ_t was used as the tolerance. Since α_{d1}, α_{d2}, $\bar{S}_{BB,0}(t)$, and $\bar{S}_{BB,90}(t)$ were all measured results, simultaneous equations were applied to learning the phase amplitude $\omega_{LR,d}\tau_d$ caused by the Doppler signal:

$$\omega_{LR,d}\tau_d = \left(\sin\alpha_{d2} - \sin\alpha_{d1}\right)^{-1} \times \left(\bar{S}_{BB,0}\left(t\right)|_{|S_{ref}(t)-\alpha_{d1}|<\theta_t} - \bar{S}_{BB,0}\left(t\right)|_{|S_{ref}(t)-\alpha_{d2}|<\theta_t}\right) \tag{9.19}$$

Substituting (9.19) to the expressions of $\bar{S}_{BB,0}(t)$ and $\bar{S}_{BB,90}(t)$ would obtain the phase offset caused by clutter signals under different phase delays [94]. Such a calculation provides a precise trajectory of the metal plate. The measure RMS error was only 0.71 mm when the displacement of the metal plate was 5 cm; while the displacement was 0.25 cm, the RMS error was 0.03 mm. The above results show that the QSIL radar has not only overcome the nonlinear issue inherent in the SIL mechanism but also provides excellent sensitivity.

Additionally, Yu *et al.* proposed a phase-canceling SIL (PCSIL) ultrasonic radar [95]. Figure 9.18 shows the architecture of the PCSIL radar, the digital SILO of which was composed of a resonator and a comparator. This oscillator is also known as the relay-feedback oscillator [96]. A square wave was used as the driving signal and injection signal of the ultrasonic component, an easy implementation on the FPGA without the need for an ADC chip. A controller, coupled with an FM demodulator, was used to monitor the phase delays caused by the Doppler signal and could control the adjustable digital delay unit to form a feedback network. Therefore,

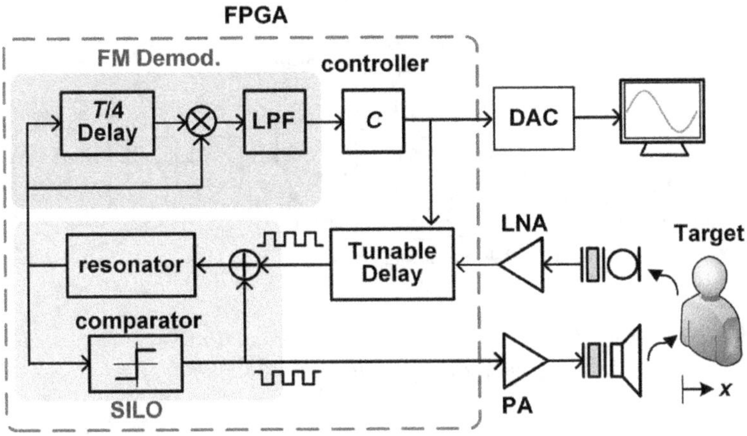

Figure 9.18 Block diagram of the PCSIL radar

the injection phase could maintain a high linearity. By dynamically adjusting the time delay of the feedback circuit to cancel the Doppler phase difference caused by motions of the subject, the output frequency of the SILO could be maintained to further eliminate the EMI issue. The displacement information of the subject was to be obtained from the digital adjustable delay, and the radar continued to operate at the optimum point of monitoring. Figure 9.19 shows a PCSIL radar adopting a 40-kHz ultrasonic wave to monitor displacements of the human chests caused by vital signs. The antennal was 30 cm away from the subject. Benefitted from its high linearity and sensitivity, the PCSIL radar can effectively retrieve subjects' RRs and heartbeat rates with an error lower than 1 bpm.

The characteristics of the PCSIL radar are as follows: (1) The transmitting source, a 40-KHz ultrasonic wave, can be categorized as the millimeter level at room temperature, and the PCSIL mechanism allows the radar to operate at the optimum point to provide better sensitivity and dynamic range. The experiment

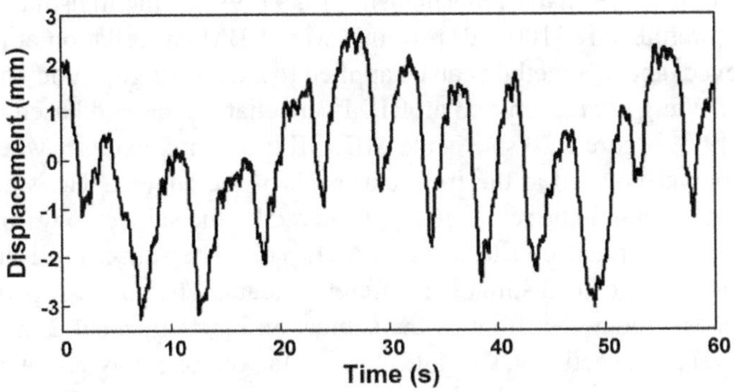

Figure 9.19 Displacement measurement of the subject's cardiopulmonary activities using PCSIL radar

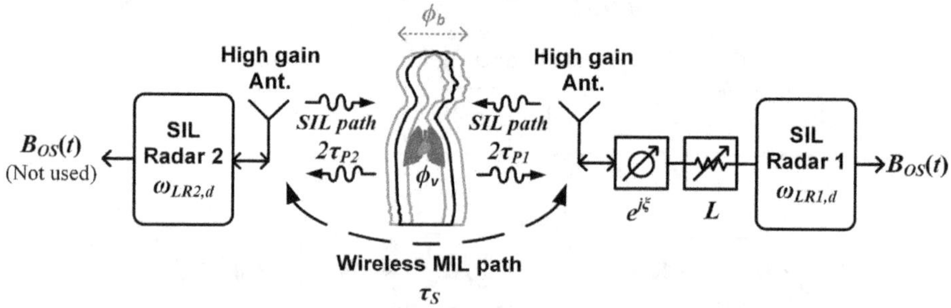

Figure 9.20 SIL-radar-based system for RBM cancellation using two MIL-SIL radars placed on the opposite side of the body

indicated that the maximum measurable displacement was 120 mm, which was approximately the length of 14 wavelength; (2) The phase cancelling demodulation solution is highly tolerant of the non-ideality of circuits. The system's accuracy can be maintained as long as the feedback loop remains stable; (3) Not constrained by the frequency locking range of the PLL circuit, the PCSIL radar is capable of having a larger operating bandwidth; and (4) This design can be incorporated into sonar or microwave radar if the transceiver is altered or the frequency converting mechanism is increased [95].

9.3.3 Random body motion cancellation

The amplitude of the subject's RBM is usually far bigger than the chest displacement caused by breaths or heartbeats. The wave fluctuations caused by the former can easily saturate the baseband amplifier of the CW radar or the ADC module. Although we can use multiple radars with digital signal processing to cancel the body movements, the maximum displacement would still be limited to under 1 cm. The method of RBM cancellation under the SIL architecture is implemented by the mutual-injection-locked (MIL) mechanism [88, 97–99] or the transmit and retransmit (T&RT) architecture [100, 101]. In this way, RBM cancellation at the RF side can be achieved, and this method can be applied to a wide range of body movements.

In 2011, Wang *et al.* proposed an RBM cancellation method based on the MIL mechanism [97]. Figure 9.20 shows the MIL-SIL radar architecture, where two sets of SIL radars were placed at the front and back of the subject. The signal sent by each unit would not only be reflected and received by the subject to form a SIL path but also by another radar unit to form an MIL path. The subscripts 1 and 2 of the parameters were used to distinguish different radars. The round-trip propagation delays of the SIL radar 1 in front of the subject and the SIL radar 2 in the back of the subject were, respectively, $2\tau_{p1}$ and $2\tau_{p2}$. τ_s was the one-way propagation delay between the SIL radars 1 and 2. The frequency domain in Figure 9.21 was derived to yield phases, $\phi_{s1}(s)$ and $\phi_{s2}(s)$, of output signals under the MIL state of the two SILOs:

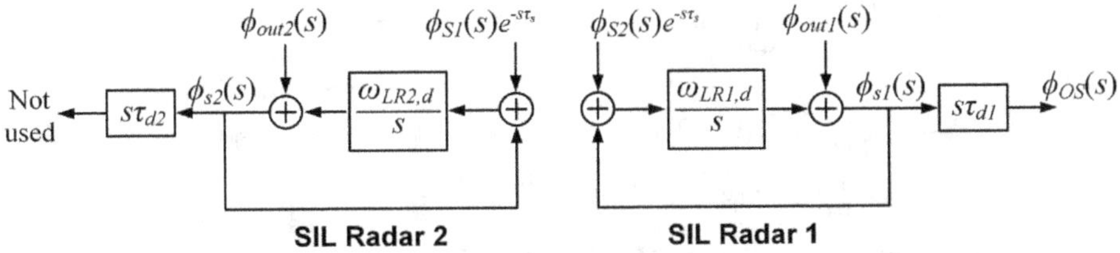

Figure 9.21 Frequency-domain model of Figure 9.20

$$\begin{cases} \phi_{s1}\left(s\right) = \dfrac{\omega_{LR1,d}}{s+\omega_{LR1,d}} \cdot \phi_{s2}\left(s\right) e^{-s\tau_s} + \dfrac{s}{s+\omega_{LR1,d}} \cdot \phi_{out1}\left(s\right) \\ \phi_{s2}\left(s\right) = \dfrac{\omega_{LR2,d}}{s+\omega_{LR2,d}} \cdot \phi_{s1}\left(s\right) e^{-s\tau_s} + \dfrac{s}{s+\omega_{LR2,d}} \cdot \phi_{out2}\left(s\right) \end{cases} \tag{9.20}$$

where

$$\begin{cases} \phi_{out1}\left(s\right) = H_{m1} \cdot \left(\phi_v - \phi_b\right) + H_{o1} \cdot \phi_{n,osc1} \\ \phi_{out2}\left(s\right) = H_{m2} \cdot \left(\phi_v + \phi_b\right) + H_{o2} \cdot \phi_{n,osc2} \end{cases} \tag{9.21}$$

$\phi_{out1}(s)$ and $\phi_{out2}(s)$ are the SILO output phases, the form of which is the same as formula (9.13), under the SIL condition. The only difference is that the Doppler signal ϕ_m would be modified as $\phi_v + \phi_b$ or $\phi_v - \phi_b$. ϕ_v and ϕ_b are phase modulation components caused by vital signals and RBMs. Since vital signs are associated with an inward-and-outward chest motion, where the body moves back and forth, The demodulated phases of the two SIL radars were the sum and difference of ϕ_v and ϕ_b. As deduced by Reference [88], it was assumed that τ_s was much smaller than 1. Substituting (9.21) into (9.20) yields $\phi_{s1}(s)$:

$$\begin{aligned} \phi_{s1}\left(s\right) \approx & \frac{\omega_{LR1,d}H_{m2}\left(s\right) + \omega_{LR2,d}H_{m1}\left(s\right)}{\omega_{LR1,d} + \omega_{LR2,d} + 2\omega_{LR1,d}\omega_{LR2,d}\tau_s}\phi_v\left(s\right) \\ & + \frac{\omega_{LR1,d}H_{m2}\left(s\right) - \omega_{LR2,d}H_{m1}\left(s\right)}{\omega_{LR1,d} + \omega_{LR2,d} + 2\omega_{LR1,d}\omega_{LR2,d}\tau_s}\phi_b\left(s\right) \\ & + \frac{\omega_{LR1,d}H_{o2}\left(s\right)\phi_{n2}\left(s\right) + \omega_{LR2,d}H_{o1}\left(s\right)\phi_{n1}\left(s\right)}{\omega_{LR1,d} + \omega_{LR2,d} + 2\omega_{LR1,d}\omega_{LR2,d}\tau_s}, \text{ for } s\tau_s \ll 1 \end{aligned} \tag{9.22}$$

When the numerator of the second term on the right side of (9.22) is zero, that is, when $\omega_{LR1,d}H_{m2}(s)=\omega_{LR2,d}H_{m1}(s)$ is established, the effect of RBM cancellation is achieved. SIL radars 1 and 2 may have different locking ranges or different amplitudes of Doppler phases caused by different relative positions to the human bodies. As indicated in Figure 9.20, in practice, one of the radars can be added with a tunable phase shifter (TPS) and attenuator L to achieve the cancelling condition. Since the baseband signals of the two radars are the same, system complexity can be reduced by using only an IQ delay discriminator. To validate the effect of body movement cancelling, subjects were asked to jog on treadmills during the experiment. The two units of the MIL-SIL radar system were placed 2 m apart, and the subject was at the center between the two radar units. One SIL radar unit was shut down to be compared with a conventional SIL radar. Figure 9.22 shows the experimental results,

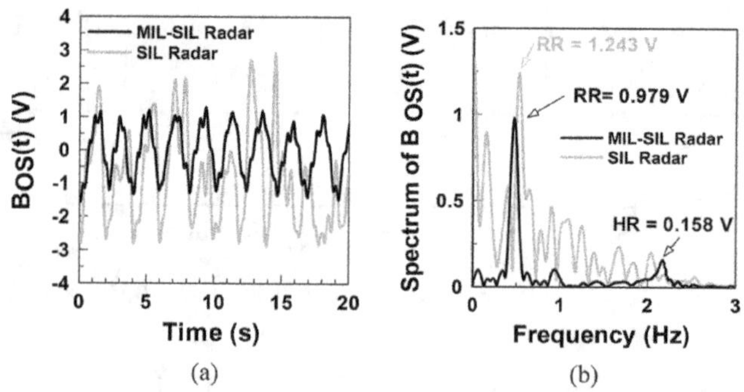

Figure 9.22 Using the MIL-SIL radar and SIL radar to monitor the vital signs of subjects jogging on treadmill. (a) Time-domain waveform. (b) Spectra of (a).

where black and gray lines were, respectively, the results of the MIL-SIL radar and the SIL radar. The output time-domain waveform $B_{OS}(t)$ is plotted in Figure 9.22a. This waveform is apparently more regular by virtue of the RBM cancellation ability provided by the MIL-SIL radar. Processed by Fourier transform, the spectra are shown in Figure 9.22b. Despite the fact that breath information was provided by the SIL radar, the data could hardly distinguish heartbeat signals from RBM signals. In contrast, MIL-SIL radar successfully distinguished breath signals from heartbeat signals, which were 30 and 130 heartbeats/min, during jogs, and the results were consistent with the measurements of the medical instrument.

Based on the same principle, the two radars were placed on the same side of the body. RBM cancellation could be achieved only if the two radars operate at the MIL-SIL state and the two injection signals possess the same amplitude but opposite phases [98]. However, such a setting may simultaneously cancel RBM and vital sign information, and therefore a special antenna positioning is required. As shown in Figure 9.23a, the first setting of this experiment was an up–down positioning consisting of a low-gain antenna and a high-gain antenna. The beam of the low-gain antenna was perpendicular to the chest, while that of the high-gain antenna was also perpendicular to the chest but with an incline of θ_b [99].

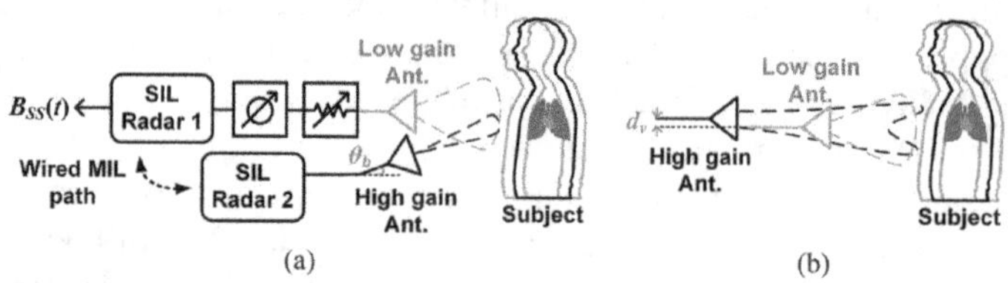

Figure 9.23 SIL-radar-based system for RBM cancellation using MIL-SIL radar placed on the same side of the body. (a) Up–down and (b) Front–rear configuration.

Human chest is not an even surface. If an EM wave radiates to a human chest and is reflected, the injection signals received by the two antennas consist of different Doppler phases caused by either body movements or vital signs. θ_b was used to adjust the radar cross section of the subject, and doing this allowed the researchers to achieve RBM cancellation while keeping the vital sign information. The second setting was depicted in Figure 9.23b, where a low-gain antenna was placed in front of the high-gain antenna. The low-gain antenna was placed in front of the high-gain antenna toward the subject. Since the antennas were made of metal, the radiation of the high-gain antenna could not be aimed directly at the chest, resulting in weak Doppler information. As a result, the height d_v of the two antennas was appropriately adjusted so that the two received injection signal phases possessed enough body movement amplitudes and different vital signs. The adjustment parameters θ_b and d_v in both settings were used to change the radar cross section of the human chest to the high-gain antenna. In fact, there are individual differences concerning the evenness of the human chest surface, and parameters for the individual subjects can just be modified accordingly [99].

The ideal correction condition of an MIL-SIL radar is as follows: (1) Identical MIL paths $\omega_{LR1,d} = \omega_{LR2,d} = \omega_{LR,m}$ and (2) Doppler phases received by the two radars have the same amplitude of body movements and different amplitudes of vital signs, meaning $\phi_{b1}(t) \approx \phi_{b2}(t) = \phi_b(t)$ and $\phi_{v1}(t) \neq \phi_{v2}(t)$. Considering a general operating circumstance, the signal strength used for mutual injection locking is lower. The system's output carrier frequency was approximately the average of the inherent oscillation frequencies of the two SILOs. It was assumed that SILO 1 and 2 had the same characteristics including the same operating frequencies, the same locking ranges, and the same SIL paths. In addition, the chest movements caused by vital signs were much smaller than the operating wavelength and the displacements caused by random movements: $x_{v1,2,\mathrm{RMS}} \ll \lambda_{osc}$ and $x_{v1,2,\mathrm{RMS}} \ll x_{b,RMS}$. The output frequency shift of the MIL-SIL radar could be approximated as

$$\omega_m(t) \approx \left(\omega_{LR1} \cdot \frac{2\pi\left(x_{v1}(t) - x_{v2}(t)\right)}{\lambda_{osc}}\right) \cdot \left(1 - 2\sin^2\left(\frac{2\pi x_b}{\lambda_{osc}}\right)\right) \tag{9.23}$$

The first and second terms on the right side of (9.23) are, respectively, target vital sign and nonlinear distortion. To avoid the effect of the latter, the range of body movements should be restrained from $-\lambda_{osc}/8$ to $\lambda_{osc}/8$. For example, when adopting an operating frequency of 2.4 GHz, an optimum sensing effect can only be achieved if the maximum displacement of body movements is restricted to under 3 cm. To overcome the range restriction for RBM cancellation of the MIL-SIL architecture, the T&RT SIL radar was proposed in 2017 [100]. The subscripts 1 and 2 of the parameters were used to distinguish different radars for transmission and reception. As shown in Figure 9.24, this architecture adopted only one SILO and the antennas were placed in the front and back of the subject. Antenna 1 was responsible for radiating the transmit signals of the SILO to the subject's body and for receiving echo signals from the front. Through circulators 1 and 2, the signals were transmitted to

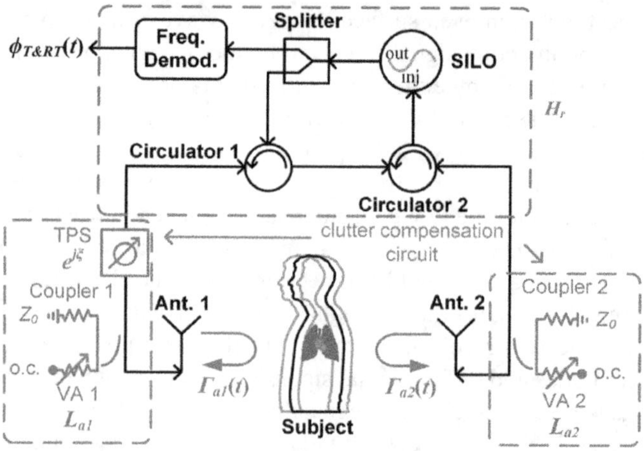

Figure 9.24 T&RT SIL radar architecture

antenna 2 and were sent again to the body. The echo signals reflected from the back were injected into the SILO through antenna 2 and circulator 2 to form an SIL path.

Vital signs caused by omnidirectional movements were kept, and signals concerning body movements were canceled. It is noteworthy that subjects might be seen at any given location in the environment, and therefore the strengths of the Doppler signals and clutter signals received by the two antennas might be different. To avoid the effect of cancelling RBM signals, as indicated by the red dotted lines, a clutter compensation circuit, which was composed of a TPS, a VA, and a coupler, was created to alter the clutter components of the system. The two channels of the coupler were used to propagate radar signals. One was terminated to complete impedance matching, and the other was a terminal open circuit that created a reflected signal with no phase modulation. The amplitude and phase of the reflected signal were controlled by VA and TPS, and its transmission coefficients were, respectively, $L_{a1,2}$ and $e^{j\xi}$. With proper modification, this system is now capable of cancelling subject's RBMs under an environment with various clutters.

The transfer function H_{sys} of this system was made up of three components: $\Gamma_{a1,2}$ were reflection coefficients of signals reflected from human bodies to the two antennas, and H_r was the transmission coefficient of the radar system:

$$H_{sys}(\mathrm{t}) = \Gamma_{a1}\left(t\right) \cdot H_r \cdot \Gamma_{a2}\left(t\right) \tag{9.24}$$

where

$$\begin{cases} H_r = T_0 e^{j(2\xi - \phi_0)} \\ \Gamma_{a1}\left(t\right) = D_1 e^{j(\phi_v(t) + \phi_b(t) - \phi_{d1})} + C_1'\left(L_{a1}\right) \cdot e^{-j\phi_{c1}'(L_{a1})} \\ \Gamma_{a2}\left(t\right) = D_2 e^{j(\phi_v(t) - \phi_b(t) - \phi_{d2})} + C_2'\left(L_{a2}\right) \cdot e^{-j\phi_{c2}'(L_{a2})} \end{cases} \tag{9.25}$$

where T_0 and ϕ_0, respectively, represent the amplitude and phase of H_r, and $D_{1,2}$ and $\phi_{d1,2}$ represent the amplitude and phase of the Doppler signal's reflection coefficient.

Considering the contribution of the clutter compensation circuit, $C'_{1,2}(L_{a1,2})$ and $\phi'_{c1,2}(L_{a1,2})$ are, respectively, the amplitude and phase of the clutter signal. Based on a reasonable assumption $\phi_v(t) \ll \phi_b(t)$, a cancellation condition [101] was

$$\begin{cases} \dfrac{D_1}{C_1'(L_{a1})} = \dfrac{D_2}{C_2'(L_{a2})} \\ \xi = \dfrac{\phi_{d1} + \phi_{d2} + \phi_{c1}'(L_{a1}) + \phi_{c2}'(L_{a2})}{4} + \dfrac{\phi_0}{2} - \left(n - \dfrac{1}{2}\right) \cdot \dfrac{\pi}{2} \end{cases} \tag{9.26}$$

The system's output phase was

$$\phi_{T\&RT}(s) \propto \omega_{LR} \sin(2\phi_v(t)) \tag{9.27}$$

The phase of RBM was cancelled, and the vital sign signals were doubled.

Reference [101] proposed a calibration method using a periodically moving metal plate to find unknown parameters, in order to satisfy the cancellation condition by adjusting VAs and TPS. The procedures were as follows:

Step 1: TPS was set to 0°, and VAs to 10 dB. Circuits of Ant.1 and 2 were left open alternately ($\Gamma_{a1} = 1$ or $\Gamma_{a2} = 1$). When the displacement of the metal plate was half the length of the operating wavelength of the radar, a frequency demodulator was used to obtain the signal's semi-amplitude, meaning the Doppler amplitude D_2 or D_1. Next, an actuator was held stationary and the position of the metal plate was fine-tuned. Phase delay of the TPS, with a single-antenna setup, was swept from 0° to 180°. The maximum semi-amplitude output voltage by the frequency demodulator represented the summation of the Doppler echo and stationary clutter amplitudes, $C'_1 + D_1$ or $C'_2 + D_2$.

Step 2: Step 1 was followed, but the attenuation of VAs was adjusted from 0 to 10 dB. D_1/C'_1 and D_2/C'_2 were measured to satisfy (9.26).

Step 3: Settings of VAs were identical to those in Step 2. The displacement of the metal plate was set to over one wavelength to modify the phase delay provided by TPS. The phase cancellation condition is approximately satisfied when the amplitude of the output signals is minimal.

It is worth noting that if the initial position of the metal plate is different, the settings of VAs and TPS are also different.

Figure 9.25 shows an experiment setting for RBM cancellation. To compare different effects, an additional conventional CW radar was employed. The T&RT SIL radar operated at 2.41 GHz, while the CW radar operated at 2.46 GHz to prevent frequency interference. During the experiment, the two antennas were placed 3 m from each other, and the subject was seated 1 m from antenna 1. The observation time was 5 min. Window length and overlap were set at 60 and 55 s for time–frequency analysis. The gray dotted lines and circle symbol in Figure 9.25b indicated that the rms value of the subject's body movements measured by the CW radar was approximately 1–185 mm. The blue triangles were breath signals measured by the T&RT SIL and were approximately 22–24 breaths/min. The blue rhombuses and red stars were heartbeat signals measured by the T&RT radar and the medical instrument, and the two symbols were consistent with values of 54–64 beats/min. Although the

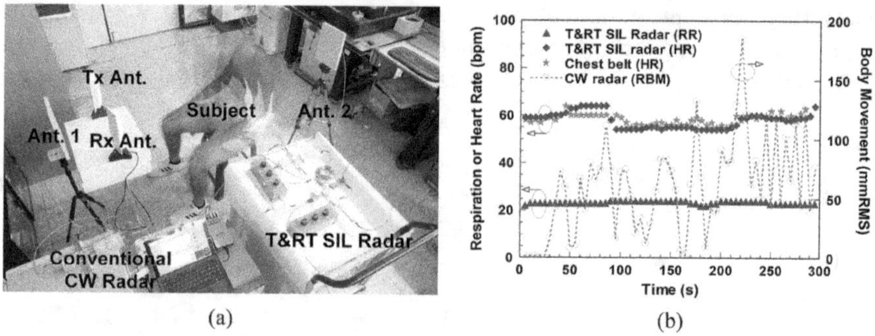

Figure 9.25 *Experimental setup and results for monitoring the vital signs using T&RT SIL radar and conventional CW radar. (a) Photograph of the experiment. (b) A 5-min experiment results of Figure 9.25a.*

subject's maximum body movement displacements were as much as 20 cm (1.6 λ_T), still the T&RT SIL radar could provide accurate vital sign information.

9.3.4 Wearable SIL radar

The architecture of the SIL radar can be roughly compartmented in two independently operating parts, a SILO and a frequency demodulator. These two parts both satisfy the requirements of clutter resistance, sensitivity, and penetrability. All kinds of wearable devices have been proposed to be attached to cloths, seatbelts, and chair back. This design not only reduces the possibility of forgetting to carry the device with you but also prevents negative influences such as interferences of multipath and moving objects. Additionally, the design enhances the system's SNR and provides abundant information for clinical and medical diagnosis. Important research results are to be provided as follows.

Wang *et al.* proposed a bistatic SIL radar system in 2015, which attached the radar antenna to the chest [102] or wrist [103] to monitor subjects' vital signs at all times. As shown in Figure 9.26, the SILO tag is composed of SILO and antenna to monitor vital signs and transmit wireless signals to injection locked oscillator (ILO)-based frequency demodulators at various locations. It should be noted that if the oscillator inside the frequency demodulator is under the SIL state, the phase

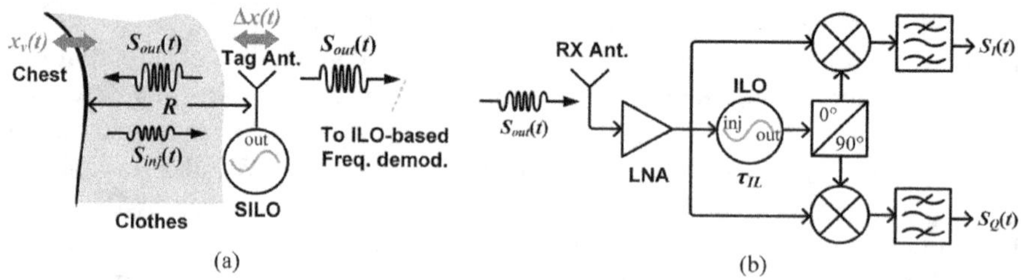

Figure 9.26 *Block diagram of the bistatic SIL radar system*

modulation information of the input signals can be duplicated. The tag exhibits outstanding frequency selectivity; its output power is the inherent oscillation power of the ILO and can be used simultaneously as a band-pass filter and an amplifier. In addition, by adjusting the differences between the injection frequency and the ILO free-running frequency, we are allowed to adjust the transmission time delay of the ILO. Thus, this design has been used to replace coaxial cable, LC filter, or surface acoustic wave filter in an RF delay unit. The replacement significantly reduces system complexity [39]. This mechanism, which allows external signals to be injected into the locking oscillator, can be applied to monitoring environmental spectra [104–107]. The subsequently proposed ILO-based frequency demodulator is often used as a frequency demodulator for passive radar or radar gesture recognition [108–113]. The architecture of the SILO tag worn on the subject is extremely simple that battery time can be significantly increased. The output frequency is affected by the relative instantaneous displacement $\Delta x(t)$ of the tag and the human chest. $\Delta x(t)$ is represented as

$$\Delta x\left(t\right) = x_v\left(t\right) + m \cdot a_b\left(t\right)/k_s \tag{9.28}$$

Parameters on the right side of the equation, $x_v(t)$, m, $a_b(t)$, and k_s, respectively, represent the instantaneous displacement caused by vital signs, mass of the SILO tag, acceleration of the body movement, and spring constant of the clothes [40]. As indicated in (28), any movements of the subject's body, such as step rates of walking and jogging, would be reflected to $a_b(t)$ to be detected by the system.

The experiment setting is shown in Figure 9.27a. The receiver was placed on the table in a stationary position. The subject walked in the place and took some deep breaths. The output IQ signals were DC offset calibrated and arctangent demodulated, and the baseband signals were plotted in Figure 9.27b. Window length and refresh rate were set to 50 and 0.1 s. Figure 9.27c is the result of time–frequency

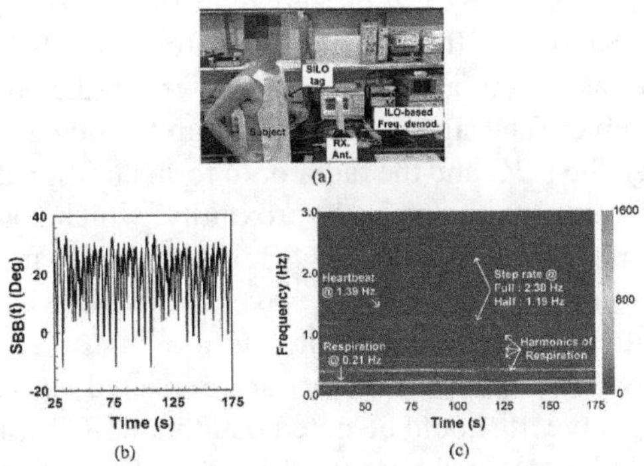

Figure 9.27 Experimental setup and results of the bistatic SIL radar system. (a) Photography of a man stepping with the SILO tag. (b) Baseband signal with the setup shown in Figure 9.27(a). (c) Spectrogram of Figure 9.27b.

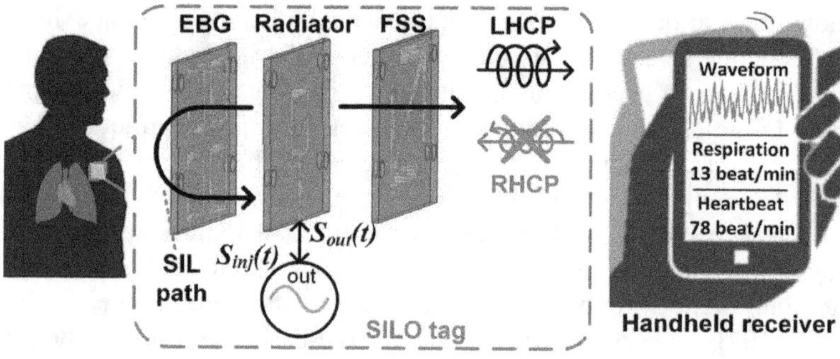

Figure 9.28 SILO tag with EBG and FSS for wearable detection

analysis. Walking would cause a regular backward–forward motion of one side of the chest, and therefore the full-step frequency was twice the half-step frequency.

The SILO tag monitored the acceleration caused by body movements and could distinctively differentiate frequency changes caused by various motions, as seen in Figure 9.27c, even if chest displacements caused by walking were very different from those caused by cardiopulmonary activities. The differentiation process first detected the dominant frequency of breaths within the range of 0.1–0.75 Hz, and then the full-step frequency was detected within the range of 1.75–5 Hz. Finally, after removing the harmonics and half-step frequency signals of breaths, the heartbeat frequency was retrieved within the range of 0.7–1.75 Hz. The subject's breaths, heartbeats, full step frequency, and half step frequency were, respectively, 0.21, 1.39, 2.38, and 1.19 Hz. This setting was later validated for monitoring the vital signs and step rate of the subject before and after the run [40]. In addition, Tang *et al.* proposed a hybrid mode SIL-CW radar that detects the fluctuations of wrist and chest. The peak time difference between heartbeat and pulse is highly correlated to pulse transmit time measured using ECG and PPG sensors and can be used to estimate blood pressure [114].

As shown in Figure 9.28, Arif *et al.* added an EM band-gap and a frequency-selective surface (FSS) to the SILO tag. The radiator in the SILO generated linear polarized (LP) waves as the output signals, which were radiated to the human body and the handheld receiver. The signals reflected from the human body were injected into the SILO through the EBG and the radiator to form an SIL path. The EBG can be used as an isolator to prevent human body proximity, which compromised the performance of the antennas [115, 116]. The FSS was primarily used as a polarization transformer, which converted the input LP wave to a left-hand circular polarization (LHCP) wave. The LHCP wave was radiated to moving objects (such as walking pedestrians or a moving handheld receiver) and reflected as a right-hand circular polarization (RHCP) wave. It should be noted that this RHCP wave, after converted by the FSS, could not be received because the output EM wave was orthogonal to the polarized direction of the EM wave created by the radiator. Polarization diversity has been applied to reducing the influence of moving clutters, which may cover the subject's vital signs, in the environment [117]. To verify the FSS's suppressibility against moving clutters, around the SILO tag in [115], an actuator was employed

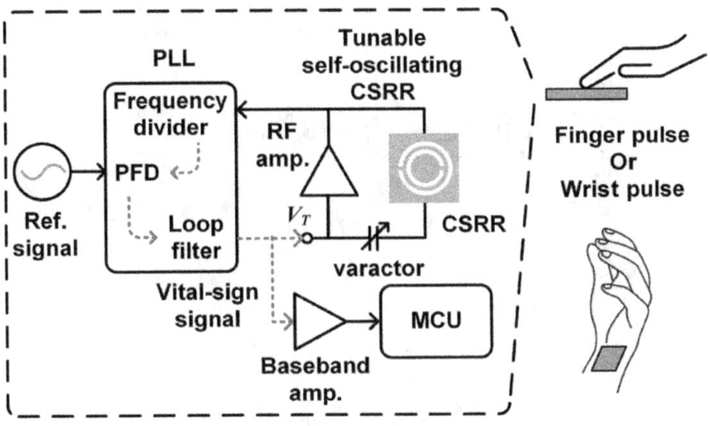

Figure 9.29 System block diagram of the PPIL with CSRR SILO tag

and wavering to simulate moving clutters. This architecture could suppress 90% or more against moving clutters.

Tseng *et al.* designed an SIL active integrated antenna (AIA), where the resonator of the SILO was replaced by a circular patch antenna. This alternative allows us to determine the frequency of the free-running oscillator, to receive and transmit EM waves, and to reduce the overall circuit size [89]. However, if the patch antenna is applied to wearable devices for monitoring, the radiation efficiency would be reduced and system SNR deteriorated [115]. In 2020, a complementary slot ring resonator (CSRR) that concentrates EM energy to a tiny region was applied to the resonator design in an SIL AIA [118]. As shown in Figure 9.29, when the subject's finger was placed on the CSSR for near-field sensing, the relaxation and contraction of capillaries caused the skin surface to move. According to the perturbation theory, the SILO's output frequency could modulate signals and provide excellent SRN gain. After performing demodulation by the backend PLL, this architecture not only provided high-quality heartbeat signals but also reduced the interference of breath harmonics [119]. Using a similar concept, Arif *et al.* combined a CSRR with a circular polarized antenna to improve the near-field sensing efficiency of the SILO tag worn on the chest [120]. The black and red lines in Figure 9.30 were two references, respectively, for the ECG measurements and the tag measurements. Regions B-A and C-C of the tag measurements corresponded, respectively, to regions Q-T and R-R of the ECG measurements. The correspondence was used to determine the time of relaxation and contraction of the subject's ventricle, while the measurements of the inter-beat interval were considered as critical information for the clinical diagnosis of cardiovascular diseases.

9.3.5 Multiple subject detection using SIL radar

In sections 9.3.1–9.3.4, we introduced methods for addressing the technical bottlenecks of SIL radars when only a single subject is present. In this section, an SIL

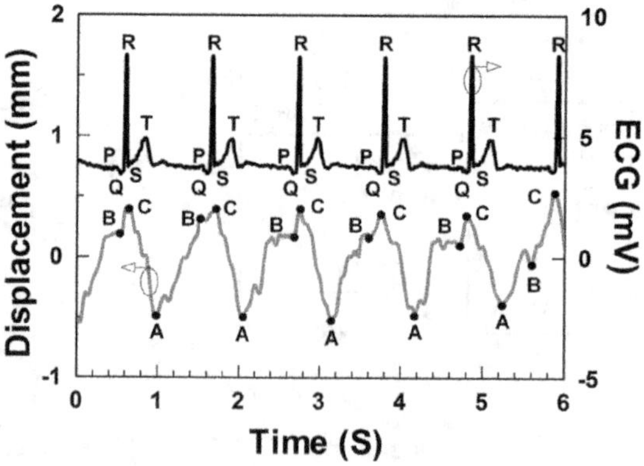

Figure 9.30 Measured heartbeat signal versus reference ECG signal

radar framework for detecting multiple subjects is introduced for scenarios in which a driver and multiple passengers are in a vehicle. Regarding the use of SIL CW radars, Juan *et al.* employed an SP3T RF switch to activate three high-gain antennas in sequence, each pointing in a different direction, and performed coherent sampling to acquire detection results at different locations. When the radar and the paper box were 50 cm apart, this approach detected the cardiopulmonary activities of mice remaining still and could therefore be applied to preventing infestation in storage areas [121]. Yuan *et al.* used the metamaterial leaky-wave antenna (LWA) technique. Because the main lobe azimuth angle of an LWA is affected by the operating frequency, the frequency tuning voltage of the SILO can be used to perform scanning in the beam direction and detect the vital sign of different subjects within the same space [122]. When the SILO is operated under FSK modulation mode, additional distance information can be obtained to distinguish between subjects in neighboring areas. Hsu *et al.* proposed an SIL-MIL technique, in which each unit of the front-end circuit is composed of an oscillator, a hybrid coupler, two power dividers, and a dual-port CP antenna. An output signal from the oscillator passes through the hybrid coupler and power dividers to generate four signals of equal amplitude; the phase of two of the signals and that of the other two signals differ by 90°. The two signals orthogonal to each other are input into the CP antenna to generate an LHCP wave, which is received along with the RHCP wave reflected from the subject by the CP antenna. The received signals then serve as the injection signals of the oscillator. This architecture effectively reduces the effect of the clutter signal and improves system sensitivity. When a unit connects with neighboring units through the power dividers and exhibits a transmit frequency similar to those of the neighboring units, the oscillator simultaneously operates under SIL and MIL states and exhibits the same output frequency. Under this scenario, the frequency tuning voltage of the oscillator can be configured to modify the phase difference of each unit, thereby changing the antenna pattern [123]. A radar system based on this architecture can detect the vital-sign information of three adults sitting next to each other, shoulder to shoulder, and determine the azimuth angle between each and the antenna.

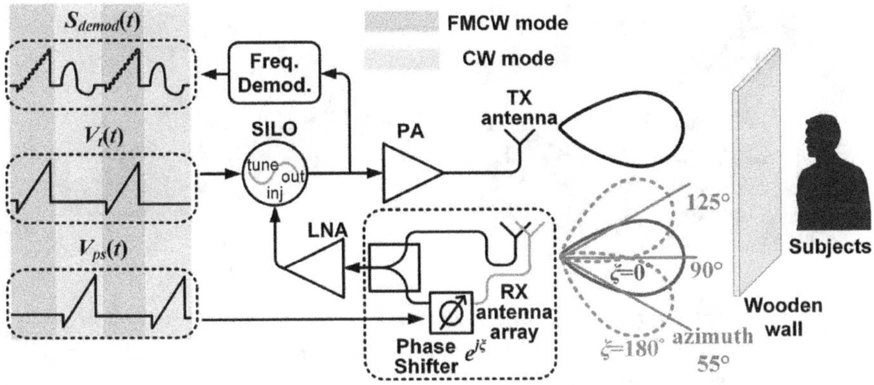

Figure 9.31 Block diagram of the FMSIL for STW detection

In the aforesaid literature [121–123], subjects must be at different azimuth angles for the detection process to work, resulting in application limitations. In 2012, Wang *et al.* proposed a frequency-modulated SIL (FMSIL) radar for see-through-wall (STW) detection. In this radar, the first set of fast-time spectrum is subtracted from all sets of fast-time spectra to distinguish between subjects and other still objects, and changes in the distance to the subject can be monitored continuously [124]. The radar can be used in tandem with a 1T2R antenna to determine the azimuth angles of multiple subjects and generate a 2-D STW image [125]. Figure 9.31 is a block diagram of an FMSIL radar system. The system has two types of voltage signals, $V_t(t)$ and $V_{ps}(t)$, which, respectively, control the SILO frequency tuning voltage and the phase delay ξ of the phase shifter in the RX antenna array. When $V_t(t)$ and $V_{ps}(t)$ are ramp and constant, respectively (labeled as blue regions), the radar operates under FMCW mode to detect the distances and vital signs of multiple subjects, and the beam direction of the RX antenna remains constant. By contrast, the tan-colored regions indicate that the radar is operating under CW mode, and the azimuth angle of the RX antenna changes with $V_{ps}(t)$, allowing the detection of the azimuth angle and physiological information of subjects; the range of detection is 55°–125°. Because the system has only two antennas, vital signs determined using the FMCW mode are used as reference signals to compare the amplitude ratio of the modulation signals contributed by different subjects when the CW mode is operated under sum–difference pattern. This greatly increases the azimuth angular resolution and enables the matching of vital signs with 2-D locations, providing high-quality STW images.

The basic principles of the system are as follows. When the system is in FMCW mode, the input signal is the sum of multiple return signals from different targets. Therefore, (9.1) is rearranged into

$$\omega_{out}(t) = \omega_{osc}(t) + 2\pi K_v V_t(t) - \sum_i \omega_{LR,i} \tau_d \sin \alpha_i(t) \quad (9.29)$$

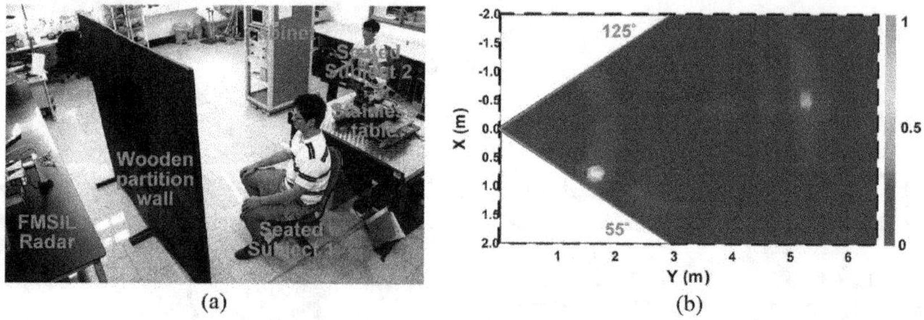

Figure 9.32 (a) Photograph of the experimental setup for STW detection of two seated subjects. (b) STW image constructed using the experimental setup in Figure 9.32a.

where the subscript i corresponds to the ith object in the environment and K_v denotes the tuning sensitivity of the oscillator. The phase difference $\alpha_i(t)$ is affected by the frequency of the free-running oscillator, internal circuit delay τ_{int}, and SIL path transmission delay $\tau_{p,i}(t)$. Accordingly, $\alpha_i(t) = \omega_{osc}(t) \times (\tau_{int} + \tau_{p,i}(t))$. After passing through a low-pass filter and undergoing frequency demodulation, the demodulated signal is expressed as follows:

$$S_{demod}(t) = -\sum_i \omega_{LR,i} \tau_d \sin \alpha_i(t) \tag{9.30}$$

Because $\alpha_i(t)$ changes when it is affected by $\omega_{osc}(t)$, the demodulated signal consists of components from different frequencies. After Fourier transformation, multiple tones are obtained on the spectrum, with each tone representing different objects. The distance information of each object can be determined from the known experimental parameters. Notably, the framework benefits from the outstanding SNR performance of the SIL mechanism and can therefore be used to continuously monitor changes in chest displacement due to respiration and heart motion without relying on the 2-D FFT introduced in section 9.2. In Figure 9.32a, two subjects are seated in front of a radar, and a wooden partition wall is placed between the subjects and the radar. Within the radar detection range, the subjects remain seated and breathe regularly. Clutter signals are caused by various objects present in the environment. The distance from the radar to subjects A and B is 0.83 and 5.28 m, respectively, and the azimuth angle is 65° and 95°, respectively. Figure 9.32b indicates that the locations of the spots displayed in the resulting STW image are consistent with those of the actual situation. Because chest displacement caused by breathing is greater than that of heart motion, the colors of the spots flicker with the breathing patterns of the subjects. In addition, because the returned echo signal of the front subject is stronger, the corresponding spot is more visible.

When frequency scanning is performed directly on the SILO, bifurcation phenomena might cause instability in the SILO, resulting in irregular changes in the output frequency [84]. This problem is more noticeable when the scanning range is large. This presents difficulty in increasing the range resolution of an FMSIL radar by increasing the modulation bandwidth. Su *et al.* proposed a single-conversion

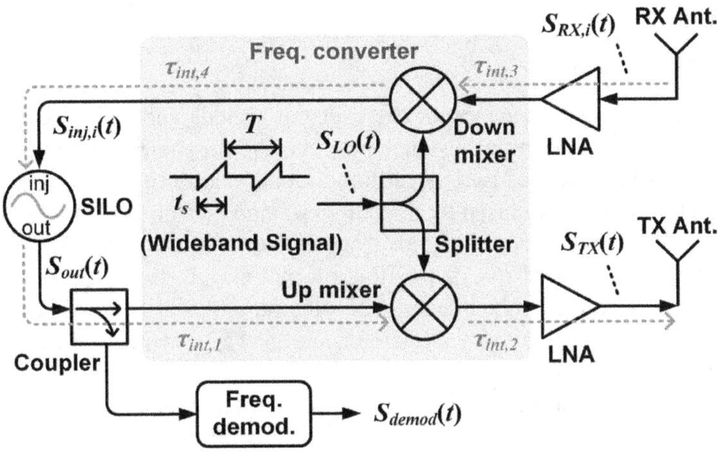

Figure 9.33 Block diagram of single-conversion SIL

stepped-frequency continuous-wave (SCSFCW) radar architecture. As shown in Figure 9.33, the output and injection ports of the SILO are connected to an up/down mixer. $S_{out}(t)$ and $S_{LO}(t)$ are mixed to increase the transmit frequency $\omega_{TX}(t)$.

Accordingly, $\omega_{TX}(t) = \omega_{out}(t) + \omega_{LO}(t)$. $S_{LO}(t)$ is compatible with various bandwidth modulation methods, including FSK, FMCW, and SFCW. After the echo signal $S_{RX,i}(t)$ is mixed with $S_{LO}(t)$, the input signal $S_{inj}(t)$ is generated and fed into the SILO, enabling the oscillator to operate under the SIL state. In this architecture, the SILO serves as the intermediate frequency signal source, and a frequency demodulator is used to determine the changes in the SILO frequency. This reduces flicker noise and PRF and enables outstanding Doppler sensitivity. In Reference [126], 5.5–6.5-GHz stepped chirp signals were used in an SCSFCW radar to increase the frequency of the 2.45-GHz SILO. The transmit frequency was 7.95–8.95 GHz, and a horn antenna was used to transmit and receive signals. To reduce energy consumption during low-frequency conversion, subharmonic up/down mixers can be used to perform frequency conversion [127].

FMSIL and SCSFCW radars provide information on the distance between the radar and the subject. However, the range resolution is limited by the bandwidth. On the basis of the framework presented in Figure 9.33, Wang *et al.* used the output signal of a VCO as $S_{LO}(t)$; the center frequency is ω_{LO}, and the frequency tuning voltage $V_t(t)$ is adjusted for the oscillator to perform scanning or maintain the voltage at a constant value. This enables the system to operate under FMCW and CW modes. In Reference [128], the operating frequency of the single-conversion SIL (SCSIL) hybrid mode radar was 5.725–5.875 GHz.

In this architecture, $\alpha_i(t)$ in (9.29) is expressed using (9.31):

$$\alpha_i(t) \approx \alpha_{IF} + (\omega_{TX} + 2\pi K_v V_T(t)) \times \left(\tau_{int,2} + \tau_{int,3} + \frac{2}{c}(R_i + x_i(t))\right) \tag{9.31}$$

where

$$\alpha_{IF} = \omega_{osc} \times \left(\tau_{int,1} + \tau_{int,4}\right) \tag{9.32}$$

ω_{TX} denotes the center frequency of the transmit signals and can be expressed as $\omega_{osc}+\omega_{LO}$. R_i denotes the initial distance between the ith subject and the radar, and $x_i(t)$ denotes displacement caused by cardiopulmonary activities. $\tau_{int,1}$ and $\tau_{int,4}$ denote the internal circuit delays caused by the output of signals from the SILO output port to the up-mixer and the input of signals from the down-mixer to the SILO injection port, respectively. $\tau_{int,2}$ and $\tau_{int,3}$ denote the internal circuit delays caused by signal transmission from the up-mixer to the TX antenna and that from the RX antenna to the down-mixer, respectively. In Reference 128, $\tau_{int,1-4}$ values were measured as 1.03, 2.24, 2.24, and 8.76 ns, respectively.

The beat signal is subject to Fourier transformation and expressed as follows:

$$\begin{aligned} S_{demod}(\omega) &= \mathbb{F}\left(w(t)\, S_{demod}(t)\right) \\ &= \frac{\tau_d}{2}\sum_i \omega_{LR,i} \times \left(W\left(\omega - \omega_{s,i}\right) e^{j\phi_{s,i}} - W\left(\omega + \omega_{s,i}\right) e^{-\left(j\phi_{s,i}-\pi\right)}\right) \end{aligned} \tag{9.33}$$

where $w(t)$ and $W(\omega)$ are the Fourier transform pairs of the window function. The center frequency and phase of each tone on the spectrum are calculated as

$$\begin{cases} \omega_{s,i} = \dfrac{\omega_B}{T_s}\left(\tau_{int,2} + \tau_{int,3} + \dfrac{2}{c}R_i\right) \\ \phi_{s,i} = 2\pi \cdot \mathrm{Mod}\left(\alpha_{IF} + \omega_{TX}\frac{\omega_{s,i}}{\omega_B}T_s + \frac{\pi}{2}, 2\pi\right) \end{cases} \tag{9.34}$$

where ω_B and T_s are the sweeping bandwidth and period, respectively. As demonstrated by the frequency–phase relationship expressed in (9.34), the frequency and phase of each tone exhibit a corresponding relationship when the circuit parameters are known. However, (9.34) becomes invalid when two objects are present within a range resolution of $c/(2\omega_B)$. Therefore, this equation can be used to identify the presence of range resolution problems and limit the computation scope of the super-resolution algorithm, thereby improving the system efficiency and accuracy. In addition, conventional super-resolution algorithms, such as MUSIC, assume that the input signal contains Gaussian noise. However, this assumption violates the principles of the SIL radar system [88]. Consequently, MUSIC cannot be used to increase the distance resolution of SCSIL radars [129].

To address the aforesaid problem, a frequency estimation algorithm (FEA) for SIL radar was proposed [128]. Figure 9.34 presents the FEA process. According to (9.30), the demodulation model y_n for multiple-subject detection can be simplified as follows:

$$y_n = \sum_{k=1}^{K} g_k e^{j\omega_k n} + \varepsilon_n \tag{9.35}$$

where g_k and ω_k are the amplitude and frequency of sinusoidal waves, respectively. In coarse frequency estimation, FFT is used to acquire the original spectral data in the preset frequency range. Then, fine frequency estimation is performed by

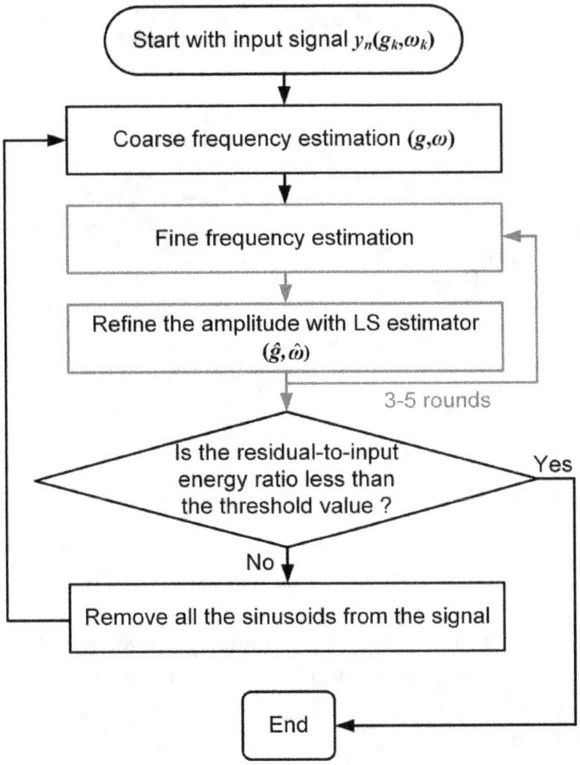

Figure 9.34 Flowchart of the proposed FEA

subtracting the wave component with the largest amplitude in the original spectral data from the input signal, followed by calculating the remaining signals. This process is repeated until the amplitude and frequency of all components are determined. Because the original spectral data provided by FFT are inaccurate, components with smaller amplitudes determined from the first iteration are subtracted from the original signal.

The least-squares estimator is used to recalculate the data of the dominant sinusoidal wave, as shown in (9.36):

$$(\hat{g}, \hat{\omega}) = \underset{g,\omega}{\operatorname{argmin}} \left\| y_n - \sum_{k=1}^{K} g_k e^{j\omega_k n} \right\|^2 \tag{9.36}$$

Next, all wave components are ranked based on their amplitude. The amplitude refinement process is repeated for three to five rounds, preventing mutual interferences between the sinusoidal waves due to insufficient resolution. In consideration of the limited SNR of the input signal and the need for computation efficiency, the computation process is terminated when the ratio of the residual energy to the input energy is smaller than the preset threshold [128]. When this method is applied, an

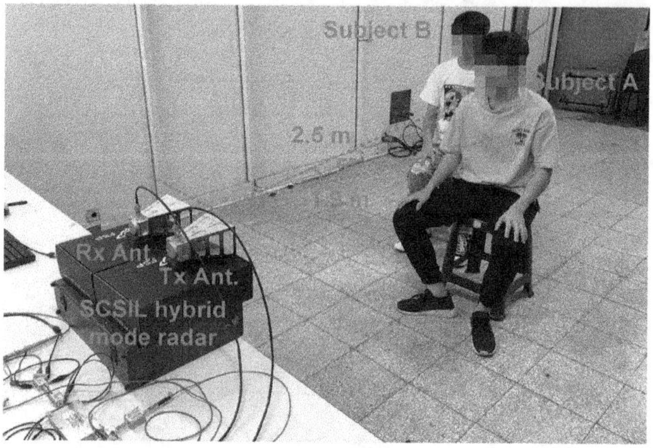

Figure 9.35 *Photograph of the environment in the two-subject experiment SCSIL hybrid radar*

SIL CW radar can simultaneously detect the vital signs of three adults sitting shoulder to shoulder and achieve an accuracy rate of 85.3% [129].

Figure 9.35 presents the configuration of an experiment using the SCSIL hybrid mode radar to detect two subjects 1.9 and 2.5 m from the antenna and with an azimuth angle of 5°. Because the positions of the subjects nearly overlapped, most energy in the echo signal came from the subject located closer to the antenna. Figure 9.36a presents the spectral results under FMCW mode; the black solid and dashed lines represent the FFT amplitude and phase, respectively. In the FFT spectrum, the peak frequency appears at 294 Hz; the ideal phase is at 344.1°, but the measured phase was at 133.8°.

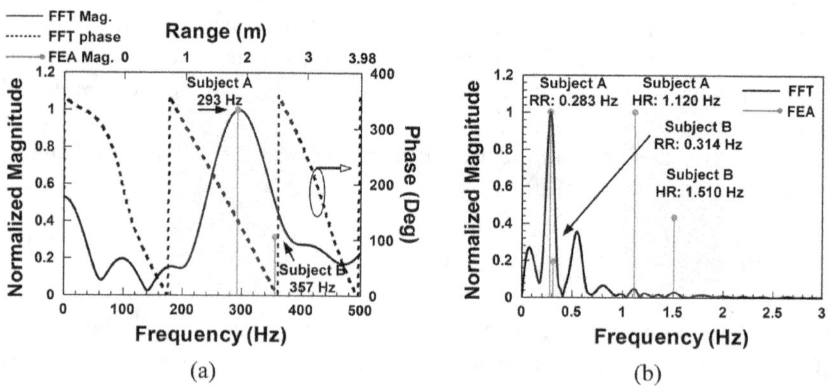

Figure 9.36 *(a) Range detection results with the experimental setup shown in Figure 9.35. (b) Results of vital sign detection using conventional FFT and FEA.*

This indicated that the frequencies of the two tones were overly close. To address this problem, the FEA computation range was limited to within ±1 frequency resolution from the peak frequency ($1/t_s$; in Reference [128], this value was 100 Hz), as indicated in the yellow region of Figure 9.36. The FEA computation revealed two signals (312 and 357 Hz), which are labeled using gray lines and circles in Figure 9.36; the signals corresponded to distances of 1.91 and 2.55 m, which were consistent with the actual positions of the subjects. Because the detection time was only 20 s, using FFT to distinguish between the RR and the HR of the two subjects was difficult. By contrast, the proposed FEA effectively detected the vital sign of the subjects. During the FEA computation, the RR and HR frequency ranges were set as 0.1–0.4 and 1–2 Hz, respectively. The RRs of Subjects A and B were determined to be 0.226 and 1.08 Hz, respectively, and their HRs were determined to be 0.3 and 1.5 Hz, respectively; these values were consistent with actual measurement. Equations (9.30) and (9.33) indicate that the amplitude of the demodulated signal determined using FMCW and CW modes is affected by the locking range. Therefore, the amplitudes of the two spectra in Figure 9.36 can be used to match distance and vital signs. In Reference [128], detection was repeatedly performed with two subjects changing the distance between themselves. The results revealed that when the distance between the subjects was reduced from 100 to 60 cm, the FEA error increased only from 1.8 to 4.4 cm. This verified the accuracy of the algorithm.

Su *et al.* proposed a switched phased-array (SPA)-SIL radar for acquiring azimuth angle information. The detection range of this radar reaches 120° [130]. When the frequency conversion mechanism of SCSIL is integrated with SPA-SIL, the 2-D positions and vital signs of multiple subjects in a space can be determined. As shown in Figure 9.37, a frequency converter increases the transmit frequency of a system to

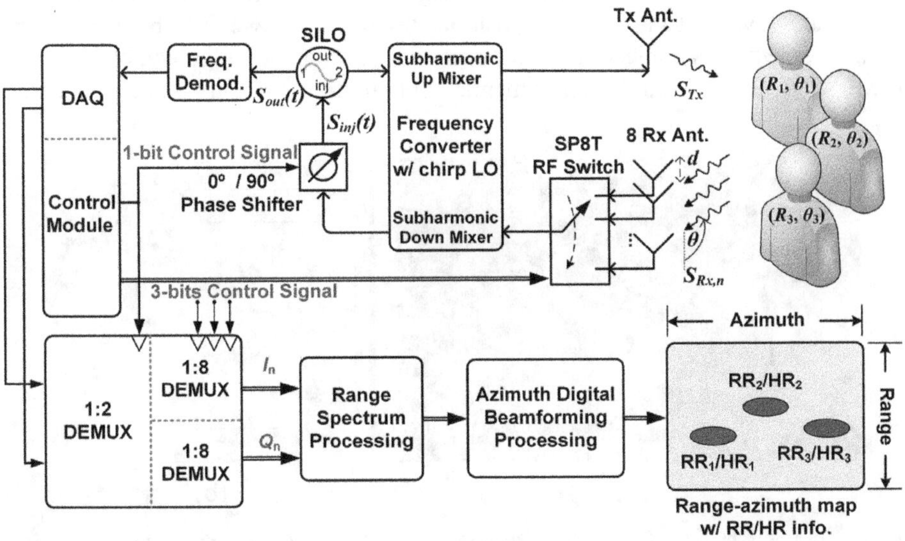

Figure 9.37 Block diagram of the 2-D SIL Doppler radar

5.8–6.8 GHz, and a 1T8R antenna array is used; the receiving antennas are arranged in a uniform linear array with 2 cm spacing. Echo signals pass through the down-mixer and are then injected into the SILO to enter the SIL state. The phase shifter enables alternate phase delays of 0° and 90° to obtain Doppler phases. After undergoing FFT, the demodulated signal is expressed as follows:

$$S_{demod,n}\left(f,\tau'\right) = a_n e^{j\phi_n} \sum_i \omega_{LR,i} e^{j\frac{4\pi f_c}{c}\left(R_i(\tau')+\frac{nd}{2}\cos\theta_i\right)} \times \operatorname{sinc}\left(\pi T\left(f-\frac{2\omega_B R_i(\tau')}{T_{SC}}\right)\right) \tag{9.37}$$

where τ', t, n, θ_i, and R_i denote slow time, fast time, nth antenna, and azimuth angle of and distance to the ith target, respectively. $a_n e^{j\phi_n}$ denotes the amplitude and phase difference caused by the nth antenna; this term must be calibrated prior to azimuth digital beamforming (ADBF) to address the errors of signals received by each antenna. The distance and azimuth angle of a subject can be determined by peaks of range spectrum and ADBF. In ADBF, the demodulated signals corresponding to the eight antennas are multiplied by $e^{j\theta}$, where θ ranges between 0° and 180°. Summing the eight modulation signals yields a peak value, which represents the subjects' azimuth angle with respect to radar. This information can be combined with the distance information to plot a range–azimuth map (RAM). Because radars have limited spatial resolutions, only one point is displayed on a RAM when multiple subjects are present in a space. To resolve this problem, the strength of the respiration signals of two subjects displayed in slow-time spectra acquired from different positions within the point space must be compared. This creates a Doppler-weight RAM to identify the positions of multiple subjects [131].

Figure 9.38 presents the configuration of an experiment in which the 2-D SIL Doppler radar illustrated in Figure 9.37 was used to detect the positions and vital signs of five subjects; the left and right images present a photograph of the experiment and a layout of the subjects' positions. Subjects 1–3 were further apart from each other, whereas Subjects 4 and 5 sat shoulder to shoulder. Figure 9.39a presents the RAM acquired from the experimental configuration in Figure 9.38. The results

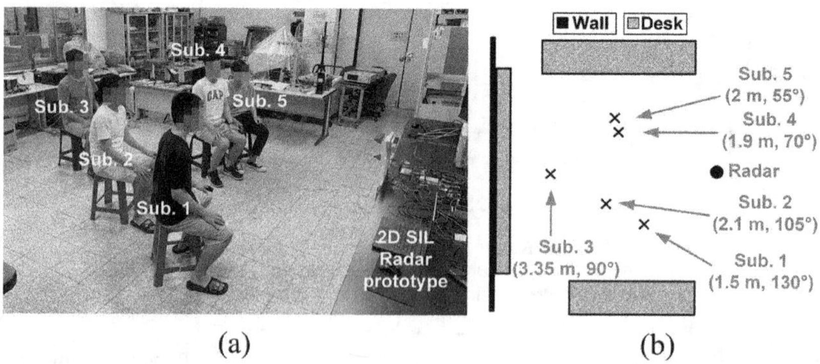

Figure 9.38 Experiment of detecting the positions and vital signs of five seated subjects with the 2-D SIL Doppler radar shown in Figure 9.37

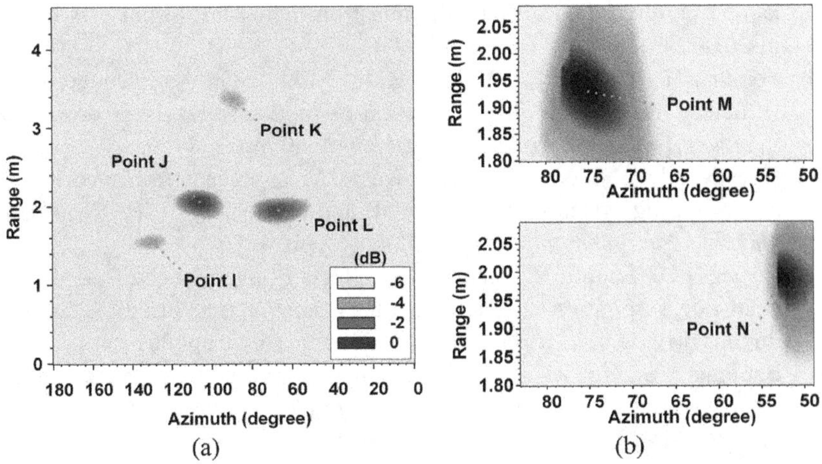

Figure 9.39 *Experiment results of the experimental setup shown in Figure 9.38. (a) Constructed RAM for the scenario shown in Figure 9.38a. (b) Constructed Doppler-weighted RAM for locating subjects 4 and 5.*

revealed only four distinguishable points (I, J, K, and L) on the map (1.54 m, 133°; 2.03 m, 104°; 3.37 m, 92°; and 1.9 m, 67°, respectively). Only one respiratory frequency was identified each from the slow-time spectra of I, J, and K, indicating that each point represented one subject (subjects 1–3). Two respiratory frequencies were identified in L, indicating that the spot was the result of combining the information of the remaining two subjects. Comparing the slow-time spectra of different positions revealed that M and N were dominated by different vital signs, indicating that these points represented subjects 4 and 5. The positioning determined using the detection method differed from that of the actual measurement by 5.12%, and the vital signs detected from each subject were consistent with those measured using medical equipment. This verified that the proposed system can be used to detect and distinguish multiple subjects in the same space.

References

[1] National Safety Council (NSC). Drivers are falling asleep behind the wheel [online]. 2020. Available from https://www.nsc.org/road/safety-topics/fatigued-driver.

[2] Sullman M.J.M. 'The expression of anger on the road'. *Safety Science*. 2015, vol. 72(2), pp. 153–9.

[3] McLaren C., Null J., Quinn J. 'Heat stress from enclosed vehicles: moderate ambient temperatures cause significant temperature rise in enclosed vehicles'. *Pediatrics*. 2005, vol. 116(1), pp. e109–12.

[4] Hall J. *KSP: 52 children died of hyperthermia from being left in hot cars in 2018 [online]*. 2018. Available from https://fox17.com/news/local/

ksp-52-children-died-of-hyperthermia-from-being-left-in-hot-cars-hot-cars-in-2018.

[5] Forsman P.M., Vila B.J., Short R.A., Mott C.G., Van Dongen H.P.A. 'Efficient driver drowsiness detection at moderate levels of drowsiness'. *Accident Analysis & Prevention*. 2013, vol. 50, pp. 341–50.

[6] Koesdwiady A., Soua R., Karray F., Kamel M.S. 'Recent trends in driver safety monitoring systems: state of the art and challenges'. *IEEE Transactions on Vehicular Technology*. 2017, vol. 66(6), pp. 4550–63.

[7] Lisper H.O., Laurell H., Stening G. 'Effects of experience of the driver on heart-rate, respiration-rate, and subsidiary reaction time in a three hours continuous driving task'. *Ergonomics*. 1973, vol. 16(4), pp. 501–6.

[8] Banning A.S., Ng G.A. 'Driving and arrhythmia: a review of scientific basis for international guidelines'. *European Heart Journal*. 2013, vol. 34(3), pp. 236–44.

[9] Thayer J.F., Yamamoto S.S., Brosschot J.F. 'The relationship of autonomic imbalance, heart rate variability and cardiovascular disease risk factors'. *International Journal of Cardiology*. 2010, vol. 141(2), pp. 122–31.

[10] Webster J.G. *Medical instrumentation application and design medical instrumentation application and design*. 4th edn. Hoboken, NJ: Wiley; 2009. pp. 147–57.

[11] Wang F., Li G., Chen J., Duan Y., Zhang D. 'Novel semi-dry electrodes for brain–computer interface applications'. *Journal of Neural Engineering*. 2016, vol. 13(4),p. 046021.

[12] Ikari T., Kurose S., Igasaki T., Kobayashi M. 'Pulse monitoring by solgsol-gel composite flexible piezoelectric sensors'. *Proceedings of IEEE International Ultrasonics Symposium*; Chicago, MA. 2014. pp. 2071–74.

[13] Kwon S., Lee J., Chung G.S., Park K.S. 'Validation of heart rate extraction through an iphone accelerometer'. *Proceedings Annual International Conference of the IEEE Engineering Medicine and Biology Society*; Boston, MA, 2011. pp. 5260–63.

[14] Hong Y., Kim S.-G., Kim B.-H., *et al.* 'Noncontact proximity vital sign sensor based on PLL for sensitivity enhancement'. *IEEE Transactions on Biomedical Circuits and Systems*. 2014, vol. 8(4), pp. 584–93.

[15] Sun Y., Thakor N. 'Photoplethysmography revisited: from contact to noncontact, from point to imaging'. *IEEE Transactions on Biomedical Engineering*. 2016, vol. 63(3), pp. 463–77.

[16] Xu X., Yu J., Chen Y., Zhu Y., Kong L., Li M. 'Breath listener: fine-grained breathing monitoring in driving environments utilizing acoustic signals'. *Proceedings of ACM International Mobile Systems, Applications, and Service Conference*; Seoul, South Korea; 2019. pp. 54–66.

[17] Abdelnasser H., Harras K.A., Youssel M. 'UbiBreathe: a ubiquitous noninvasive WiFi-based breathing estimator'. *Proceedings of ACM International Mobile Ad Hoc Networking and Computing Symposium*; Hangzhou, China; 2015. pp. 277–86.

[18] Jia W., Peng H., Ruan N., Tang Z., Zhao W. 'WiFind: driver fatigue detection with fine-grained Wi-Fi signal features'. *IEEE Transactions on Big Data*. 2020, vol. 6(2), pp. 269–82.

[19] Shi J., Sha M., Yang Z. 'DiGs: distributed graph routing and scheduling for industrial wireless sensor-actuator networks'. *Proceedings of IEEE International Distributed Computing Systems Conference*; Vienna, Austria; 2018. pp. 354–64.

[20] Lin J.C., Kiernicki J., Kiernicki M., Wollschlaeger P.B. 'Microwave apexcardiography'. *IEEE Transactions on Microwave Theory and Techniques*. 1979, vol. 27(6), pp. 618–20.

[21] Droitcour A.D., Boric-Lubecke O., Lubecke V.M., Lin J., Kovacs G.T.A. 'Range correlation and I/ Q performance benefits in single-chip silicon doppler radars for noncontact cardiopulmonary monitoring'. *IEEE Transactions on Microwave Theory and Techniques*. 2004, vol. 52(3), pp. 838–48.

[22] Park B.-K., Yamada S., Boric-Lubecke O., Lubecke V. 'Single-channel receiver limitations in Doppler radar measurements of periodic motion'. *Proceeding of IEEE Radio and Wireless Symp*; San Diego, MA; 2006. pp. 99–102.

[23] Park B.-K., Boric-Lubecke O., Lubecke V.M. 'Arctangent demodulation with DC offset compensation in quadrature Doppler radar receiver systems'. *IEEE Transactions on Microwave Theory and Techniques*. 2007, vol. 55(5), pp. 1073–9.

[24] Chou C.-C., Lai W.-C., Hsiao Y.-K., Chuang H.-R. '60-GHz CMOs Doppler radar sensor with integrated V-band power detector for clutter monitoring and automatic clutter-cancellation in noncontact vital-signs sensing'. *IEEE Transactions on Microwave Theory and Techniques*. 2018, vol. 66(3), pp. 1635–43.

[25] Zakrzewski M., Raittinen H., Vanhala J. 'Comparison of center estimation algorithms for heart and respiration monitoring with microwave Doppler radar'. *IEEE Sensors Journal*. 2012, vol. 12(3), pp. 627–34.

[26] Xu W., Gu C., Li C., Sarrafzadeh M. 'Robust Doppler radar demodulation via compressed sensing'. *Electronics Letters*. 2012, vol. 48(22), pp. 1428–30.

[27] Ahn S.J., Rauh W., Oberdorfer B. *Least squares fitting of circle and ellipse [online]*. 2020. Available from https://reurl.cc/e8o7nM.

[28] Gao X., Boric-Lubecke O. 'Radius correction technique for Doppler radar noncontact periodic displacement measurement'. *IEEE Transactions on Microwave Theory and Techniques*. 2017, vol. 65(2), pp. 621–31.

[29] T.-Y.-J K., Yan Y., Shen T.-M., Chen A.Y.-K., Lin J. 'Design and analysis of a 60-GHz CMOs Doppler micro-radar system-in-package for vital-sign and vibration detection'. *IEEE Trans. Microw. Theory Techn*. 2013, vol. 61(4), pp. 1649–59.

[30] Huang T.-Y., Hayward L.F., Lin J. 'Noninvasive measurement and analysis of laboratory rat's cardiopulmonary movement'. *IEEE Transactions on Microwave Theory and Techniques*. 2017, vol. 65(2), pp. 574–81.

[31] Mostafanezhad I., Boric-Lubecke O. 'Benefits of coherent low-IF for vital signs monitoring using Doppler radar'. *IEEE Transactions on Microwave Theory and Techniques*. 2014, vol. 62(10), pp. 2481–7.

[32] Wei C., Lin J. 'Digitally assisted low IF architecture for noncontact vital sign detection'. *Proceedings of IEEE MTT-S International Microwave Symposium*; Phoenix, USA; 2015. pp. 1–4.

[33] Ma X., Li L., Ming S., You X., Lin J. 'Envelope detection for an ADC-relaxed double-sideband low-IF CW Doppler radar'. *IEEE Transactions on Microwave Theory and Techniques*. 2018, vol. 66(12), pp. 5833–41.

[34] Mostafanezhad I., Yavari E., Boric-Lubecke O., Lubecke V.M., Mandic D.P. 'Cancellation of unwanted Doppler radar sensor motion using empirical mode decomposition'. *IEEE Sensors Journal*. 2013, vol. 13(5), pp. 1897–904.

[35] Tomii S., Ohtsuki T. 'Heartbeat detection by using doppler radar with wavelet transform based on scale factor learning'. *Proceedings of IEEE International Communication Conference*; London, 2015. pp. 483–88.

[36] Lv Q., Chen L., An K., *et al.* 'Doppler vital signs detection in the presence of large-scale random body movements'. *IEEE Transactions on Microwave Theory and Techniques*. 2018, vol. 66(9), pp. 4261–70.

[37] Yang Z.-K., Shi H., Zhao S., Huang X.-D. 'Vital sign detection during large-scale and fast body movements based on an adaptive noise cancellation algorithm using a single Doppler radar sensor'. *Sensors*. 2020, vol. 20(15), p. 4183.

[38] Kazemi S., Ghorbani A., Amindavar H., Li C. 'Cyclostationary approach to Doppler radar heart and respiration rates monitoring with body motion cancelation using radar Doppler system'. *Biomedical Signal Processing and Control*. 2014, vol. 13(4), pp. 79–88.

[39] Wang F.-K., Fang C.-H., Horng T.-S., Peng K.-C., Li J.-Y., Chen C.-C. 'Concurrent vital sign and position sensing of multiple individuals using self-injection-locked tags and injection-locked I/Q receivers with arctangent demodulation'. *IEEE Transactions on Microwave Theory and Techniques*. 2013, vol. 61(12), pp. 4689–99.

[40] Wang F.-K., Chou Y.-R., Chiu Y.-C., Horng T.-S. 'Chest-worn health monitor based on a bistatic elf-injection-locked radar'. *IEEE Transactions on Biomedical Engineering*. 2015, vol. 62(12), pp. 2931–40.

[41] Kazemi S., Ghorbani A., Amindavar H., Morgan D.R. 'Vital-sign extraction using bootstrap-based generalized warblet transform in heart and respiration monitoring radar system'. *IEEE Transactions on Instrumentation and Measurement*. 2016, vol. 65(2), pp. 255–63.

[42] Lee K.J., Park C., Lee B. 'Tracking driver's heart rate by continuous-wave Doppler radar'. *Proceedings of International Conference IEEE Engineering Medicine Biology Society*; Orlando, FL; 2016. pp. 5417–20.

[43] Li C., Lin J. 'Random body movement cancellation in doppler radar vital sign detection'. *IEEE Transactions on Microwave Theory and Techniques*. 2008, vol. 56(12), pp. 3143–52.

[44] Fletcher R., Han J. 'Low-cost differential front-end for Doppler radar vital sign monitoring'. *Proceedings of IEEE MTT-S International Microwave Symposium*; Boston, MA; 2009. pp. 1325–8.

[45] Yu X., Li C., Lin J. 'Two-dimensional noncontact vital sign detection using Doppler radar array approach'. *Proceedings of IEEE MTT-S International Microwave Symposium*; Baltimore, MD; 2011. pp. 1–4.

[46] Rahman A., Yavari E., Singh A., Lubecke V.M., Lubecke O.-B. 'A Low-IF Tag-based motion compensation technique for mobile Doppler radar life signs monitoring'. *IEEE Transactions on Microwave Theory and Techniques*. 2015, vol. 63(10), pp. 3034–41.

[47] Zhu F., Wang K., Wu K. 'A fundamental-and-harmonic dual-frequency Doppler radar system for vital signs detection enabling radar movement self-cancellation'. *IEEE Trans. Microw. Theory Techn.* 2018, vol. 66(11), pp. 5106–18.

[48] Mostafanezhad I., Boric-Lubecke O., Lubecke V., Host-Madsen A. 'Cancellation of unwanted motion in a handheld Doppler radar used for non-contact life sign monitoring'. *Proceedings of IEEE MTT-S International Microwave Symposium*; Atlanta, GA; 2008. pp. 1171–4.

[49] Gu C., Wang G., Li Y., Inoue T., Li C. 'A hybrid radar-camera sensing system with phase compensation for random body movement cancellation in Doppler vital sign detection'. *IEEE Transactions on Microwave Theory and Techniques*. 2013, vol. 61(12), pp. 4678–88.

[50] Li C., Lubecke V.M., Boric-Lubecke O., Lin J. 'A review on recent advances in Doppler radar sensors for noncontact healthcare monitoring'. *IEEE Transactions on Microwave Theory and Techniques*. 2013, vol. 61(5), pp. 2046–60.

[51] Li C., Peng Z., Huang T.-Y., *et al*. 'A review on recent progress of portable short-range noncontact microwave radar systems'. *IEEE Transactions on Microwave Theory and Techniques*. 2017, vol. 65(5), pp. 1692–706.

[52] Lee J., Li Y.-A., Hung M.-H., Huang S.-J. 'A fully-integrated 77-GHz FMCW radar transceiver in 65-nm CMOs technology'. *IEEE Journal of Solid-State Circuits*. 2010, vol. 45(12), pp. 2746–56.

[53] Park J., Ryu H., Ha K.-W., Kim J.-G., Baek D. '76–81-GHz CMOs transmitter with a phase-locked-loop-based multichirp modulator for automotive radar'. *IEEE Transactions on Microwave Theory and Techniques*. 2015, vol. 63(4), pp. 1399–408.

[54] Wang G., Gu C., Inoue T., Li C. 'A hybrid FMCW-Interferometry radar for indoor precise positioning and versatile life activity monitoring'. *IEEE Transactions on Microwave Theory and Techniques*. 2014, vol. 62(11), pp. 2812–22.

[55] Scheiblhofer S., Schuster S., Stelzer A. 'High-speed FMCW radar frequency synthesizer with DDS based linearization'. *IEEE Microwave and Wireless Components Letters*. 2007, vol. 17(5), pp. 397–9.

[56] Scheiblhofer S., Schuster S., Stelzer A. 'Signal model and linearization for nonlinear chirps in FMCW radar SAW-ID tag request'. *IEEE*

Transactions on Microwave Theory and Techniques. 2006, vol. 54(4), pp. 1477–83.

[57] Hitzler M., Saulig S., Boehm L., *et al*. 'Ultracompact 160-GHz FMCW radar MMIC with fully integrated offset synthesizer'. *IEEE Transactions on Microwave Theory and Techniques*. 2017, vol. 65(5), pp. 1682–91.

[58] Balon S., Mouthaan K., Heng C.-H., Chen Z.N. 'A C-band FMCW SAR transmitter with 2-GHz bandwidth using injection-locking and synthetic bandwidth techniques'. *IEEE Transactions on Microwave Theory and Techniques*. 2018, vol. 66(11), pp. 5095–105.

[59] Wang K., Gu J.-F., Ren F., Wu K. 'A multitarget active backscattering 2-D positioning system with super resolution time series post-processing technique'. *IEEE Transactions on Microwave Theory and Techniques*. 2017, vol. 65(5), pp. 1751–66.

[60] Baral A.B., Torlak M. 'Impact of number of noise eigenvectors used on the resolution probability of music'. *IEEE Access*. 2019, vol. 7, pp. 20023–39.

[61] Yang J., Pyo G., Kim C.-Y., Hong S. 'A 24-GHz CMOs UWB radar transmitter with compressed pulses'. *IEEE Transactions on Microwave Theory and Techniques*. 2012, vol. 60(4), pp. 1117–25.

[62] Jain V., Tzeng F., Zhou L., Heydari P. 'A single-chip dual-band 22–29-GHz/77–81-GHz BiCMOS transceiver for automotive radars'. *IEEE Journal of Solid-State Circuits*. 2009, vol. 44(12), pp. 3469–85.

[63] Xu R., Jin Y., Nguyen C. 'Power-efficient switching-based CMOs UWB transmitters for UWB communications and radar systems'. *IEEE Trans. Microw. Theory Techn*. 2006, vol. 54(8), pp. 3271–7.

[64] Arbabian A., Callender S., Kang S., Rangwala M., Niknejad A.M. 'A 94 GHz mm-wave-to-baseband pulsed-radar transceiver with applications in imaging and gesture recognition'. *IEEE Journal of Solid-State Circuits*. 2013, vol. 48(4), pp. 1055–71.

[65] Ren L., Koo Y.S., Wang H., Wang Y., Liu Q., Fathy A.E. 'Noncontact multiple heartbeats detection and subject localization using UWB impulse Doppler radar'. *IEEE Microwave and Wireless Components Letters*. 2015, vol. 25(10), pp. 690–2.

[66] Ren L., Wang H., Naishadham K., Kilic O., Fathy A.E. 'Phase-based methods for heart rate detection using UWB impulse Doppler radar'. *IEEE Transactions on Microwave Theory and Techniques*. 2016, vol. 64(10), pp. 3319–31.

[67] Wang Y., Liu Q., Fathy A.E. 'CW and pulse-Doppler radar processing based on FPGA for human sensing applications'. *IEEE Trans. Microw. Theory Techn*. 2013, vol. 51(5), pp. 3097–107.

[68] Wang G.R., Han H.G., Kim S.Y., Kim T.W. 'Wireless vital sign monitoring using penetrating impulses'. *IEEE Microwave and Wireless Components Letters*. 2017, vol. 27(1), pp. 94–6.

[69] Lim S., Lee S., Jung J., Kim S.-C. 'Detection and localization of people inside vehicle using impulse radio ultra-wideband radar sensor'. *IEEE Sensors Journal*. 2020, vol. 20(7), pp. 3892–901.

[70] Leem S.K., Khan F., Cho S.H. 'Vital sign monitoring and mobile phone usage detection using IR-UWB radar for intended use in car crash prevention'. *Sensors*. 2017, vol. 17(6), p. 1240.

[71] Yang Z., Bocca M., Jain V., Mohapatra P. 'Contactless breathing rate monitoring in vehicle using UWB radar'. *Proceeding of International Workshop ACM Real-World Embedded Wireless Systems and Networks*; Shenzhen, China; 2018. pp. 13–18.

[72] Schires E., Georgiou P., Lande T.S. 'Vital sign monitoring through the back using an UWB impulse radar with body coupled antennas'. *IEEE Transactions on Biomedical Circuits and Systems*. 2018, vol. 12(2), pp. 292–302.

[73] Yang M., Yang X., Li L., Zhang L. 'In-car multiple targets vital sign monitoring using location-based VMD algorithm'. *Proceedings of International Conference IEEE Wireless Communications Signal Process*; Hangzhou, China; 2018. pp. 1–6.

[74] Zheng T., Chen Z., Cai C., Luo J., Zhang X. 'V2Fi: in-vehicle vital sign monitoring via compact RF sensing'. *Proceedings of ACM Interactive Mobile Wearable Ubiquitous Technology*. 2020, vol. 4(2), pp. 1–27.

[75] Dragomiretskiy K., Zosso D. 'Variational mode decomposition'. *IEEE Transactions on Signal Processing*. 2014, vol. 62(3), pp. 531–44.

[76] Schleicher B., Nasr I., Trasser A., Schumacher H. 'IR-UWB radar demonstrator for ultra-fine movement detection and vital-sign monitoring'. *IEEE Transactions on Microwave Theory and Techniques*. 2013, vol. 61(5), pp. 2076–85.

[77] Li C.-J., Hsiao C.-H., Wang F.-K., Horng T.-S., Peng K.-C. 'A rigorous analysis of local oscillator pulling in frequency and discrete-time domain'. *IEEE Radio Frequency Integrated Circuit Symposium Digest*; Boston, MA, 2009. pp. 409–12.

[78] Li C.-J., Hsiao C.-H., Wang F.-K., Horng T.-S., Peng K.-C. 'A rigorous analysis of A phase-locked oscillator under injection'. *IEEE Transaction Microwave Theory Techniques*. 2010, vol. 58(5), pp. 1391–400.

[79] Wang F.-K., C.-J L., Hsiao C.-H. 'An injection-locked detector for concurrent spectrum and vital sign sensing'. *IEEE MTT-S International Microwave Symposium Digest*; Anaheim, CA; 2010. pp. 768–71.

[80] Wang F.-K., Wu C.-T.M., Horng T.-S., *et al.* 'Review of self-injection-locked radar systems for noncontact detection of vital signs'. *IEEE Journal of Electromagnetics, RF and Microwaves in Medicine and Biology*. 2020, vol. 4(4), pp. 294–307.

[81] Adler R. 'A study of locking phenomena in oscillators'. *Proceedings of the IRE*. 1946, vol. 34(6), pp. 351–7.

[82] Hsiao C.-H., Li C.-J., Wang F.-K., Horng T.-S., Peng K.-C. 'Study of direct conversion transmitter pulling effects in constant envelope modulation

systems'. *IEEE MTT-S International Microwave Symposium Digest*; Anaheim, CA, 2010. pp. 1174–77.

[83] Hsiao C.-H., Li C.-J., Wang F.-K., Horng T.-S., Peng K.-C. 'Analysis and improvement of direct-conversion transmitter pulling effects in constant envelope modulation systems'. *IEEE Transactions on Microwave Theory and Techniques*. 2010, vol. 58(12), pp. 4137–46.

[84] Pontón M., Suárez A. 'Wireless injection locking of oscillator circuits'. *IEEE Transactions on Microwave Theory and Techniques*. 2016, vol. 64(12), pp. 4646–59.

[85] Wang F.-K., Horng T.-S.J., Shih J.-Y., Hsu Z.-J., Su W.-C., Juan P.-H. 'See-through-wall (STW) life detector using self-injection-locked (SIL) technology'. *Proceedings of IEEE Asia-Pacific Microwave Conference*; Hong Kong, China; 2020. pp. 496–8.

[86] Wang F.-K., Li C.-J., Hsiao C.-H, *et al.* 'A novel vital-sign sensor based on a self-injection-locked oscillator'. *IEEE Transactions on Microwave Theory and Techniques*. 2010, vol. 58(12), pp. 4112–20.

[87] Lu K.-C., Wang F.-K., Horng T.-S. 'Ultralow phase noise and wide-band CMOs VCO using symmetrical body-bias PMOS varactors'. *IEEE Microwave and Wireless Components Letter*. 2013, vol. 23(2), pp. 90–2.

[88] Wang F.-K., Horng T.-S., Peng K.-C., Jau J.-K., Li J.-Y., Chen C.-C. 'Single-antenna Doppler radars using self and mutual injection locking for vital sign detection with random body movement cancellation'. *IEEE Transactions on Microwave Theory and Techniques*. 2011, vol. 59(12), pp. 3577–87.

[89] Tseng C.-H., Yu L.-T., Huang J.-K., Chang C.-L. 'A wearable self-injection-locked sensor with active integrated antenna and differentiator-based envelope detector for vital-sign detection from chest wall and wrist'. *IEEE Transactions on Microwave Theory and Techniques*. 2018, vol. 66(5), pp. 2511–21.

[90] Wu P.-H., Jau J.-K., Li C.-J., Horng T.-S., Hsu P. 'Phase- and self-injection-locked radar for detecting vital signs with efficient elimination of DC offset and null points'. *IEEE Transactions on Microwave Theory Techniques*. 2013, vol. 61(1), pp. 685–95.

[91] Juan P.-H., Chen K.-H., Wang F.-K. 'Frequency-offset self-injection-locked (FOSIL) radar for noncontact vital sign monitoring'. *IEEE MTT-S International Microwave Symposium Digest*; Los Angeles, CA; 2020. pp. 643–6.

[92] Juan P.-H., Chen K.-H., Wang F.-K. 'Frequency-offset self-injection-locked radar with digital frequency demodulation for SNR improvement, elimination of EMI issue, and DC offset calibration'. *IEEE Transactions on Microwave Theory and Techniques*. 2021, vol. 69(1), pp. 1149–60.

[93] Wang F.-K., Su S.-C., Tang M.-C., Horng T.-S. 'Displacement monitoring system based on a quadrature self-injection-locked radar technology'. *IEEE MTT-S International Microwave Symposium Digest*; 2017. pp. 1–4.

[94] Wang F.-K., Juan P.-H., Su S.-C., Tang M.-C., Horng T.-S. 'Monitoring displacement by a quadrature self-injection-locked radar with measurement- and

differential-based offset calibration methods'. *IEEE Sensors Journal*. 2019, vol. 19(5), pp. 1905–16.

[95] Yu S.-H., Horng T.-S. 'Highly linear phase-canceling self-injection-locked ultrasonic radar for non-contact monitoring of respiration and heartbeat'. *IEEE Transactions on Biomedical Circuits and Systems*. 2020, vol. 14(1), pp. 75–90.

[96] S.-H Y., P.-H W. 'Two kinds of self-oscillating circuits mechanically demonstrated'. *International Journal of Electronics and Communication Engineering*. 2014, vol. 8(7), pp. 1334–8.

[97] Wang F.-K., Horng T.-S., Peng K.-C., Jau J.-K., Li J.-Y., Chen C.-C. 'Mutual injection-locked SIL sensor array for vital sign detection with random body movement cancellation'. *IEEE MTT-S International Microwave Symposium Digest*; Baltimore, MD, Section TH2C-4; 2011. pp. 1–4.

[98] Tang M.-C., Kuo C.-Y., Wun D.-C., Wang F.-K., Horng T.-S. 'Same side dual SIL-radar system for real-time vital sign monitoring with random body movement cancellation'. *IEEE MTT-S International Microwave Symposium Dig*est; San Francisco, CA, Section WE1A-1, May; 2016. pp. 1–4.

[99] Tang M.-C., Kuo C.-Y., Wun D.-C., Wang F.-K., Horng T.-S. 'A self- and mutually injection-locked radar system for monitoring vital signs in real time with random body movement cancellation'. *IEEE Transactions on Microwave Theory and Techniques*. 2016, vol. 64(12), pp. 4812–22.

[100] Tang M.-C., Wang F.-K., Horng T.-S. 'A single radar-based vital sign monitoring system with resistance to large body motion'. *IEEE MTT-S International Microwave Symposium Digest*; Honololu, HI; 2017. pp. 1–4.

[101] Tang M.-C., Wang F.-K., Horng T.-S. 'Single self-injection-locked radar with two antennas for monitoring vital signs with large body movement cancellation'. *IEEE Transactions on Microwave Theory and Techniques*. 2017, vol. 65(12), pp. 5324–33.

[102] Wang F.-K., Chou Y.-R., Chiu Y.-C., Tang M.-C., Horng T.-S. 'Chest-worn health monitor based on a bistatic self-injection-locked radar'. *IEEE MTT-S International Microwave Symposium*; Phoenix, AZ; 2015. pp. 1–4.

[103] Wang F.-K., Tang M.-C., Su S.-C., Horng T.-S. 'Wrist pulse rate monitor using self-injection-locked radar technology'. *Biosensors*. 2016, vol. 6(4), pp. 54–65.

[104] C.-J L., Wang F.-K., Horng T.-S., Peng K.-C. 'A novel RF sensing circuit using injection locking and frequency demodulation for cognitive radio applications'. *IEEE Trans Microwave Theory Techniques*. 2009, vol. 57(12), pp. 3143–52.

[105] C.-J L., Wang F.-K., Horng T.-S., Peng K.-C. 'A novel RF sensing circuit using injection locking and frequency demodulation for cognitive radio applications'. *IEEE MTT-S International Microwave Symposium Digest*; Boston, MA; 2009. pp. 1165–8.

[106] Wang F.-K., Chen C.-T., Tsai J.-R., *et al.* 'A CMOS spectrum sensor using injection locking of two voltage-controlled oscillators for cognitive radio

system'. *IEEE Radio Frequency Integrated Circuit Symposium Digest*; Baltimore, MD, Section RMO3A-3; 2011. pp. 1–4.

[107] Wang F.-K., Horng T.-S., Peng K.-C., J.-Y L., Chen C.-C. 'Combining SIL tag and IL receiver for concurrent vital sign and position sensing'. *IEEE MTT-S International Microwave Symposium Digest*; Seattle, WA, Section TU2B-2; 2013. pp. 1–4.

[108] Chiu Y.-C., Wang F.-K., Chou Y.-R., Horng T.-S. 'Wearable doppler radar health monitor with gesture control'. *Proceedings of IEEE Asia-Pacific Microwave Conference*; Sendai, Japan; 2014. pp. 944–6.

[109] Tang M.-C., Wang F.-K., Horng T.-S. 'Human gesture sensor using ambient wireless signals based on passive radar technology'. *IEEE MTT-S International Microwave Symposium Digest*; Phoenix, AZ, Section WE1G-1; 2015. pp. 1–4.

[110] Tang M.-C., Wang F.-K., Horng T.-S. 'Vital-sign detection based on a passive WiFi radar'. *IEEE MTT-S International Microwave Workshop Series, RF and Wireless Technology for Biomedical and Healthcare Application*; Taipei, Taiwan; 2015. pp. 74–5.

[111] Wang F.-K., Tang M.-C., Chiu Y.-C., Horng T.-S. 'Gesture sensing using retransmitted wireless communication signals based on Doppler radar technology'. *IEEE Transactions on Microwave Theory and Techniques*. 2015, vol. 63(12), pp. 4592–602.

[112] Tang M.-C., Chen C.-L., Lin M.-H., Wang F.-K., Yeh C.-H., Horng T.-S. 'A hybrid computer vision and Wi-Fi Doppler radar system for capturing the 3-D hand gesture trajectory with a smartphone'. *IEEE MTT-S International Microwave Symposium Digest*; Honolulu, HI, Section WEIF1-21; 2017. pp. 1–4.

[113] Lai Y.-C., Chou C.-C., Tang M.-C., Horng T.-S., Wang F.-K. 'Finger gesture sensing and recognition using a Wi-Fi-based passive radar'. *IEEE MTT-S International Microwave Symposium Digest*; Boston, MA, Section Tu3G-1; 2019. pp. 1–4.

[114] Tang M.-C., Liao C.-M., Wang F.-K., Horng T.-S. 'Noncontact pulse transit time measurement using a single-frequency continuous-wave radar'. *IEEE MTT-S International Microwave Symposium Digest*; Philadelphia, PA; 2018. pp. 1409–12.

[115] Arif R.E., Tang M.-C., W.-C S., Horng T.-S., Wang F.-K., Tseng C.-H. 'Designing a metasurface-based tag antenna for wearable vital sign sensors'. *IEEE MTT-S International Microwave Symposium Digest*; Boston, MA; 2019. pp. 373–6.

[116] Juan P.-H., Tsai H.-C., Wang F.-K. 'Wearable vital sign sensing radar antenna with moving clutter immunity'. *IEEE International Radio-Frequency Integration Technology Symposium Digest*; Hiroshima, Tokyo; 2020. pp. 247–9.

[117] Hsu C.-Y., Hwang L.-T., Wang F.-K., Horng T.-S. 'Wearable vital sign sensor using a single-input multiple-output self-injection-locked oscillator tag'. *IEEE MTT-S International Microwave Symposium Digest*; Philadelphia, PA; 2018. pp. 248–51.

[118] C.-Z W., Tseng C.-H. 'A perturbation-injection-locked sensor with self-oscillating active CSRR for vital-sign detection from fingertip'. *IEEE MTT-S International Microwave Symposium Digest*; Boston, MA; 2019. pp. 369–72.

[119] Tseng C.-H., Wu C.-Z. 'A novel microwave phased- and perturbation-injection-locked sensor with self-oscillating complementary split-ring resonator for finger and wrist pulse detection'. *IEEE Transactions on Microwave Theory and Techniques*. 2020, vol. 68(5), pp. 1933–42.

[120] Arif R.E., W.-C S., Tang M.-C., Horng T.-S., Wang F.-K. 'Chest-worn self-injection-locked oscillator tag for monitoring heart rate variability'. *IEEE MTT-S International Microwave Symposium Digest*; Los Angeles, CA, 2020. pp. 512–15.

[121] Juan P.-H., Wang F.-K., Tzeng Y.-T. 'SIL-radar-based rat detector for warehouse management system'. *Proceedings of IEEE International Microwave Biomedical Conference*; Nanjing, China, Section TA2B-2; 2019. pp. 1–4.

[122] Yuan Y., Lu C., Chen A.Y.-K., Tseng C.-H., Wu C.-T.M. 'Multi-target concurrent vital sign and location detection using metamaterial-integrated self-injection-locked quadrature radar sensor'. *IEEE Transactions on Microwave Theory and Techniques*. 2019, vol. 67(12), pp. 5429–37.

[123] Hsu C.-Y., Chuang C.-Y., Wang F.-K., Horng T.-S., Hwang L.-T. 'Detection of vital signs for multiple subjects by using self-injection-locked beam scanning array'. *IEEE MTT-S International Microwave Symposium Digest*; Honolulu, HI; 2017. pp. 1–4.

[124] Wang F.-K., Horng T.-S., Peng K.-C., Jau J.-K., J.-Y L., Chen C.-C. 'Seeing through walls with a self-injection-locked radar to detect hidden people'. *IEEE MTT-S International Microwave Symposium Digest*; Montreal, Canada, Section THPE-4; 2012. pp. 1–3.

[125] Wang F.-K., Horng T.-S., Peng K.-C., Jau J.-K., Li J.-Y., Chen C.-C. 'Detection of concealed individuals based on their vital signs by using a see-through-wall imaging system with a self-injection-locked radar'. *IEEE Transactions on Microwave Theory and Techniques*. 2013, vol. 61(1), pp. 696–704.

[126] W.-C S., Tang M.-C., Arif R.E., Horng T.-S., Wang F.-K. 'Single conversion stepped-frequency continuous-wave radar using self-injection-locking technology'. *IEEE MTT-S International Microwave Symposium Digest*; Boston, MA; 2019. pp. 1–4.

[127] W.-C S., Tang M.-C., Arif R.E., Horng T.-S., Wang F.-K. 'Stepped-frequency continuous-wave radar with self-injection-locking technology for monitoring multiple human vital signs'. *IEEE Transactions on Microwave Theory and Techniques*. 2019, vol. 67(12), pp. 5396–405.

[128] Wang F.-K., Juan P.-H., Chian D.-M., Wen C.-K. 'Multiple range and vital sign detection based on single-conversion self-injection-locked hybrid mode radar with a novel frequency estimation algorithm'. *IEEE Transactions on Microwave Theory and Techniques*. 2020, vol. 68(5), pp. 1908–20.

[129] Chian D.-M., Wen C.-K., Wang F.-K., Wong K.-K. 'Signal separation and tracking algorithm for multi-person vital signs by using doppler radar'. *IEEE Transactions on Biomedical Circuits and Systems*. 2020, vol. 14(6), pp. 1346–61.

[130] W.-C S., Juan P.-H., Chian D.-M., Horng T.-S., Wen C.-K., Wang F.-K. 'Human tracking and vital sign monitoring with a switched phased-array self-injection-locked radar'. *IEEE MTT-S International Microwave Symposium Digest*; Los Angeles, CA; 2020. pp. 659–62.

[131] W.-C S., Juan P.-H., Chian D.-M., Horng T.-S., Wen C.-K., Wang F.-K. '2D self-injection-locked doppler radar for locating multiple people and monitoring their vital signs'. *IEEE Transactions on Microwave Theory and Techniques*. 2021, vol. 69(1), pp. 1016–26.

Chapter 10
Conclusion

Zhengyu Peng[1], Changzhi Li[2], and Faruk Uysal[3]

Thanks to the advances of embedded signal processing, high-frequency integrated circuit (IC), and monolithic microwave IC, miniature radar systems can be realized on a printed circuit board or even on an IC [1–5]. Applications of radar systems have been extended to commercial areas, such as through-the-wall detection [6–9], indoor localization [10–13], biomedical applications [14, 15], and driver assistance [16, 17]. Radars are the key technology in the safety system of a modern vehicle. Automotive radars are the critical sensors in advanced driver-assistance systems (ADAS), which are used for adaptive cruise control, collision avoidance, blind spot detection, lane change assistance, and parking assistance. Evolving from the driver assistance system, a fully autonomous car has much higher demands for its sensors, especially for automotive radars. The topics of this book mainly focus on modern radar technology and its applications on automotive vehicles. The fundamentals of modern automotive radars, including radar system architecture, radar signal processing, noise modeling, and basic detection theorems, are introduced at the beginning by using the frequency-modulated continuous-wave (FMCW) waveform as the example. Due to the scope of this book, details of some theoretical analysis are not discussed in this book. Interested readers may refer to Reference [18] for more theoretical contents. In addition to the radar fundamentals, this book also includes the contents of advanced technologies that have been used in modern automotive radar systems, such as multiple-input and multiple-output (MIMO), sensor fusion, target classification, as well as interference detection and mitigation. These technologies are the keys for supporting ADAS and future fully autonomous vehicles.

In this book, we also recognize that the future of automotive radars should not only address conventional exterior applications but also play important roles for interior applications, such as gesture sensing for human–vehicle interaction, driver/passenger vital signs and presence monitoring, etc. The readers could refer

[1]Aptiv Corporation, Indiana Tech Center (ITC), Fort Wayne, IN, USA
[2]Department of Electrical and Computer Engineering, Texas Tech University, Lubbock, TX, USA
[3]Radar Technology Department, Netherlands Organisation for Applied Scientific Research (TNO), The Netherlands

to Reference [19] for additional contents on short-range micromotion sensing with radar technology.

References

[1] Droitcour A.D., Boric-Lubecke O., Lubecke V.M., Lin J., Kovacs G.T.A. range correlation and I/Q performance benefits in single-chip silicon Doppler radars for noncontact cardiopulmonary monitoring '. *IEEE Transactions on Microwave Theory and Techniques*. 2014, vol. 52(3), pp. 838–48.

[2] Yeap S.B., Qing X., Chen Z.N. ' 77-ghz dual-layer transmit-array for automotive radar applications '. *IEEE Trans Antennas Propag*. 2015, vol. 63(6), pp. 2833–37.

[3] Luo T.-N., Wu C.-H.E., Chen Y.-J.E. 'A 77-ghz CMOS automotive radar transceiver with anti-interference function'. *IEEE Transactions on Circuits and Systems I: Regular Papers*. 2013, vol. 60(12), pp. 3247–55.

[4] Lee J., Li Y.-A., Hung M.-H., Huang S.-J. 'A fully-integrated 77-ghz FMCW radar transceiver in 65-nm CMOS technology'. *IEEE Journal of Solid-State Circuits*. 2010, vol. 45(12), pp. 2746–56.

[5] Li C., Yu X., Lee C.M., Li D., Ran L., Lin J. 'High- sensitivity software-configurable 5.8- ghz radar sensor receiver chip in 0.13-μm CMOS for non-contact vital sign detection'.*IEEE Transactions on Microwave Theory and Techniques*. 2010, vol. 58(5), pp. 1410–19.

[6] Ahmad F., Amin M. 'Noncoherent approach to through-the-wall radar localization'. *IEEE Transactions on Aerospace and Electronic Systems*. 2006, vol. 42(4), pp. 1405–19.

[7] Lubecke V.M., Boric-Lubecke O., Host-Madsen A., Fathy A.E. 'Through-the-wall radar life detection and monitoring'. *IEEE/MTT-S International Microwave Symposium*. 2007, pp. 769–72.

[8] Debes C., Amin M.G., Zoubir A.M. 'Target detection in single- and multiple-view through-the-wall multiple-view through-the-wall radar imaging'. *IEEE Trans Geosci Remote Sens*. 2009, vol. 47(5), pp. 1349–61.

[9] Yoon Y.-S., Amin M.G. 'High-resolution through-the-wall resolution through-the-wall radar imaging using beamspace MUSIC'. *IEEE Trans Antennas Propag*. 2008, vol. 56(6), pp. 1763–74.

[10] Vossiek M., Wiebking L., Gulden P., Wieghardt J., Hoffmann C., Heide P. 'Wireless local positioning'. *IEEE Microwave Magazine*. 2003, vol. 4(4), pp. 77–86.

[11] Peng Z., Ran L., Li C. 'A 24-ghz low-cost continuous beam steering phased array for indoor smart radar'. *IEEE 58th International Midwest Symposium on Circuits and Systems (MWSCAS). IEEE*. 2015, pp. 1–4.

[12] Peng Z., Li C. 'A portable 24-ghz FMCW radar based on six-port for short-range human tracking". *2015 IEEE MTT-S 2015 International Microwave Workshop Series on RF and Wireless Technologies for Biomedical and Healthcare Applications (IMWS-BIO)*. 2015, pp. 81–82.

[13] Peng Z., Li C. 'A portable K-bband 3-D MIMO radar with nonuniformly spaced array for short-range localization'. *IEEE Transactions on Microwave Theory and Techniques*. 2018, vol. 66(11), pp. 5075–86.

[14] Li C., Lubecke V.M., Boric-Lubecke O., Lin J. 'A review on recent advances in doppler radar sensors for noncontact healthcare monitoring'. *IEEE Transactions on Microwave Theory and Techniques*. 2013, vol. 61(5), pp. 2046–60.

[15] Li C., Peng Z., Huang T.-Y, *et al.* 'A review on recent progress of portable short-range noncontact microwave radar systems'. *IEEE Transactions on Microwave Theory and Techniques*. 2017, vol. 65(5), pp. 1692–706.

[16] Hasch J., Topak E., Schnabel R., Zwick T., Weigel R., Waldschmidt C. 'Millimeter-wave technology for automotive radar sensors in the 77 ghz frequency band'. *IEEE Transactions on Microwave Theory and Techniques*. 2012, vol. 60(3), pp. 845–60.

[17] Patole S.M., Torlak M., Wang D., Ali M. 'Automotive radars: a review of signal processing techniques'. *IEEE Signal Processing Magazine*. 2017, vol. 34(2), pp. 22–35.

[18] Richards M.A. *Fundamentals of radar signal processing fundamentals of radar signal processing*. New York, NY: McGraw-Hill Education; 2014.

[19] *Short-range micro-motion sensing with radar technology [online]. IET digital library*. Available from https://digital-library.theiet.org/content/books/ce/pbce125e

Index

www.ingramcontent.com/pod-product-compliance
Lightning Source LLC
LaVergne TN
LVHW011847200926
848484LV00002B/413